CAMBRIDGE SOUTH ASIAN STUDIES

PEASANTS & IMPERIAL RULE

CAMBRIDGE SOUTH ASIAN STUDIES

These monographs are published by the Syndics of Cambridge University Press in association with the Cambridge University Centre for South Asian Studies. The following books have been published in this series:

1 S. Gopal: *British Policy in India, 1858–1905*
2 J. A. B. Palmer: *The Mutiny Outbreak at Meerut in 1857*
3 A. Das Gupta: *Malabar in Asian Trade, 1740–1800*
4 G. Obeyesekere: *Land Tenure in Village Ceylon*
5 H. L. Erdman: *The Swatantra Party and Indian Conservatism*
6 S. N. Mukherjee: *Sir William Jones: A Study in Eighteenth-Century British Attitudes to India*
7 Abdul Majed Khan: *The Transition in Bengal, 1756–1775: A Study of Saiyid Muhammad Reza Khan*
8 Radhe Shyam Rungta: *The Rise of Business Corporations in India, 1851–1900*
9 Pamela Nightingale: *Trade and Empire in Western India, 1784–1806*
10 Amiya Kumar Bagchi: *Private Investment in India, 1900–1939*
11 Judith M. Brown: *Gandhi's Rise to Power: Indian Politics, 1915–1922*
12 Mary C. Carras: *The Dynamics of Indian Political Factions*
13 P. Hardy: *The Muslims of British India*
14 Gordon Johnson: *Provincial Politics and Indian Nationalism*
15 Marguerite S. Robinson: *Political Structure in a Changing Sinhalese Village*
16 Francis Robinson: *Separatism among Indian Muslims: The Politics of the United Provinces' Muslims, 1860–1923*
17 Christopher John Baker: *The Politics of South India, 1920–1937*
18 David Washbrook: *The Emergence of Provincial Politics: The Madras Presidency, 1870–1920*
19 Deepak Nayyar: *India's Exports and Export Policies in the 1960s*
20 Mark Holmstrom: *South Indian Factory Workers: Their Life and Their World*
21 S. Ambirajan: *Classical Political Economy and British Policy in India*
22 M. M. Islam: *Bengal Agriculture 1920–1946: A Quantitative Study*
23 Eric Stokes: *The Peasant and the Raj: Studies in Agrarian Society and Peasant Rebellion in Colonial India*
24 Michael Roberts: *Caste Conflict and Elite Formation: The Rise of a Karava Elite in Sri Lanka, 1500–1931*
25 John Toye: *Public Expenditure and Indian Development Policy 1960–1970*
26 Rashid Amjad: *Private Industrial Investment in Pakistan 1960–1970*
27 Arjun Appadurai: *Worship and Conflict under Colonial Rule: A South Indian Case*
28 C.A. Bayly: *Rulers, Townsmen and Bazaars: North Indian Society in the Age of British Expansion, 1770–1870*
29 Ian Stone: *Canal Irrigation in British India: Perspectives on Technological Change in a Peasant Economy*
30 Rosalind O'Hanlon: *Caste, Conflict and Ideology: Mahatma Jotirao Phule and Low Caste Protest in 19th Century Western India*
31 Ayesha Jalal: *The Sole Spokesman: Jinnah, the Muslim League and the Demand for Pakistan*
32 Neil Charlesworth: *Peasants and Imperial Rule: Agriculture and Agrarian Society in the Bombay Presidency, 1850–1935*
33 Claude Markovits: *Indian Business and Nationalist Politics 1931–9: The Indigenous Capitalist Class and the Rise of the Congress Party*

PEASANTS AND IMPERIAL RULE

Agriculture and Agrarian Society in the Bombay Presidency, 1850–1935

NEIL CHARLESWORTH

Lecturer in Economic History, University of Glasgow

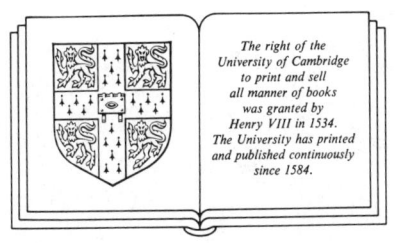

CAMBRIDGE UNIVERSITY PRESS
CAMBRIDGE
LONDON NEW YORK NEW ROCHELLE
MELBOURNE SYDNEY

Published by the Press Syndicate of the University of Cambridge
The Pitt Building, Trumpington Street, Cambridge CB2 1RP
32 East 57th Street, New York, NY 10022, USA
10 Stamford Road, Oakleigh, Melbourne 3166, Australia

© Cambridge University Press 1985

First published 1985

Printed in Great Britain at the University Press, Cambridge

Library of Congress catalogue card number: 84–9603

British Library cataloguing in publication data

Charlesworth, Neil
Peasants and imperial rule. –
(Cambridge South Asian studies; 32)
1. Peasantry – India – Bombay (State)
– History 2. Bombay (India: state) –
Rural conditions
I. Title
305.5′63   HD1339.I5

ISBN 0 521 23206 6

To Rex and Lavinia Charlesworth

CONTENTS

List of maps and tables		page ix
Preface		x
Note on technical terms and references		xi
Maps		xii

1 **Introduction: the peasant in India and Bombay Presidency**   1
    The Bombay Presidency, 1850–1935   10

2 **The village in 1850: land tenure, social structure and revenue policy**   17
    The Deccan and the Southern Maratha Country   19
    The Konkan   30
    Gujarat   34
    British revenue policy and the village   40
    The ryotwari system and its enemies   47
    Revenue policy and social structure in the Bombay village   65

3 **The village in 1850: land and agriculture**   70
    Credit and rural trade   82

4 **Indebtedness and the Deccan Riots of 1875**   95
    The official response   115

5 **Continuity and change in the rural economy, 1850–1900**   125
    Expansion and contraction, 1850–1880   134
    A renewed expansionary cycle, 1880–1896   142
    The famine problem, 1896–1900   155
    Continuity and change in the agrarian economy, 1850–1900   159

6 **The Bombay peasantry, 1850–1900: social stability or social stratification?**   162
    The evidence: deductions from patterns of economic change   165
    The evidence: land tenure and social structure   174
    Identifying the rich peasantry   192

7	**The agricultural economy, 1900–1935: the critical watershed?**	204
	Before the depression: expansion and inflation	210
	The agricultural depression and its legacy	225
	The problem of land ownership and organisation	231
8	**The impact of government policy, 1880–1935**	239
	The policy debates, 1880–1935	240
	The impact of the Deccan Agriculturists' Relief Act	252
	The impact of administrative action	260
	The influence of official policy	267
9	**The peasant and politics in the early twentieth century**	268
	'Poor peasant politics': tenant protest in the Konkan	271
	'Rich peasant politics' in the Deccan and Gujarat	276
	The Bardoli campaign of 1928: the middle peasant in politics?	283
	The peasantry and politics in the Bombay Presidency	289
10	**Conclusions: the problem of differential commercialisation**	292
	Glossary	301
	Bibliography	303
	Index	315

MAPS

1	The Bombay Presidency in the British period	page xv
2	Gujarat, showing British administrative divisions	xvi
3	The Deccan, showing British administrative divisions	xvii
4	The Southern Maratha Country, showing British administrative divisions	xviii
5	The Konkan, showing British administrative divisions	xix

TABLES

1	Ownership of land by Marwaris in certain villages of Poona District, 1874–5	105
2	Prices of jowar in three taluks, 1866–1910	132
3	Increases in numbers of carts in Satara District, 1860s–1890s	146
4	Estimated landownership by 'non-agriculturists': the land transfer enquiry, late 1890s	194–5
5	Estimated landownership by 'non-agriculturists': the record of rights enquiry in the Deccan, 1910s	197
6	Estimated landownership by 'non-agriculturists': the completed record of rights enquiry, 1917	197
7	Foodgrain price trends, 1900–22	209
8	Changes in district patterns of cotton cultivation, 1914/15–1924/5	216
9	Increases in land revenue in districts of Bombay during the late nineteenth century	265

PREFACE

This book has been a long time in the making. My work on the subject of the Bombay peasantry in the British period began when I was a research student at Cambridge in the early 1970s and has continued since then during my time at Glasgow. Over this period I have incurred many debts. Financially, I have been assisted by the Social Science Research Council, which sponsored my original research, Cambridge University, through the award of the Holland Rose Studentship, and Glasgow University, with a number of additional grants. I am grateful to librarians and archivists, in particular in four institutions: the Maharashtra State Archives in Bombay, the India Office Library and Records and the University Libraries of Cambridge and Glasgow.

Academically, I have benefited enormously from the fertile climate of debate in South Asian history which existed at Cambridge while I was there and also, since, from numerous valuable discussions with colleagues and friends at Glasgow. It would be invidious to name names but many may notice here the influence of their ideas and I am grateful to them. One academic debt is, however, so fundamental that it must be specifically recorded. The late Professor Eric Stokes first suggested the subject as an area of research to me in 1969 and thereafter as supervisor and friend constantly supported the work's development until his untimely death in February 1981. It is a source of deep regret to me that this book was not published during his lifetime. At least, I can record here the very great debt which I owe to him and to the example of scholarship and academic comradeship which he set.

I am very grateful to Mrs Blythe O'Driscoll, our departmental secretary, who typed the manuscript with notable calmness and efficiency. My wife and daughters, once again, provided essential support throughout and a congenial domestic atmosphere for writing. However, this book is dedicated to my parents, without whose consistent help and encouragement I would never have enjoyed an academic career.

Glasgow
December 1983

NOTE ON TECHNICAL TERMS AND REFERENCES

I have not followed any dogmatic scheme in the presentation of Indian language words and technical terms. I have, however, consistently sought to use the spelling and terminology which would seem most easily intelligible to modern readers in both India and Britain. Thus I have, for example, preferred the more natural 'jowar' and 'taluk' to the 'jowari' and 'taluka' used in the official literature. The same practice has been followed with all proper names.

All references are given in full at their first presentation. The following are the most common abbreviations used in reference:

IOL India Office Library and Records, London
MARD Maharashtra State Archives, Bombay, Revenue Department Papers
NS New Series
SBG *Selections from the Records of the Bombay Government*

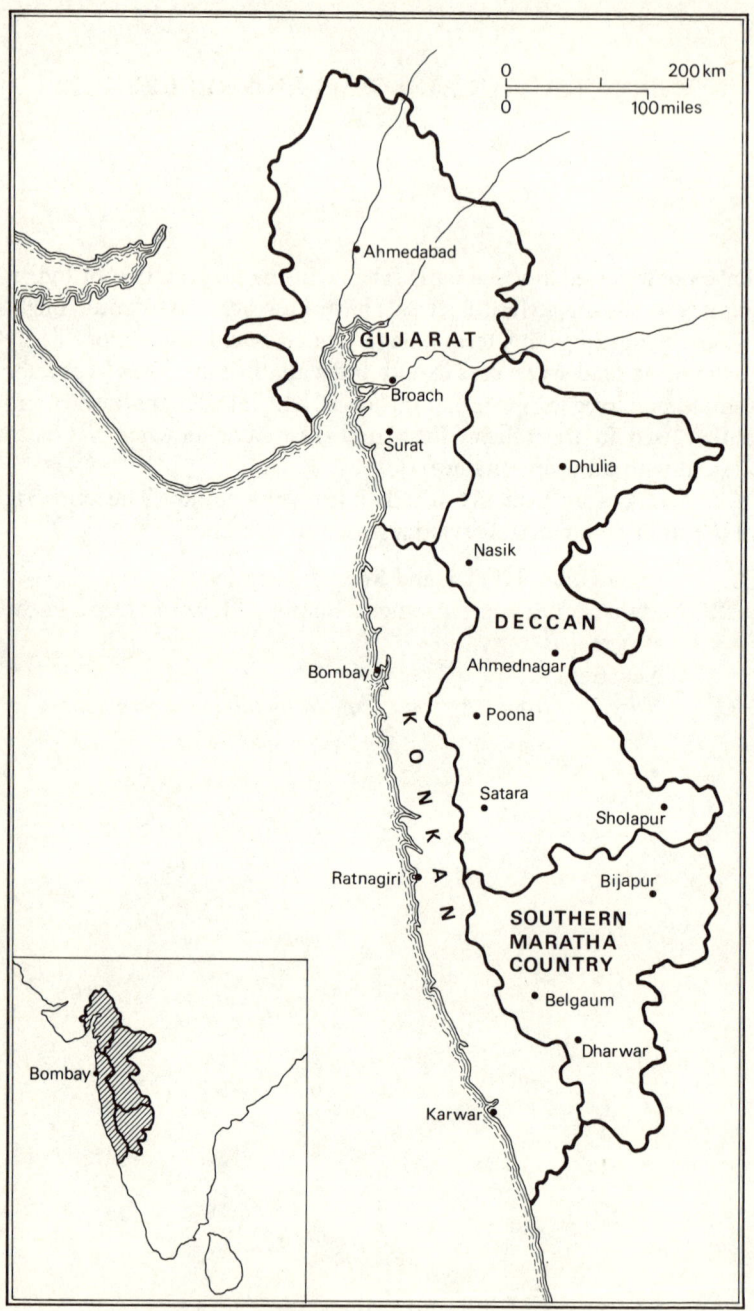

Map 1 The Bombay Presidency in the British period

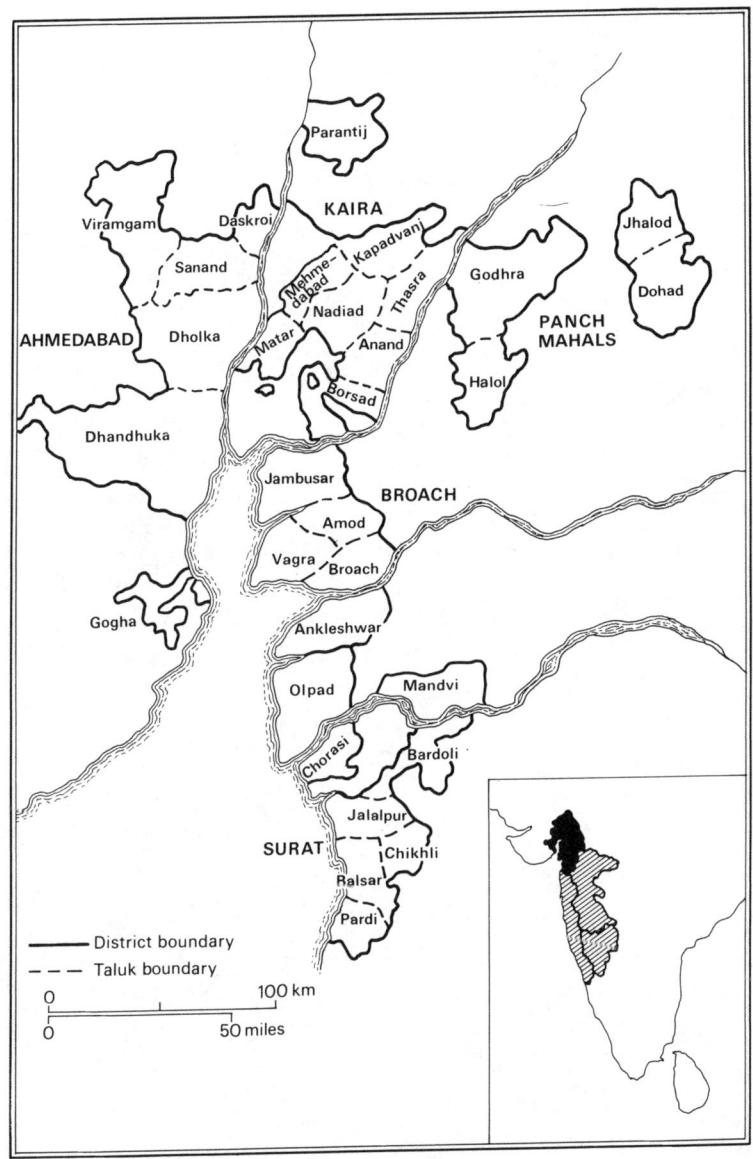

Map 2 Gujarat, showing British administrative divisions

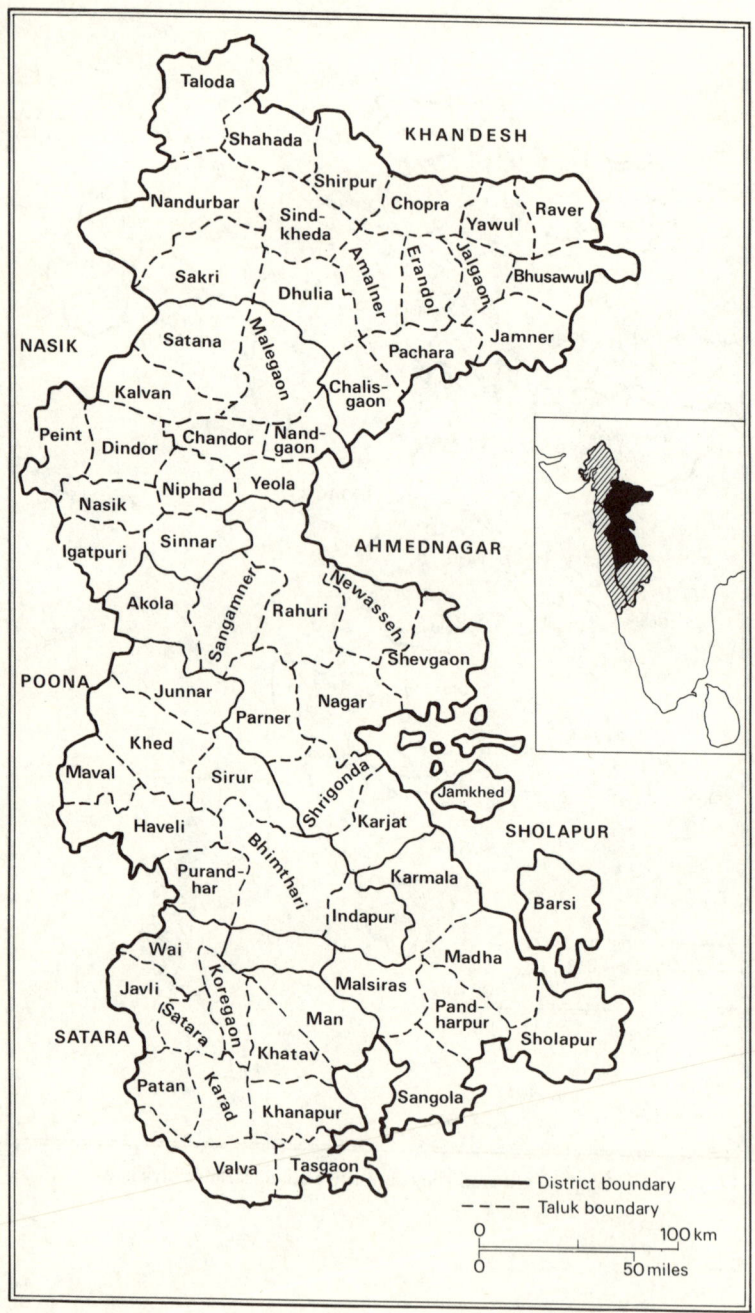

Map 3 The Deccan, showing British administrative divisions

Map 4   The Southern Maratha Country, showing British administrative divisions

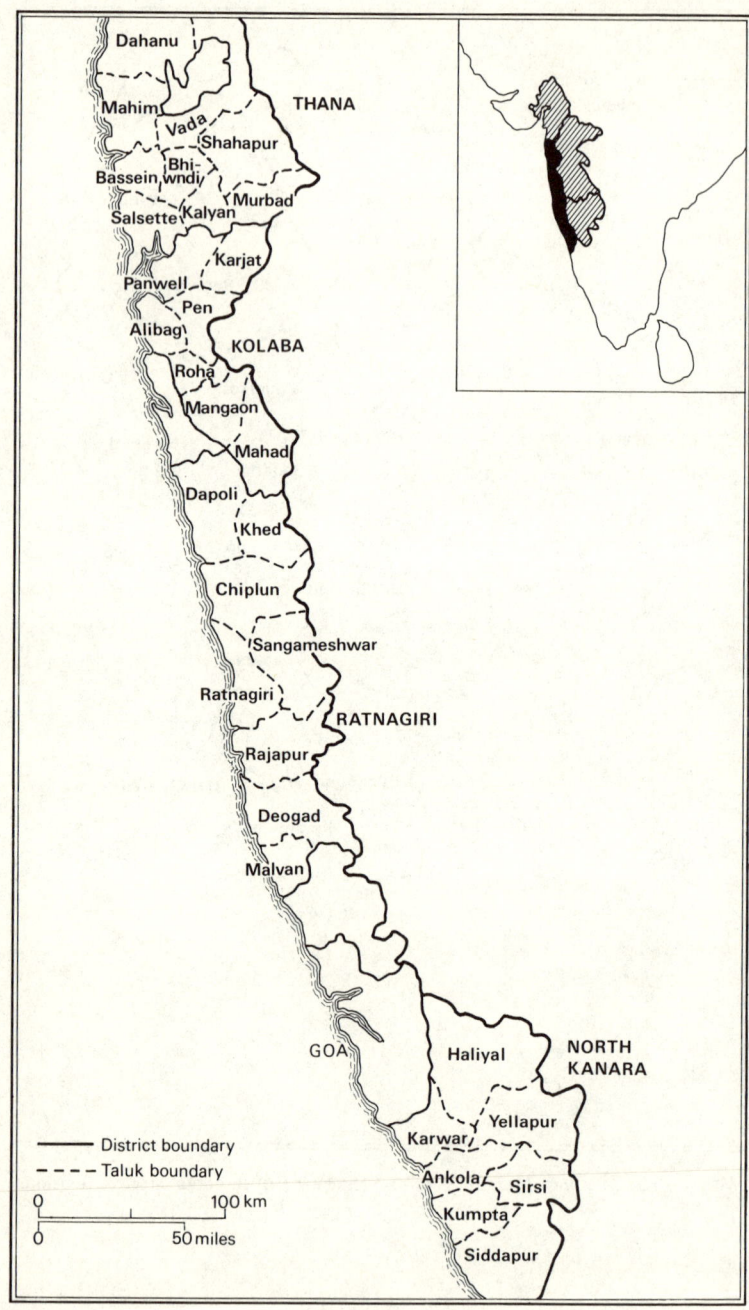

Map 5 The Konkan, showing British administrative divisions

1
Introduction: The peasant in India and Bombay Presidency

The peasantry is not just 'the awkward class' but also the typical class. Historically, most men's occupation has been within small-scale agriculture and examining change and development in this context is therefore vital to any realistic understanding of the evolution of the modern world. That is the root justification for this study of the agrarian economy and society of one major province of British India, the Bombay Presidency, in the period between the mid-nineteenth century and the 1930s. As a methodology, in the face of the daunting potential scope of the subject, three distinct areas of enquiry will predominate. Firstly, we need to investigate the economics of agriculture and changes within it; the financing of operations, the types of crops grown, the methods employed and, so far as possible, the levels and trends in output performance. Yet agriculture does not operate in a vacuum. Land tenure and the structure of power and status in the village determine its context. In turn, agricultural developments can create agrarian change: extending commercialisation of agriculture, for example, might prove an engine of revolution within traditional patterns of landownership, land tenure and peasant social relationships. Thirdly, in British India, the imperial rulers, dependent on the village for revenue and the mass acquiescence which guaranteed their political security, were always intimately concerned with rural issues. So our study must also, in part, be a study of British agrarian and revenue policy and its effects. From these subjects, hopefully, might emerge some answers about the dominant themes of the Bombay countryside's history: the genuine occurrence of economic diversification and change without, in developmental terms, the revolutionary consequences evident in many Western societies.

All the particular issues provide wide scope for speculation and debate. They were central to the polemics of the 'great tradition' of British Indian economic and social history – that of Marx, Curzon and Dutt – which painted contrary sweeping images of the beneficent diffusion of advanced economic techniques and civilisation or the impoverishment of India through a parasitic imperialism. If we are to 'judge' British rule, then the rural economy and society because of its vast numerical predominance must be the true test. Thus when Morris argued the thesis of 'rather substantial increases in total real output in the Indian economy in the nineteenth century'[1] he was primarily advancing an argument about agricultural performance. Yet assessing the balance of that performance is extremely difficult. Morris' case rests on a wide range of indicators: the steady expansion of the cultivated acreage in most provinces, the extensive development of commercial crops, notably cotton and jute, for export, the marked, sometimes spectacular, rises in crop and land prices. Further, it would be difficult to argue that the state siphoned off in taxation a disproportionately high share of any increment. Land revenue levels, even if excessive in the early years of British rule, probably failed to keep up with general price increases during the late nineteenth century – self-evidently so in permanently settled Bengal – and were then further cut down, in real terms, by the more rapid inflation of the period 1900–20.[2]

Any agricultural expansion in the nineteenth century, too, took place against a background of only slow aggregate growth in population. As late as the period between 1891 and 1921 the population within the boundaries of the present-day state of India increased merely from 235.9 million to 248.1 million, a rise of just 5.17 per cent over the thirty years.[3] Unspectacular increases in total overall output seemed, therefore, required to improve per capita performance perceptibly. This general pattern, though, provides a false illusion of stability. It masked quite swift short-term increases in population accompanied by periodic bursts of high mortality, caused by famine and disease. Bombay Presidency was strikingly subject to this process:[4] its population was sharply cut back by serious famine in the late 1870s and late 1890s and by the influenza epidemic at the end of the First World War. As Klein has argued,[5] it is hard to see buoyancy in this situation. Yet nineteenth-century famine may have been primarily the product of commercial and social change – for example, the creation of a newly vulnerable minority – and not an inevitable indicator of decline in aggregate output. Sluggish overall population growth, substantially created by occasional high mortality, is also the story of late

---

1 Morris D. Morris, 'Towards a Reinterpretation of Nineteenth Century Indian Economic History', *The Indian Economic and Social History Review*, 5, 1, March 1968, p. 11.
2 D. A. Washbrook, 'Law, State and Agrarian Society in India', *Modern Asian Studies*, 15, 3, July 1981, p. 671.
3 See *Census of India, 1951* (Delhi, 1953), Vol. 1, India, Part 1A, Report, p. 122.
4 In the Visarias' 'West Zone' the population growth rate was notably below the all-India norm between 1871 and 1941. Leela and Pravin Visaria, 'Population', in Dharma Kumar, ed., *The Cambridge Economic History of India*, Vol. 2 (Cambridge, 1983), p. 490.
5 Ira Klein, 'Population and Agriculture in Northern India, 1872–1971', *Modern Asian Studies*, 8, 2, April 1974, pp. 191–216.

Tokugawa Japan and here many have drawn from this data optimistic conclusions about long-term agricultural trends.[6]

After about 1920, however, Indian population started to grow much more quickly and consistently. Output and earnings now, apparently, required rapid expansion even to maintain per capita levels and in the late 1920s the onset of depression struck a severe blow at commercial agriculture. Even so, in the inter-war years there was no major famine in India, no fundamental crisis of subsistence. So Morris' case might be extended over the whole British period, not just the nineteenth century. Indeed one could argue that, by the sort of 'test' which is often applied to agriculture's role in Japanese or British industrialisation, agriculture in British India played a stimulatory role within the economy. It provided government with a major revenue raiser in the form of the land tax. Its cash-crop exports were a vital earner of foreign exchange. Further, the famine evidence of the late British period does not suggest automatic decline in per capita foodgrain availability.[7]

However, a less sanguine interpretation of events can be presented. Understanding the meaning of population changes and famine incidence is undoubtedly a difficult matter. Population increase, for example, seems to have begun to accelerate first in Bengal, the very province where pressure on land was, in places, already acute and where expansions of the cultivated acreage, by the late nineteenth century, were much less extensive. A marked deterioration of the land–labour ratio apparently occurred here after 1870 at a time when, we are told, productivity was not

---

6 Writers such as T. C. Smith, Kozo Yamamura, Susan Hanley and E. S. Crawcour have created a new orthodoxy on Tokugawa Japan that change and growth in agriculture throughout the Tokugawa period (1600–1868) firmly laid the foundation for Meiji development. Few would dispute the thesis of expansion in the first half of the period, but between the 1720s and the mid-nineteenth century the evidence is complex. Unlike in the seventeenth century, cultivated acreage was now expanding very little. Overall population growth, too, became minimal, partly, the writers argue, because of conscious restrictions on size of family but also, clearly, through the persistence of famine down to the 1830s. The argument, which might also be applied to British India, remains that, even with slowly improving productivity, the low population growth in aggregate terms meant rising per capita output, whilst social change and extending commercialisation diversified the products of agriculture. For this interpretation, see particularly T. C. Smith, *The Agrarian Origins of Modern Japan* (Stanford, 1959), and Susan B. Hanley and Kozo Yamamura, 'Population Trends and Economic Growth in Pre-industrial Japan', in D. V. Glass and Roger Revelle, eds., *Population and Social Change* (London, 1972), pp. 451–99. However, for a challenge, see also Seymour Broadbridge, 'Economic and Social Trends in Tokugawa Japan', *Modern Asian Studies*, 8, 3, July 1974, pp. 347–72.

7 For a fuller statement of the general argument that agriculture's role in British India's economic development was less inhibitive than traditionally assumed, see my *British Rule and the Indian Economy, 1800–1914* (London, 1982).

improving.[8] Yet Bengal, too, avoided famine until the disaster of 1943 complicated by the dislocations of the war: can its peasants genuinely have avoided a decline in per capita mass consumption? On the other hand, nineteenth-century famine was no respecter of traditionally more prosperous regions. One possible pointer to agricultural decline was the striking of serious crises in areas hitherto apparently immune. In western India, for example, the famines of the late 1890s affected not just the 'famine belt' east Deccan, where rainfall was always unreliable, but 'secure' parts of Gujarat, the Konkan and the Southern Maratha Country. This may reflect particular climatic or ecological problems or, equally, could suggest fatal flaws in the process of agricultural expansion. Much of the new land brought under the plough during the British period may have been marginal. Officials sometimes commented that capital and labour spent on the clearing of waste land might have been more profitably invested in better techniques applied to land already cultivated.

One problem with the argument for agricultural buoyancy lies in explaining the impetus behind any supposed increase in productivity. The technology of most Indian agriculture changed little. The provision of new canal irrigation facilities was the greatest source of improvement but, except in the Punjab and parts of the United Provinces, was too scantily spread to revolutionise peasant production. Attempts to generate more capital in agriculture through co-operative societies or government action against moneylenders apparently enjoyed patchy success. Yet the pessimistic conclusions of Blyn's quantitative study,[9] based on the agricultural statistics from the 1890s to 1947, have provided no definitive conclusion to the argument. The deficiencies of the output figures are serious. The basic mechanism of collection was open to serious abuse and miscalculation and subjective assessments, as on 'quality of season', which so shaped many provinces' statistics, may have been significantly influenced by political pressures on local administrations.[10] At the broadest all-India level, statistical enquiry has borne some fruits, but

---

8 Rajat Ray claims that in Bengal between 1901 and 1921 cultivated acreage actually decreased whilst the population it was required to support went up by 10.8 per cent. See Rajat Ray, 'The Crisis of Bengal Agriculture 1870–1927 – the Dynamics of Immobility', *The Indian Economic and Social History Review*, 10, 3, September 1973, pp. 244–79.
9 George Blyn, *Agricultural Trends in India, 1891–1947: Output, Availability and Productivity* (Philadelphia, 1966).
10 On the deficiencies of the collection procedure, see Clive Dewey, 'Patwari and Chaukidar: Subordinate Officials and the Reliability of India's Agricultural Statistics', in Clive Dewey and A. G. Hopkins, eds., *The Imperial Impact. Studies in the Economic History of Africa and India* (London, 1978), pp. 280–314. For the influence of subjective factors on the Bombay agricultural statistics see Alan W. Heston, 'Official Yields Per Acre in India, 1886–1947: Some Questions of Interpretation', *The Indian Economic and Social History Review*, 10, 4, December 1973, pp. 303–32.

detailed provincial disaggregation for the nineteenth century remains intractable.[11] Deduction from the weight of empirical evidence, then, rather than dependence on statistics will be the main methodology of our examination of agricultural trends.

Agrarian change – shifts in patterns of landownership and tenure and its consequences – has equally proved the subject of radically different interpretations. Current orthodoxy now typically stresses continuity: that, whatever the illusion of tenurial turmoil, landowning elites and peasant cultivators in most localities came from substantially the same groups in 1947 as in 1850. The Sahotas of Kessinger's Vilyatpur,[12] Stokes' peasant elite[13] remain throughout the dominant figures in their local agriculture. In the U.P., again, 'the stability of the great estates in the post-1860 period' is 'striking',[14] whilst Dharma Kumar's Madras exhibits 'very little change in the degree of inequality between 1853–54 and 1945–46'.[15] Change of personnel, of course, occurred, sometimes abruptly. Undeniably ownership of many estates in Bengal changed hands following the Permanent Settlement of 1793 but even here, apparently, 'many of the land transfers were made to relatives, dependents and former employees of the old zamindars'.[16] There was, then, no revolution in the real substance of agrarian social structure.

Undoubtedly this interpretation has substantial attractions. Against it the traditional recital of alleged social change on the land – growing peasant indebtedness and dispossession, land transfer to 'new' proprietary groups, the creation of a pauperised agrarian proletariat – looks simplistic. The latter drew its authority from the literature created by British officialdom but, arguably, much of this mistakenly accepted institutional

---

11 Thus it seems clear now that overall per capita national income rose, to some degree, in the century before 1920. For a survey of the assessments, see A. Heston, 'National Income', in Kumar, ed., *The Cambridge Economic History of India, Vol. 2*, pp. 376–462. At the same time, however, *The Cambridge Economic History* conspicuously fails to provide any detailed statistical coverage of long-term agricultural trends in the provinces.
12 Tom G. Kessinger, *Vilyatpur 1848–1968* (Berkeley, Los Angeles and London, 1974).
13 Eric Stokes, 'The Structure of Landholding in Uttar Pradesh, 1860–1948'. *The Indian Economic and Social History Review*, 12, 2, April–June 1975, pp. 113–32. Reprinted as Ch. 9 of *The Peasant and the Raj. Studies in Agrarian Society and Peasant Rebellion in Colonial India* (Cambridge, 1978).
14 Ibid., p. 123.
15 Dharma Kumar, 'Landownership and Inequality in Madras Presidency: 1853–54 to 1946–47', *The Indian Economic and Social History Review*, 12, 3, July–September 1975, p. 242.
16 John R. McLane, 'Revenue Farming and the Zamindari System in Eighteenth Century Bengal', in R. E. Frykenberg, ed., *Land Tenure and Peasant in South Asia* (New Delhi, 1977), p. 29.

change at its face value and failed to see the subtle adaptation to continuity underneath. Hence the alien intruder into landownership, the spectre raised by such as Thorburn, has now been largely replaced in the literature by the successful representative of established peasant groups. Further, even where the latter is seen as expanding ownership, a revisionist view of the economic and social consequences is still possible. Physical eviction of poor peasants was by no means automatic. Often, where a peasant 'lost' legal proprietorship of his holding, he remained its cultivator: the landlord was not going to take to the plough himself. Many 'land transfers' were really the conversion of a long-established creditor-debtor relationship into one of landlord and tenant. This may be simple adaptation to the Westernised legal forms introduced by British rule, enabling the creditor to protect his interests more effectively in the new institutional climate. If so, there is clearly continuity in the economic relationship: the cultivator remains the same person and he continues to render a proportion of his proceeds to a figure who, according to taste, can be seen as patron or exploiter.

The new orthodoxy of agrarian continuity also accords with political realities and developments. On late Tsarist Russia Shanin has argued that the interpretation, founded by Lenin, of social revolution on the land little accords with observed historical events.[17] If Russian agrarian society had really become stratified – 'depeasantised' Lenin had called it – into a commercial elite and a mass depressed proletariat, then why was there not political evidence of class tension, why did the poor peasant masses not rise up against their oppressors? In India too, it is hard to correlate the complexion of peasant political behaviour with, say, the argument for the creation of a large new class of landless labourers. Rural campaigns became a vital adjunct to nationalist activity after Gandhi's rise, but they hardly suggested cross-currents and tensions based on violent economic differentiation. Unlike perhaps in China, there was never political breakdown in the countryside, never the development of widespread peasant alienation from the social and political system. Transitions in political power, such as Reeves' demise of the traditional landlords in the U.P.,[18] tended to be accomplished smoothly. In sum, the wider history of India's rural society suggests the continued involvement and control of established groups rather than the creation of a new peasant proletariat which would have had nothing to lose in turbulence and revolution.

17 Teodor Shanin, 'Socio-economic Mobility and the Rural History of Russia, 1905–1930', *Soviet Studies*, 23, 1971–2, pp. 222–35, and *The Awkward Class* (Oxford, 1972).
18 P. D. Reeves, 'Landlords and Party Politics in the United Provinces, 1934–37', in D. A. Low, ed., *Soundings in Modern South Asian History* (London, 1968), pp. 261–93.

Continuity, therefore, makes much sense. Yet one might wonder whether social historians' praiseworthy zeal to stress the adaptability of established structures and groups now requires such prominence: some emphasis might return to the developments which called forth the need to adapt. The economic historian is bound to be struck by the extensive and powerful innovations of British rule: the expansion of the cultivated acreage and of cash-crop agriculture, the improvements in communications and the rising value of and demand for land. Many of these economic forces had widespread consequences. One specific example can be given: by the early twentieth century the size of units of cultivation indisputably had fallen in many regions. Despite first appearances, this did not necessarily mean that overall land availability per cultivator was deteriorating, for the most striking factor – at any rate in Bombay – appears to be the growing division of peasant holdings into a larger number of individual plots.[19] This very process of fragmentation, however, must have had considerable effects on agriculture and its organisation. It apparently limited the scope for technological change involving economies of scale. On the other hand, small plots which could have made only a marginal contribution to family subsistence needs may have encouraged trends towards cash-cropping and provided a powerful force for the fuller incorporation of the peasant within the market economy.

The Indian rural economy and society, therefore, have to be seen, whatever their underlying stability, as entities undergoing constant economic flux and change. One obvious example can be given of the rapidity with which changes can occur in quite basic agricultural patterns: under the impact of the American Civil War, total cotton acreage in the Bombay Presidency leapt up from just 1.00 million acres in 1860–1 to 1.98 million acres by 1869–70 only to decline again during the 1870s.[20] Here alone is an important story. The mechanics of the developments must have been complex: the financing of the expansion, the degree of commitment to the market and its organisation and, later, the process of reducing the scale of cotton cultivation. These issues themselves suggest another, more fundamental, question. Why did the whirlpool of flux, such as that created by the American War cotton boom, fail to produce revolutionary long-term structural change? Manifestly the Indian agriculturist during the British period was neither depeasantised nor simply converted into a large capitalist farmer. For the Indian countryside, too, cash-crop export was not the route to substantial economic development

19 For a fuller discussion of this phenomenon, see Chapter 7.
20 See my 'Agrarian Society and British Administration in Western India, 1847–1920' (University of Cambridge, unpublished Ph. D. thesis, 1974), Ch. 2.

which it provided for the white dominions of the British Empire. At the same time, there was so little question of mass exodus from the countryside that industrial historians can sometimes speak of labour supply as an important problem of Indian industrialisation. Flux and change, then, as the modern orthodoxy has it, was accommodated to the rural environment.

What did British policy contribute to the pattern? The traditional historiography of Indian history, whatever the radical differences of interpretation, was based on a common assumption about the power of the imperial rulers. Whether the results were good or bad, British policy decisions were seen as automatically seminal to developments in the indigenous economy and society. Modern scholarship has delighted in undermining these assumptions: from all-powerful pro-consuls of European civilisation and capitalism, British officials have become baffled victims of a complex society. The extreme, reached perhaps in Frykenberg's *Guntur District*,[21] is to view particularly revenue policy as the plaything of powerful local forces, indeed not as a 'policy' at all in the sense of some organised centrally directed momentum. Decisions, it is argued, had to be taken entirely pragmatically if the vital land revenue which paid for British rule was to emerge from the 'silent village'.[22] Even then, it only emerged fitfully and on conditions dictated by local society.

This view is right to stress the weaknesses of British power. The most obvious is the simple numerical position, which left the small number of Europeans in the countryside as overlords of a large Indian-staffed bureaucracy which could evolve its own compromises on the ground. Policy was clearly adapted greatly by interaction with local conditions. However, it is one thing to stress the problems policy faced and its shaping by rural reality in implementation and quite another to imply the absence of any active role. Policy evolution and its complex effects should remain one of the fundamental organising principles of British Indian history. On this, as on the vitality of economic and social change, the pendulum is ready – and needs – to swing back somewhat.[23]

The argument can be amplified by looking at two broad areas. Nobody now would see Indian society at the British takeover as some empty canvas to be filled by the new rulers or as so riven with 'anarchy' that

---

21 See R. E. Frykenberg, *Guntur District, 1788–1848: a History of Local Influence and Central Authority in South India* (Oxford, 1965).
22 The phrase is Frykenberg's. See ibid., p. 230.
23 There are some signs of this occurring in the sections on 'Agrarian Relations' in Kumar, ed., *The Cambridge Economic History of India, Vol. 2.*

traditional social forms and structures had dissolved. The British inherited a sophisticated rural economy and society and the stream of continuity in their development continued across the changeovers of power. Nevertheless, the annexations of each of the major provinces provided that crucial moment in their histories – equivalent to 1789 in France and 1917 in Russia – when, with a radical change of government, the new rulers, for all their weakness and lack of local power and knowledge, possess the self-confident vitality to try to impose their ideals. The formation of the land revenue 'systems' was a product of this. However much they were altered or reduced to a lowest common denominator of pragmatic tax gathering on the ground, a momentum of applied principles still remained. It is the interaction between that momentum and local conditions which provides one of the dominant influences on Indian agrarian development.

Again, Indian governments have been characterised frequently as laissez-faire in their economic and social policy. Yet, compared with other imperial societies and the Third World in general in the nineteenth century, the intervention of government in rural affairs is striking. This was the product not only of the security fears of foreign rulers but also of the inundation of official reports and commissions which brought problems to light and of the ideology of administrators who needed to try to 'improve' society to justify their existence. Bombay officialdom made its own unique contribution here. In 1879 the Bombay government enacted the Deccan Agriculturists' Relief Act: this attempted to ameliorate alleged exploitation of poor peasants by regulating credit and debt transactions between moneylenders and agriculturists in the Deccan districts, a direct interference in village relationships. Of course, the effects of such agrarian legislation in India were complex. Occasionally it proved so totally untuned to the actualities of rural life that the impact was superficial. Invariably the results were far from what the rulers expected. Even so, the chemistry of reaction between official activity and local reality typically produced a blend of forces which could prove highly influential in shaping agrarian society. For this reason, government policy will provide an important element in our story.

This brief review will have served its purpose if it introduces to the reader the broad issues which underlie this study. Already he may have realised that the empirical basis for many of the arguments is still not very substantial. It may be that, granted the inadequacies of Indian statistics, definitive statements on long-term provincial economic trends will never be safely made. Again, local variation, particularly in social structure and change, may be so intense as to invalidate much generalisation even at regional level. However, one way forward, arguably, is for more provincial case studies of economic and social development in the countryside. This is an attempt to provide one.

The Bombay Presidency, 1850–1935

The choice of region for study reflects the belief in the importance of British government and its processes, for the Bombay Presidency had no obvious overriding geographical or cultural unity. It was an administrative entity, solely the creation of British rule and replaced since Independence by more logically constructed, linguistically defined states.[24] In fact Bombay was arguably the most heterogeneous of all the British Indian provinces, the territories it administered including Aden and the province of Sind, a major region of modern Pakistan. However, the area discussed in this book is the 'Bombay Presidency proper', that is excluding Sind and the overseas territories but including all parts of the British Presidency contained within the modern state of India.

This region was formed exclusively by the pattern of conquest. A British stake in western India stemmed back to Charles II's marriage treaty and the early trading posts at Salsette Island and Surat remained of commercial importance throughout the eighteenth century. However, the bulk of the Bombay Presidency was created, relatively swiftly, by two major early nineteenth-century blocks of conquest: the districts of Gujarat won in 1802–3 and the Deccan and Konkan regions acquired at the fall of the Poona Peshwa in 1818.[25] The unit so formed comprised most of the western coastline of the sub-continent from the wastes of Kutch to beyond Goa, plus the inland territory whose natural administrative centre was Poona. It never wandered more than 200 miles from the coast and the boundaries did not follow any notable geographical features. In addition, the Presidency's misshapen form was compounded by the existence of several large princely states like Kolhapur and Baroda eating into the British-ruled districts: in 1872 the total area and population of the Bombay princely states amounted to well over half the totals for the British Presidency.[26] Within the Bombay Presidency, only a common basic religion – nearly four-fifths of the population were Hindus in

---

24 The modern state which grew out of the British Bombay Presidency was Maharashtra, designed to encompass all major Marathi-speaking areas including some not previously in Bombay. In 1960 Gujarat was separated from Maharashtra and constituted as a distinct state, thereby recognising the fundamental cultural and linguistic divide within western India.
25 Some additions were made later; for example, Satara was one of Dalhousie's annexations.
26 See *Census of the Bombay Presidency, 1872* (Bombay, 1875), Part 2, para. 627.

1872[27] – papered over deep cultural and linguistic differences. The fundamental distinction was between the Marathi-speaking regions of the Deccan and Konkan, and Gujarat with its own language and traditions. Why, then, examine so diverse and apparently ill-integrated a region? The unity imposed by a common administration in the British period is not the sole justification: the political history of the different districts had been interlinked long before. Shivaji's empire had brought much of the region under one administration and the influence of its officials, mainly 'the Brahman and Prabhu castes of the Maratha desh', had been 'diffused throughout the area of Maratha power'.[28] Chitpavan Brahmins, the most ubiquitous and powerful of these groups, could be found as administrators in the eighteenth century in districts as far-ranging as Baroda and Kolhapur.[29] At the same time, traders too migrated between Gujarat and Maharashtra.[30] By 1850 the ubiquity of the Gujarati Vania or, from further north, the Marwari as moneylender and dealer in many a Deccan village matched that of the Chitpavans and the Deshastha Brahmins as civil servants. These were regions, then, with a tradition of inter-change and association. To this was added some social affinity between the established proprietary and cultivating groups on the land over much of the Presidency. The vast Kunbi caste in most areas epitomised the western Indian peasant farmer. The 1881 Census enumerated over 52 per cent of all the Presidency's agriculturists as 'Kunbis' and the allied Gujarati 'Kanbis' provided a further nearly 5 per cent.[31] Underneath this broad umbrella, it is true, the status and claims of the sub-units in the different localities varied subtly. In the central Deccan the 'Maratha' Kunbis represented the generality of proprietary cultivators; the 'Patidar' Kunbis of Gujarat comprised a more exclusive elite.[32] Even so, the extent of the landholdings of those who called themselves some type of 'Kunbi' gave the pattern of ownership throughout most of western India some semblance of homogeneity.

This, then, is the justification for studying the province: common political traditions – though the force and impact of Maratha rule varied widely from locality to locality and provides a major reason for variations in social history – and some social cohesion, completed, of course, by the

---

27 Ibid., para. 228.
28 John Roberts, 'The Movement of Elites in Western India under Early British Rule', *The Historical Journal*, 14, 2, June 1971, p. 243.
29 Gordon Johnson, *Provincial Politics and Indian Nationalism. Bombay and the Indian National Congress 1880–1915* (Cambridge, 1973), p. 58.
30 Roberts, 'The Movement of Elites in Western India', p. 242.
31 *Census of India, 1881* (Bombay, 1882), Bombay, Vol. 2, Tables, Appendix C, Table 3.
32 On the Patidars, see David F. Pocock, *Kanbi and Patidar. A Study of the Patidar Community of Gujarat* (Oxford, 1972).

British creation of one united political and administrative entity. Otherwise, there is diversity and the opportunity to examine the impact of a common administrative and revenue organisation on regions different geographically and agriculturally. Like Washbrook's Madras, the Bombay Presidency contained distinct areas of wet-crop and dry-crop agriculture.[33] Throughout the southern bulk of the Presidency the division was created by the backbone ridge of the Western Ghat mountains running north–south some fifty miles from the Arabian Sea. The coastal strip and the Ghat spurs invited rain but, in so doing, they robbed the regions further east. As a rule in Bombay, the further east the less it rained. At Ratnagiri on the coast an average of over 101 inches of rain fell in each of the 28 years ending in 1878.[34] At Sholapur in the East Deccan, in contrast, the average annual rainfall for the period 1873–82 stood at just over 31 inches.[35] These climatic variations ensured fundamental differences in agricultural and social systems. On the coast, even though grain cultivation was widespread, wet-crop agriculture and a relatively large population could be sustained. The further east one went, progressively lower population density, larger units of landownership and greater dependence on millet foodgrains were necessary for peasant agriculture to survive.

The progression of this change was evident from taluk to taluk, occasionally even from village to village. Nevertheless, the geographer can generalise and firmly point to four basic regions within the Bombay Presidency. The largest was the great plateau-land to the east of the Ghat spurs, the Deccan. In the British period the Deccan comprised six administrative districts: Khandesh and Nasik in the north, Ahmednagar, Poona, Satara and Sholapur. Here the low rainfall of typically 20 to 40 inches per year was matched by soil conditions of variable quality. As a result, the Deccan was supremely a millet-growing area: higher quality cereals like wheat were less common than in north India and 'probably no region of the sub-continent, comparable in size, has so little rice'.[36] Jowar and bajra, the most important varieties of local millet, therefore

---

33 Washbrook has made the distinctions between the areas of wet-crop and dry-crop agriculture a crucial determinant of differential patterns of social structure and change in his accounts of the Madras Presidency. See D. A. Washbrook, 'Country Politics: Madras 1880–1930', *Modern Asian Studies*, 7, 3, July 1973, pp. 155–211, and *The Emergence of Provincial Politics. The Madras Presidency 1870–1920* (Cambridge, 1976).
34 *Gazetteer of the Bombay Presidency* (hereafter *BG*), Vol. 10, *Ratnagiri and Savantvadi* (Bombay, 1880), p. 23.
35 *BG*, Vol. 20, *Sholapur* (Bombay, 1884), p. 6.
36 O. H. K. Spate, *India and Pakistan. A General and Regional Geography* (London, 1954), p. 648.

dominated the fields of the Deccan. In Sholapur District, for example, in the early 1880s, of just under 1.6 million acres under tillage jowar occupied 950,000 acres and bajra 208,000 acres.[37] In an east Deccan district like Sholapur, too, the low population density relative to other regions of the Presidency was especially marked: Sholapur's figure published in its *Gazetteer* of 1884 was 128.84 persons per square mile[38] compared with contemporary norms of over 200 for the coastal districts and Gujarat.

Unreliable climate and, frequently, poor soil presented constant problems for Deccan agriculture. The east, where rainfall was lowest, ran a continual risk of famine. Yet the region was traditionally no downtrodden backwater. Politically it was the heartland of the old Mahratta empire and in the British period too, its natural administrative centre of Poona provided a more logical lynchpin for western India than peripheral, cosmopolitan Bombay. Socially, the peasant elite who staffed Shivaji's army had built up administrative power and privilege for themselves in the village at home. Agriculturally, too, low population density meant, at least, the potentiality for relatively large peasant holdings. The *Gazetteer* enquiries quoted what seem astonishingly large average areas for holdings in Deccan districts: 48 acres in Sholapur District was the extreme example[39] but even in the less marginal Ahmednagar District 59 per cent of holdings were categorised as being over 10 acres in size.[40] Whatever the unreliability of these statistics,[41] it is indisputable that land was, in absolute terms, hardly under great pressure from population in the Deccan of the early British period.[42] Typically, land for cultivation was freely available, particularly so in the frontier northern district of Khandesh. To this extent, the Deccan of 1850 was a region of some agricultural opportunity, though at existing levels of technology a rapid upswing in population growth might, in most localities, soon press on the far from limitless resources of the region.

These characteristics – of standing in 1850 at some sort of economic crossroads – also affected the Presidency's second region, the Karnatak or

---

37 *BG, Vol. 20, Sholapur*, p. 229.
38 Ibid., p. 23.
39 Ibid., p. 218.
40 *BG, Vol. 17, Ahmednagar* (Bombay, 1884), pp. 244.
41 Not only are the absolute sizes of holdings quoted sometimes difficult to believe, but also the differences from district to district. One suspects that the appearance of drastic decline in average sizes of holding over the late nineteenth and early twentieth centuries is partly created by over-estimation at the beginning of the period.
42 Fukazawa speaks of the 'abundance' of land in pre-British Maharashtra. H. Fukazawa, 'Agrarian Relations: Western India', in Kumar, ed., *The Cambridge Economic History of India, Vol. 2*, p. 179.

'Southern Maratha Country'. As the latter name implied, this was the region immediately to the south of the Deccan and it comprised the three districts of Belgaum, Dharwar and Bijapur. Geographically it provided 'a transitional zone between Maharashtra and Mysore'.[43] Politically and commercially, the Southern Maratha Country had to look firmly northwards or southwards for its administrative centre and trading outlets, and the links had developed with Maharashtra. Poona and then Bombay became the region's political overseer and the ports of Ratnagiri District were by 1850 long-established as the destination for exports of produce. Agriculturally, the Southern Maratha Country mirrored Deccani patterns with, perhaps, the contrasts painted more starkly. The region shared the problems of living in the Ghats' rain shadow and the eastern district of Bijapur, like the areas of the east Deccan, risked serious famine when rainfall levels fell below the norm. Yet west of Bijapur climate and soil conditions were much more favourable. Both Belgaum and Dharwar districts were suitable for widespread cultivation of cotton and in 1850 the Dharwar variety was among India's most successful.

If the Deccan and the Southern Maratha Country share common characteristics, then the long, narrow coastal strip of the Konkan region provides a radically different case. In 1850 there were four districts in the British territory here – Thana, Salsette, Kolaba and Ratnagiri – and these were joined by a fifth in 1862 when the southern district of North Kanara was transferred from the Madras Presidency. In a number of senses the Konkan was an 'old' region. Its land was already in 1850 intensively cultivated, its Brahmin elites were well accustomed to political and administrative power throughout western India and its commercial routes, both coastal and across the Arabian Sea, were long-established. In the south of the region, grain cultivation, as in the rest of the Presidency, was important, though the major crops here were the coarse grains like harik rather than the millets of the Deccan.[44] North of Bombay City, however, patterns of wet-crop cultivation predominated: in Thana District 63.9 per cent of the cultivated acreage was growing rice in the late 1870s.[45] Throughout the region its intensive agriculture had become capable of and accustomed to supporting large numbers by the mid-nineteenth century. Ratnagiri's figure for population density – 269 persons per square mile[46] – was the largest revealed by the *Bombay Gazetteer* enquiries of the 1870s and 1880s. This meant, typically, small, well-

---

43 Spate, *India and Pakistan*, p. 659.
44 Harik was the major crop of Ratnagiri District. See *BG, Vol. 10, Ratnagiri and Savantvadi*, p. 146.
45 *BG, Vol. 13, Thana*, Part 1 (Bombay, 1882), p. 286.
46 *BG, Vol. 10, Ratnagiri and Savantvadi*, p. 2.

worked holdings: in both Ratnagiri and Thana over half were of not more than 5 acres in the late 1870s.[47] Productivity in Konkani agriculture was traditionally relatively high but the region faced a problem, severe by Bombay standards, of relative scarcity of land for expansion.

The three regions described so far all had, between themselves, some natural geographical unity. Gujarat in the north was different. Geologically the alluvial and deltaic characteristics of much of its land linked it more with Sind and the north rather than the bulk of Peninsula India. These conditions often provided, particularly in the valleys of the Tapti and Narbada rivers, relatively rich dark soil which enabled dry-crop agriculture to support a large population without an especially favourable climate. Gujarat boasted near-Konkani levels of population density with a rainfall typical of many parts of the Deccan. Ahmedabad District's *Gazetteer* population density figure slightly exceeded, at 215 persons per square mile,[48] that of Thana District. At the same time an average of just 32 inches of rain a year fell on Ahmedabad City between 1870 and 1878.[49] However, the greater potentiality of the soil was shown by the cropping patterns: in Ahmedabad District jowar and bajra came second to wheat in extent of cultivation.[50] In 1850 cash-crop agriculture was probably pursued more extensively in Gujarat than in any other region of the Bombay Presidency. Cotton was, as in the south, the most important commercial crop: criticism from British cotton interests of the East India Company's alleged failure to develop the region, particularly its communications, reflected a strong mid-nineteenth-century belief that Gujarat could become one of the major suppliers of raw cotton to British industry. Other commercial crops suitable for cultivation in Gujarat included tobacco, grown in Kaira District, and sugarcane produced in Surat. The four British districts of central Gujarat – Ahmedabad, Kaira (or Kheda), Surat and Broach[51] – therefore boasted in 1850 a mixed agriculture, potentially vulnerable in the peripheries to famine but more diversified in production than the other regions of western India.

Overall, a recurring relationship is evident between the regions' different geographies, ecologies and climates and their agriculture in the early British period. Some equilibrium had long been struck and up to 1850 was usually maintained between what a region could support and its

---

47 See ibid., p. 145, and *BG, Vol. 13, Thana*, Part 1, p. 284.
48 *BG, Vol. 4, Ahmedabad* (Bombay, 1879), p. 2.
49 Ibid., p. 20.
50 Wheat covered 26–37% of the total cultivated acreage as against 24–26% and 18–21% for jowar and bajra respectively. See ibid., p. 54.
51 A fifth district, the outlying Panch Mahals District, was added in 1861.

types of crop, its levels of population and size of holding. To make a sweeping generalisation, in 1850, although there were everywhere extremes of relative wealth and poverty, average total output per typical unit of peasant ownership probably did not vary very substantially throughout the Presidency. Yet this did not mean that average living standards were necessarily the same nor, especially, that subsequent historical development was likely to maintain intact the various levels of equilibria. In the economic conditions of British rule, promising improved communications and the release of new international demand for Indian agricultural produce, radically different advantages and disadvantages of the regions' various characteristics might be revealed: would it prove, for example, 'better' to boast, like the Deccan, more extensive supplies of virgin land or, like Gujarat, a more extensive traditional commitment to commercial agriculture? Also, the performance of British imperialism, in, for example, supplying communications and irrigation facilities according to regional needs would clearly affect the equation considerably. Hence an examination of the Bombay Presidency enables us to study what will be a constant theme of this book; the markedly differential patterns of continuity, change and development in the different types of region contained within one Indian province.

The varying ecological and climatic conditions in the regions had important consequences in another way. They greatly influenced the local social and tenurial structure and patterns of land control. How much and what commodities the land could produce helped to determine who owned it and the form that ownership took. The total factors controlling this situation, however, were very complex, for political influences were also vital: the degree of power and capacity for change exerted on the ground by the previous Mahratta administration, and, then, the impact on landownership and tenure of the British rulers' early policies in their quest for land revenue. In the following chapter we enter this labyrinth. Our concern here will be with the village's formal organisation: proprietorship over land, the holding of village office and the mechanism of 'settlement' of land revenue. To obtain a balanced view, we will then subsequently need to examine the more informal economic structures, for example, the power and control within the village exercised through the vast local structure of credit and debt or through regulation of marketing and commercial arrangements. However, the institutional social structure supplied the basic framework for village life and, most important, provided the first decisive point of interaction between local society in India and imperial rule. So this requires delineation first.

2
The village in 1850: land tenure, social structure and revenue policy

Traditionally, the Bombay Presidency occupies a simple place in generalised descriptions of land tenure and social structure in British India. It is seen, par excellence, as the land of the small peasant. Not here, according to orthodoxy, was either the rentier landlord, the overlord chieftain or, alternatively, the sense of complex interdependence between cultivators which produced the co-sharing principles of the 'village community'. Peasant proprietorship and the raising of land revenue directly from the smallholding cultivator was always the declared objective of British revenue administration here and 'from the first the Collectors tried to reach behind the headmen to the individual peasant'.[1] By 1850 the cultivator was genuinely established as de iure and de facto owner of much of the land.

Yet the generalisation masks much. In many areas various types of superior right to the land had established themselves. Elsewhere, cultivators had evolved conditions of collective responsibility or degrees of leasing out and sub-infeudation which complicated the simple pattern of the tiller owning the land. These processes had in many ways been considerably stimulated during the first half of the nineteenth century, despite the British authorities' general belief in ryotwari arrangements. A leading feature of the two or three decades which followed the overthrow of the Poona Peshwa in 1818 was widespread agricultural depression. Partly this was an all-Indian phenomenon, the product of price decline caused by damage to commerce created by the British conquests and monetary contraction in the wake of India's new balance of payments obligations.[2] In the Bombay Presidency, though, these economic difficulties were undoubtedly exacerbated by the effects of the land revenue system, particularly the Pringle settlements of the late 1820s and 1830s. Pringle unrealistically attempted to base revenue charges on estimates of

---

1 Kenneth Ballhatchet, *Social Policy and Social Change in Western India 1817–1830* (London, 1957), p. 116.
2 Asiya Siddiqi, *Agrarian Change in a Northern Indian State. Uttar Pradesh 1819–1833* (Oxford, 1973), p. 186.

gross returns from cultivation.³ In addition, these and other early settlements erred in assuming that the formal demands of the later Peshwas and their revenue farmers, in fact hoped-for objectives in most areas, represented reliable indicators for British revenue charges. The result was, often, intense and long-lasting over-assessment. Bunkapur taluk of Dharwar District in the south, for example, had, one official commented in 1846, 'been suffering from over-assessment during the whole course of our administration'.⁴ Depression and over-assessment in turn inhibited cultivation. As we have noted, land in the Deccan seemed abundant relative to population levels, but now new large amounts of cultivable land lay waste because of low demand and the revenue cost of bringing them under the plough. Cultivation in Bunkapur, a potentially prosperous cotton-growing tract, was in 1846 'more limited than at any former period of our rule, instead of exhibiting the increase that might naturally have been looked for from 30 years of peace and security'.⁵ At the same time, early British revenue demands were often inequitable as well as harsh. In Bunkapur one piece of garden land was rated at Rs. 40 per acre and as a result 'the holder is well-nigh ruined',⁶ but elsewhere in the taluk large landowners paid next to nothing, having clearly bribed classifying officers. Repeated examples in the diaries for the late 1830s of one leading settlement officer, Wingate, suggest that Bunkapur's experience was not untypical.⁷

These features may have had important social consequences. Over-assessment could have weakened or destroyed established landholding groups at the same time as ready availability of land for cultivation, with much potentially cultivable land waste, permitted the entry of new men. Alternatively and more subtly, many landowners may in fact, like the

3 For two different interpretations which, nevertheless, concur on the basic problems of the Pringle settlements, see Eric Stokes, *The English Utilitarians and India* (Oxford, 1959), pp. 99–103, and Ira Klein, 'Utilitarianism and Agrarian Progress in Western India', *Economic History Review*, 18, 3, 1965, pp. 576–97.
4 *(British) P(arliamentary) P(apers)*, 1852–3, Vol. 75, Official Correspondence on the System of Revenue Survey and Assessment in the Bombay Presidency, Settlement Report on Bunkapur Taluk, Dharwar District, by G. Wingate, 29 September 1846, para. 20. Fukazawa talks of over-assessment throughout the Deccan from 1818. H. Fukazawa, 'Agrarian Relations. 3. Western India', in Kumar, ed., *The Cambridge Economic History of India, Vol. 2*, p. 184.
5 *PP*, 1852–3, Vol. 75, Official Correspondence on the System of Revenue Survey and Assessment in the Bombay Presidency, Settlement Report on Bunkapur Taluk, Dharwar District, by G. Wingate, 29 September 1846, para. 20.
6 Ibid., para. 30.
7 See University of Durham, School of Oriental Studies, Sudan Archive, Papers of Sir George Wingate. For example, Wingate notes of one part of Sholapur District: 'the wretched condition of most of the villages about here is strong evidence of the oppressive nature of the assessment, while the marked differences in their condition speaks equally strongly of its inequality'. Wingate Papers, Box 293/1, Diary, 16 November 1838.

Bunkapur tax evaders, have turned the situation to their own advantage by reacting positively to over-assessment with defensive institutional arrangements. For example, much land throughout western India was, as we shall see, held free of land revenue and clearly this right would become far more valuable in conditions of over-assessment. The propagation, then, of claims to such rights and to any other forms of superior tenurial status was an obvious response, particularly when and where government seemed weak.

Apposite to this last point are the geographical as well as the qualitative limitations of early British revenue administration in Bombay. The surveys and experiments of the Pringle period were largely confined to the Deccan. Throughout the rest of the Presidency early revenue settlements were cobbled up with merely sporadic detailed investigations of local circumstance. Only Wingate's appointment in 1851 as Commissioner of Revenue Surveys presaged the first attempt to organise systematic land surveys throughout Gujarat, the Konkan and the north Deccan and create, for the first time, a coherent blueprint for the settlement of the whole Presidency.[8] Subsequent local reports by Wingate like those on Ratnagiri District in 1851 and Khandesh District in 1852 at last mapped out the principles of revenue administration in such areas.[9] The sheer unevenness of the pressure of early British administration on the Bombay countryside was, therefore, very marked and this, again, might encourage tenurial diversity to flourish. The trends of the period before 1850, in sum, did not work towards the establishment of a single ryotwari monopoly – or any sort of overall uniformity – in land tenure and agrarian social structure.

The Deccan and the Southern Maratha Country

Even so, the ryotwari ideal did seem to fit local conditions well in most areas of the Deccan. In fact, both the political and agricultural realities of life here determined the supremacy of cultivating peasant proprietorship. Landlordism in India at root typically rested on an intensive (often wet-

---

8 Even so, Wingate's post lapsed on his retirement. Supervision over settlements was then vested in the Commissioners of the several 'divisions' but during the 1860s and 1870s these – notably W. C. Anderson and J. T. Francis – were former subordinates of Wingate, who closely followed his administrative principles. Anderson briefly headed two divisions in the late 1870s and then in 1881 the post of Survey Commissioner for the whole Presidency was re-created.

9 See S(elections from the Records of the) B(ombay) G(overnment), No. 2, Report by Capt. Wingate on Introducing a Survey and Revision of Assessment in the Ratnagiri Collectorate, 30 January 1851, and SBG, No. 1, Report by Capt. Wingate on the Survey and Assessment of Khandesh District, 29 March 1852.

crop) agriculture with two main traditional attributes: firstly, sufficient population density to make landownership a valuable and exclusive asset and, secondly, a regular and reliable surplus from agricultural production to enable the cultivators to support an overlord elite. Neither of these conditions existed in the Deccan of the late eighteenth and early nineteenth centuries: a smallholding peasantry was always the logical cultivator of its arid millet-lands. Politically, independent landlordism had the best chance to flourish in regions of weaker governmental authority. In India the natural tendency of any strong political authority to try to limit the autonomy of landlord elites had added force, because of the need to raise the bulk of government income from the land revenue of the localities. The Deccan, however, particularly the central Deccan, was always the most governed area of western India. Here the Mahratta government writ ran most effectively and British rule continued the tradition of possessing greatest power to enforce its wishes on the ground nearest to Poona. Following Pringle, for example, Bombay revenue reassessments throughout the nineteenth century always began in Indapur taluk of Poona and the central Deccan's land revenue settlements were consistently the most up to date and therefore potentially oppressive.

These deep-seated influences were reflected in the characteristics of the Deccan's social structure, as reported by early British administrators. Here farming seemed highly individualist. Villages and holdings were relatively large in size and land for new cultivation or grazing freely available. Hence British officials could 'find no traces of joint ownership, and must believe that it never existed'.[10] It was, also, soon clear that 'we have here few considerable landed proprietors or other capitalists, who would be willing to engage for the revenue of our large villages under any system calculated to preserve proprietary rights, and prevent oppression on their part'.[11] The latter caveat was prompted by the previous attempts to introduce landlord-orientated revenue settlements in the Deccan: the agreements made with revenue farmers in the late Mahratta period and during the first years of British rule. Their legendary oppression convinced mid-nineteenth-century officials that artificially to introduce intermediaries between government and ryot 'would be viewed with the utmost dismay and dissatisfaction by the whole agricultural community'.[12] So in the Deccan of 1850 ryotwari arrangements seemed mere

10 *PP*, 1852–3, Vol. 75, Official Correspondence on the System of Revenue Survey and Assessment in the Bombay Presidency, Report by H. E. Goldsmid and G. Wingate, 17 October 1840, para. 37.
11 Ibid., para. 46.
12 Ibid.

acknowledgement of practical reality and, in addition, extremely easy to enforce. The 'Joint Report' of 1847, a guide-book for settlement officers in the field, in making recommendations about the size of the 'survey field' was able to assume that almost all 'survey fields' would comprise exactly one holding. Only occasionally, it seemed, would a group of tiny holdings have to be joined together or the farm of a rich peasant sub-divided to create a convenient, fairly standardised unit for the levying of land revenue.[13]

Nevertheless, this did not imply economic and social equality. Local investigators found considerable differentiation within the peasantry, particularly in the Southern Maratha Country: thus in Dharwar District Wingate spoke of 'very substantial cultivators' each paying, without particular over-assessment, around Rs. 250 of annual land revenue.[14] There, also, seemed fundamental distinctions of tenurial status. Sykes, the Bombay government's Statistical Reporter, estimated in 1838 that, whilst 40 to 55 per cent of the Deccan population comprised registered landholders, the number of tenants and labourers brought the total agricultural population up to 75 per cent of the whole.[15] Particularly on the better quality lands, sometimes growing crops for sale, owners leased out to tenants or employed labourers. In this way Wingate found, in 1847, much of the fertile garden land of Seegaon in Dharwar District being cultivated by tenants of landlords.[16]

Some diversity of status was, of course, almost bound to exist to suit different agricultural patterns and peasant family needs. Crops varied in their degree of labour intensity and the social as well as agricultural factors, which Dharma Kumar has shown necessitated the existence of a labouring class in Madras,[17] were present in Bombay too. The real question, though, is whether differences in tenurial status in the Deccan depended on economic rationale or on a tradition of formalised social and tenurial polarisation. Some Bombay officials, interestingly in view of ryotwari predispositions, believed in the latter: that the Deccan had for-

---

13 *PP*, 1852–3, Vol. 75, Official Correspondence on the System of Revenue Survey and Assessment in the Bombay Presidency, Joint Report by H. E. Goldsmid, G. Wingate and D. Davidson, 2 August 1847 (hereafter 'Joint Report, 1847'), para. 13.
14 Wingate Papers, Box 293/2, Diary, 24 January 1846.
15 W. H. Sykes, 'Special Report on the Statistics of the Four Collectorates of Deccan under the British Government', *Report of the Seventh Meeting of the British Association for the Advancement of Science* (London, 1838) (hereafter 'Sykes, "Special Report on the Deccan" '), p. 266. The extreme vagueness and unreliability of these figures are self-evident. They merely demonstrate official acknowledgement of the existence of a significant landless peasant class.
16 Wingate Papers, Box 293/3, Diary, 2 May 1847.
17 Dharma Kumar, *Land and Caste in South India. Agricultural Labour in the Madras Presidency during the Nineteenth Century* (Cambridge, 1965).

merly been a land of large estates. Sykes, for example, argued that 'the lands of villages were divided into hereditary family estates, called Thals'[18] and the Joint Report of 1847 also referred to 'Thals'.[19] However, the 'Thals', if they had ever existed, had disintegrated long before 1850. To many contemporaries and subsequent historians, the distinction which then survived in the Deccan villages was the famous one between 'mirasdar', or hereditary landholder, and 'upri', an inferior newcomer with no proprietary claim to the soil.[20] Though both mirasdar and upri were cultivating landholders, the upri would legally have no status in the village community, no established rights to the common services of the 'watandars', the village servants. Some made much of these differences. Baden-Powell, the great late nineteenth-century authority on land revenue, rejected the contemporary official argument that joint ownership 'never existed' in the Deccan and argued that mirasdars were probably 'once members of co-sharing landlord communities'.[21]

Clearly under Mahratta rule mirasdars were cultivators with full proprietary rights whilst upris had typically come into cultivation as tenants, often of state lands.[22] Yet it is hard to believe that the distinction in practice had much significance by the first half of the nineteenth century. Mirasdar had never been an exclusive title, for even in the seventeenth century intruders could acquire mirasi rights over time by settling permanently in the village.[23] Throughout the first thirty years of British rule of the Deccan the relatively limited extent of cultivation provided the keynote of agrarian trends. Typically, only around half of the cultivable land in most Deccan districts, according to formal British estimates, was being tilled; 52,000 acres out of an estimated possible of 122,000 in Bunkapur taluk in Dharwar District in 1846,[24] 106,000 acres out of 191,000

18 *PP*, 1866, Vol. 52, W. H. Sykes, Report on the Land Tenures of the Deccan, December 1830 (hereafter 'Sykes, Report on Land Tenures'), p. 3.
19 See Joint Report, 1847, para. 11.
20 The distinction between mirasdar and upri is advanced and discussed in the nineteenth-century classic, B. H. Baden-Powell, *The Land Systems of British India* (Oxford, 1892), Vol. 3, pp. 256–8, and in the major modern study, Ravinder Kumar, *Western India in the Nineteenth Century* (London and Toronto, 1968), Ch. 1.
21 Baden-Powell, *The Land Systems of British India*, Vol. 3, p. 256.
22 Hiroshi Fukazawa, 'Land and Peasants in the Eighteenth Century Maratha Kingdom', *Hitotsubashi Journal of Economics*, 6, 1, June 1965, pp. 32–61.
23 A. R. Kulkarni, 'Village Life in the Deccan in the Seventeenth Century', *The Indian Economic and Social History Review*, 4, 1, March 1967, p. 49.
24 *PP*, 1852–3, Vol. 75, Official Correspondence on the System of Revenue Survey and Assessment in the Bombay Presidency, Settlement Report on Bunkapur Taluk, Dharwar District, by G. Wingate, 29 September 1846, para. 20. At the same time, we should note that the situation may not have been quite as marked as here implied. Official estimates of 'cultivable land' were apt to exaggerate the extent of land which could be profitably cultivated. Much of the land classified as 'cultivable waste' was always very poor. For a discussion of this point, see B. H. Farmer, *Agricultural Colonization in India since Independence* (London, 1974), pp. 29–35.

The Deccan and the Southern Maratha Country   23

in Bhimthari taluk of Poona in 1838.[25] As a result, most settlers who wished to take up land could easily become proprietary holders and in some Deccan villages the wealthiest cultivators in 1850 had relatively recently established their position.[26] Under these circumstances, a social and cultural distinction between mirasdar and upri possibly remained – the upri was the new man, without deep roots in the village, whilst the mirasdar was the inheritor of a traditional family holding – but their practical rights probably differed very little. The title of mirasdar was a boast to be worn with pride. Nevertheless, it did not need the British ryotwari settlements to confirm the former upri as effective owner of his land.

We return, then, to the original notion of a large proprietary community, roughly undivided in terms of tenurial status. However, this formal legal pattern of peasant landownership may still have disguised acute divisions of actual power and control over the land and village. The Deccan villages boasted a long-established and organised 'constitution for their internal government'.[27] Within this structure a whole series of offices existed each with its traditional honours, rewards and duties, from the Patel, the village headman, and the Kulkarni, the village accountant, to the 'balutedars', the group of artisans and menials who provided services.[28] As early as the seventeenth century the holding of a 'watan', the rights bound up in a major village office, was very valuable: 'an instrument for a political career, the only means of livelihood and an indicator of social prestige and dignity'.[29] In addition, the emoluments of the large body of village servants were considerable. The clearest sign of this was that a fair proportion of total taxation raised on the village remained within it, and the 'tunkha', or government share, was only paid when village expenses and 'huks'[30] had been met. In a typical Deccan village in the early nineteenth century Sykes found that of total tax collections of Rs. 8,522, Rs. 3,022 were paid out within the village.[31]

Village office, therefore, provided the opportunity for significant economic and social differentiation, for each of the important posts offered its own source of power. The Brahmin Kulkarni was the literate clerk of the village and his educational skills might sometimes enable him to eclipse the headman.[32] The master of the village, however, was

25 SBG, N(ew) S(eries), No. 151, Settlement Report on Bhimthari Taluk, Poona District, by A. Nash, 2 November 1838, para. 6.
26 For some examples, see my 'Rich Peasants and Poor Peasants in Late Nineteenth Century Maharashtra', in Dewey and Hopkins, eds., *The Imperial Impact*, pp. 97–113.
27 Sykes, Report on Land Tenures, p. 3.
28 For example, the kumbhar (potter) and the village watchman, normally a Mahar by caste.
29 Kulkarni, 'Village Life in the Deccan in the Seventeenth Century', p. 41.
30 Huks were the rewards paid by the villagers to their officials.
31 Sykes, 'Special Report on the Deccan', p. 320.
32 Kulkarni, 'Village Life in the Deccan', p. 44.

traditionally its Patel, usually a Maratha by caste.[33] His extensive privileges included a share of the village's revenue collections, free gifts of articles from artisans and shopkeepers, and the right of precedence on festive occasions.[34] Commentators of both Mahratta and British periods provided impressive testimony of the Patel's authority. According to the chronicler Tukaram, in the seventeenth century nobody who incurred the Patel's displeasure could remain in the village.[35] Sykes evidenced the dignity of the Patel's position by the way Mahratta princes had installed themselves in the office, 'Holkur, for instance, at Munchur; Seendeh (Sindiah) at Jamgaon'.[36] To Gooddine, the author of a definitive report on the Deccan village, the Patel in eighteenth-century Ahmednagar District was the king of rural society: 'his authority was unquestionable, and his will absolute: while his privileges of precedence on all public and social occasions formed ... an imposing picture of greatness'.[37]

Clearly the Patel's position offered unique opportunities based, as it was, on two distinct roles: he was at once the leader and spokesman of rural society and the government's representative in the village. The Patel's huks alone gave him a good basis for economic security, since often the grain-share he received from the village was enough to feed his family. At Karjat in Ahmednagar District, for example, the Patel was officially entitled to 128 seers of grain for every 120 bighas of land cultivated. In 1827 8,491 bighas were under cultivation and so the Patel was due to receive 9,057 seers, sufficient, according to Sykes, to feed 25 people for a year.[38] At the same time the role of government representative brought extensive powers. Even under the purest ryotwari arrangements, the Patel had considerable control over the way he allocated the village's revenue load and reported the best time for the ryots to pay their assessments: he could aid his friends and harm his enemies by manipulating the timing of instalments.[39]

33 Sykes wrote in the 1830s: 'originally the Patels were Mahrattas only, but sale, gift, or other causes, have extended the right to many other castes. A very great majority of Patels, however, are still Mahrattas.' Sykes, Report on Land Tenures, p. 11.
34 For an enumeration of Patel rights and emoluments, see Sykes, 'Special Report on the Deccan', p. 288.
35 Kulkarni, 'Village Life in the Deccan', p. 43.
36 Sykes, 'Special Report on the Deccan', p. 288.
37 SBG, No. 4, R. N. Gooddine, Report on the Village Communities of the Deccan, 1845 (hereafter 'Gooddine's Report'), para. 21.
38 Sykes, 'Special Report on the Deccan', p. 289.
39 In Khandesh District, for example, as in many other cases, revenue instalments were taken at different times according to which of three main harvesting times the peasants used. The Patels, Wingate commented, distributed each total instalment due 'among the Ryots according to their knowledge of the proportions of the three kinds of crop cultivated by each. This must put a great deal of power into their hands.' Wingate Papers, Box 293/4, Diary, 25 January 1852.

This evidence, then, might suggest that Deccan village society was dominated by a powerful office holding elite. Yet had the traditional honour and wealth of the village officer survived into the nineteenth century or, even, ever existed universally? Many commentators of the early British period, like Sykes and Gooddine, seemed either to be referring to an idealised golden age of Patel hegemony or simply taking the formal list of privileges and rewards entirely at face value. Clearly the Karjat Patel was potentially a wealthy man, but to realise that potential required the political task of strictly enforcing collection of dues. This was not always easy, for all village officials had their rivals and enemies. The central government had an obvious interest in cutting back their local autonomy and freedom of action independent of the state and chroniclers like Sabhasad suggest that as early as Shivaji's day the Mahratta state had ambitions to restrict the village watandar's power.[40] Sometimes the government might find tacit allies against the watandars in the peasant masses of the village. The Patel, as normally the caste fellow of the bulk of the agriculturists, was rarely actively unpopular – as often was the Kulkarni – but the peasant community had a manifest stake in keeping his huks at a minimum.

In fact, the informal alliance of state and peasantry had, it is clear, already substantially undermined the power of one office holding group by the early nineteenth century. The civil governors and collectors of the districts in the traditional Mahratta hierarchy were the 'Deshmukhs'. These men, mainly the descendants of great Maratha families – around 90 per cent were Marathas in the 1830s[41] – had exercised an ill-defined, supervisorial control over administration: undoubtedly the authority of the office had always largely depended on the strength and personality of the men who held it. Without the special village power base of the Patel, these district officers were now ill-placed to grapple with the political turmoils and social changes of the late eighteenth and early nineteenth centuries. They fared particularly badly in the Southern Maratha Country, where in several areas Tipu Sultan confiscated all their rights to independent watans. One such locality was Nuwulgond taluk of Dharwar District where the district officers British rule inherited lived 'on the pensions assigned to them by the Mahratta government and... [took]... no part in the public affairs of their districts'.[42]

The Deccan Patel was not so easy a victim. His roots were normally deep in the village and all governments and administrative systems required his co-operation. Even so, snipings by state and villagers seem to

40 Kulkarni, 'Village Life in the Deccan', p. 50.
41 Sykes, Report on Land Tenures, p. 9.
42 Wingate Papers, Box 118, W. Hart and G. Wingate to A. N. Shaw, Collector of Dharwar, 25 December 1846, para. 5.

have limited his rights, particularly in the central Deccan where Mahratta government activity could be most energetic. In many areas around Poona, Patels had lost their huks from the village and were paid entirely by the state by the onset of British rule: in Newasseh and Parner taluks of Ahmednagar, for example, where Patels' huks had been rescinded by the Peshwa's powerful official, Naroo Babajee.[43] In parts of the Southern Maratha Country, too, the attack on the financial claims of the district officers was sustained and broadened with reductions of village officials' huks. Wingate soon found, on visiting Badami and Bagalkot taluks of Belgaum District in 1851, that many huks to village as well as district officials were now virtually non-existent. In the case of one grant, 'in a very few years more the levy would be wholly discontinued without the intervention of the Survey, merely from the village officers being now without sufficient influence to enforce its collection'.[44]

In practice, then, the watandar was, by the time of the British takeover of the Deccan and Southern Maratha Country, by no means always as prosperous and powerful as his formal rewards might imply, particularly in the central Deccan districts of Poona and Ahmednagar and in Belgaum and Dharwar in the south. A crucial transformation had, in many cases, occurred. Many village officers were now substantially government employees rather than local lords with an independent power base. Even where real wealth and status survived, the very attractiveness of village office in difficult times had sometimes contrived to weaken its position. In the late Mahratta and early British periods, many village offices, particularly that of Patel, became increasingly divided up, as those seeking status and income bought their way in. By the 1840s in the central Deccan 'in most villages, the Patelship has been divided among a great number of sharers, and the increasing families of these sharers have caused it to be still further subdivided'.[45] This not only produced elaborate joint proprietorship and division of function between Patel-sharers of different caste,[46] but also greatly reduced rewards for the individual office-holder. In 1843–4 the Patels of 33 villages in the Patoda taluk of Ahmednagar were due a total remuneration of Rs. 4,207, over Rs. 100 per village.[47] However, there were 284 sharers in the Patelship of these villages – one

---

43 Gooddine's Report, para. 15.
44 Wingate Papers, Box 293/3, Diary, 11 July 1851.
45 Gooddine's Report, para. 21.
46 For some examples, see Sykes, Report on Land Tenures, p. 30.
47 Gooddine's Report, para. 36. Where the money came from was very revealing about the status of these Patels. They were clearly substantially government employees: their independent huks from the villagers totalled only just over Rs. 600 compared with Rs. 3,588 worth of government pay.

had 61 Patel-sharers[48] – so that the average receipt was under Rs. 15 per sharer, 'a very paltry remuneration'.[49]

To many early British officials, therefore, the office of Deccan Patel seemed now simply an agglomeration of minor perks, squabbled over by what Gooddine called 'a beggarly mob of claimants',[50] in contrast to its former position as the great administrative lynchpin of the village. Possibly the story was genuinely – as Gooddine thought – of a relatively recent and dramatic fall from grace. Equally, though, the power of the Patel and his fellow watandars may always have been gravely circumscribed by the political realities of agrarian life. A decline in authority in the century before 1850 does not especially correlate with the known characteristics of the period: one where political fluctuations, as we shall see, frequently gave social elites the opportunity to enhance their position. That this was not the case with the Deccan village officer suggests limits to the long-standing virility of the institutions.

If, however, specific watandar perks were not necessarily very valuable by the nineteenth century, this by no means exhausted possible means of economic and social distinction within the Deccan village. Indeed, direct payments were probably only ever the tip of the iceberg of privilege. A more effective and secure source of wealth and prestige lay in ownership and control over a piece of land, where the revenue rights had been wholly or partially 'alienated' by government. Such a privilege, originally granted as reward or bribe for services rendered, involved typically either the cultivation of a holding exempt from some or all land revenue or the rights of revenue collection over an area of land. At the British takeover, such alienations were wide-ranging throughout western India. They included, as the largest units, grants of 'surinjam' and 'jagir', normally substantial estates given originally for military service or the provision of units of soldiers.[51] In addition, major landowners held lands, often amounting to whole villages and more, in 'inam'. At a lower level and ubiquitous throughout the village lands was the 'service inam' granted to watandars or others who performed some duty for local society. Every village, too, had its 'personal inams', given to specific individuals for a myriad of reasons: at Wangi, Ahmednagar District, in 1830 one man held 15 bighas of inam land 'for reading stories at the Uchaos, or festivals of the goddess Devi', and another 15 bighas were held by 'the Charseh, or clarionet players, who daily play before the idol'.[52] As in most agrarian

48 Ibid., Table 3.
49 Ibid., para. 36.
50 Ibid., para. 21.
51 Both 'surinjam' and 'jagir' signified the 'supply' of troops or the 'provision' of military service.
52 Sykes, Report on Land Tenures, p. 8.

societies, the institutions of organised religion were prominently involved in privileged landownership: the village temple usually held its slice of alienated land.

There can be no doubt that during the late eighteenth and early nineteenth centuries there was a massive extent of alienated land and revenue throughout the Bombay Presidency. Besides the whole villages and groups of villages held within jagirs and inams, a significant proportion of the land of the typical village was alienated. An official estimate in 1850 placed total alienations in Bombay at about a third of the Presidency's gross land revenue, involving lands assessable at over Rs. 82 lakhs.[53] Certainly we can safely guess that in mid-century the government's revenue writ did not run in between 20 and 30 per cent of the Presidency's cultivated land. Of the major regions, the highest proportion was probably in Gujarat, but the Deccan and Southern Maratha Country also boasted substantial alienations. Sykes discovered in 1828 that in the four districts of Poona, Ahmednagar, Khandesh and Dharwar, nearly a fifth of the towns and villages were completely alienated and, in addition, 'almost every village has rent-free lands held by the Patel, Kulkarni and Mahrs'.[54] Within the region, alienations grew thicker the further one travelled from Poona. In the south, in Dharwar and Belgaum Districts Maratha military chiefs and their descendants held sprawling estates, comprising large groups of villages, in jagir or surinjam. Here even in the government villages inamdars controlled much of the land: in 1848 there were an estimated 60,000 'minor alienations', roughly 25 per village, in the two districts' government villages, 'the share left for government, even in these its khalsat [full revenue paying] villages, not averaging one-half' of potential revenue proceeds.[55] Badami taluk in Belgaum provides a fairly typical case study within the south. Here Wingate found in 1852 76 completely alienated villages and 151 where government possessed rights. In the latter, 42 per cent of the arable land was inam or subject merely to a quit-rent.[56]

Although many alienations stemmed originally from government grant, the authorities clearly possessed a substantial interest in limiting

---

53 *SBG*, NS, No. 30, Selection of Papers Explanatory of the Origin of the Inam Commission and its Progress (hereafter 'Inam Commission Papers'), H. E. Goldsmid, Secretary to Bombay Government, to F. J. Halliday, Secretary to Government of India, 7 January 1850, para. 4.
54 Sykes, Report on Land Tenures, p. 8.
55 Inam Commission Papers, W. Hart, Inam Commissioner, to D. A. Blane, Revenue Commissioner, 1 July 1848, para. 8.
56 See *SBG*, No. 5, Report Explanatory of the Revised Assessment Introduced into the Talookas of Badamee and Bagulkote, Belgaum District, by Capt. G. Wingate, 9 June 1852 (hereafter 'Wingate's Belgaum Report').

them, for even a small reduction would mean an immense accretion of revenue. So the Mahratta government had periodically waged a fierce war to prevent the growth, often fraudulent extensions, of land alienations. Their success, though, seems to have been small. In the Southern Maratha Country, in fact, 'the unauthorised proceedings by which land was alienated from the state became more and more frequent'.[57] The reason was not simply the 'mismanagement and anarchy' by which the British characterised late Mahratta administration. Since inam rights were very widely held and much of the village community had a vested interest in the continuation and extension of land alienations, the central government could not find the same peasant allies which had enabled it to reduce watandar privileges. Then came the British takeover of power, offering, with a foreign government so unsure of its local touch, an unparalleled opportunity for the further advance of inam claims. Not surprisingly, many had 'furtively taken advantage of the confusion ... to enlarge their Inam lands, ready-money claims, and grain-fees'.[58] By 1850 the inamdar was probably at his zenith, in extent both of his alienated landholdings and of his independence from central authority. The Bombay government's information on the subject remained, as it itself admitted, 'in almost the same imperfect state in which it was on our acquisition of the country'.[59]

Even though this was a Presidency-wide phenomenon, the extent of land alienated in the Deccan and particularly the Southern Maratha Country did reveal them as regions with the potential at least for considerable disparity of social power and distinction. Nevertheless, any elite formed by control over revenue alienations here would not appear very homogeneous or exclusive. For these rights, in the Deccan, were extremely widely distributed. In most villages all the major caste groups – the untouchable Mahrs as well as the landowning Kunbis – held some inam land and the status it conferred, therefore, cut across orthodox social ranking. In most Deccan villages in 1850 leaders existed in both economic and social distinction, but they were not necessarily a cohesive entity and kinship and habits linked them closely to the bulk of the peasantry. There was, in sum, no simple equality, but agrarian elites were not a sharply divided class and potential fluidity in social structure existed.

57 Inam Commission Papers, W. Hart, Inam Commissioner, to D. A. Blane, Revenue Commissioner, 1 July 1848, para. 2.
58 Sykes, Report on Land Tenures, p. 17.
59 Inam Commission Papers, H. E. Goldsmid, Secretary to Bombay Government, to F. J. Halliday, Secretary to Government of India, 7 January 1850, para. 4.

The Konkan

These characteristics, however, would not necessarily recur in other parts of western India. The great geographical and climatic differences between the Deccan and the Konkan coastal strip suggested the likelihood of contrasting social structures. The Konkan fulfilled our earlier conditions for the growth of landlordism: high population density and the tradition of reliable surplus from an intensive agriculture. The opportunities, particularly in the northern part of Kolaba District, for reclaiming land from the sea added another minor but significant influence. Major land reclamation in a peasant economy invariably creates landlordism, for the capital needs of the operation demand men of substance who are not likely themselves to cultivate: the extensive reclamations made by merchants in Tokugawa Japan is a case in point.[60] In northern Kolaba, therefore, reclamation had produced 'a kind of superior occupant, who lets out the lands'.[61] In addition, the earliest British administrative precedent in the region had been the creation of overlords. The British, obtaining Salsette taluk, the area immediately surrounding Bombay City, in 1772–3, had leased out many of the villages to large landowners.[62] The intention had been little more than to pass on the problems of an area then 'much depopulated and scantily cultivated',[63] but the precedent was not challenged by any subsequent attempt to introduce a ryotwari system. By the mid-nineteenth century, the holders of the leases, by then mainly Bombay City merchants and traders, were still landlords of over fifty of the villages of Salsette taluk.[64]

Nevertheless, most of the evidence, although scanty, suggests that traditional Konkani agrarian society was not unlike that of the Deccan. Originally, early British officials believed, 'a more or less complete village system' with Patels and other watandars on the Deccan model existed[65] and, tenurially, peasant proprietorship was probably the rule. Baden-Powell later claimed that the cultivating 'dharekaris' had been 'acknowledged as virtually proprietors of their holdings', with rights 'very similar' to those of the Deccan mirasdars.[66] The tenurial structure,

---

60 See C. D. Sheldon, *The Rise of the Merchant Class in Tokugawa Japan, 1600–1868* (New York, 1958), pp. 82–3.
61 Baden-Powell, *The Land Systems of British India*, Vol. 3, p. 295.
62 See *SBG*, NS, No. 180, Correspondence Relating to the Conditions on which Certain Estates are Held in the Salsette Taluk of the Thana Collectorate.
63 Ibid., Pestanji Jehangir, Alienation Settlement Officer, to L. R. Ashburner, Revenue Commissioner Northern Division, 17 July 1872, para. 3.
64 Ibid., para. 7.
65 *BG*, Vol. 10, *Ratnagiri and Savantvadi*, p. 137.
66 Baden-Powell, *The Land Systems of British India*, Vol. 3, p. 289.

though, had changed markedly by the mid-nineteenth century. The major development was the rise of village leaders called 'khots', beginning with their installation as revenue farmers and the equivalent of village headmen at the beginning of the sixteenth century.[67] By the early British period the khots, having firmly established themselves as revenue and administrative controllers, were claiming proprietary rights over the soil in large areas of the two major Konkan districts of Ratnagiri and Kolaba. The khots' rise was accompanied by an important change in personnel. The first khots were typically Marathas, like their Patel fellows east of the Ghats. However, Brahmins had 'come in by sale and mortgage, and by grant'[68] and by 1850 'a large majority' of khots were now Brahmins,[69] of the same Chitpavan, Javal and Devrukha sub-castes who provided the educational and administrative elite of the Bombay Presidency. Their accession to the khoti office signalled their tightening control over Konkani society and, in turn, they transformed the position of khot so that most khots became de facto landlords of the areas they controlled, exacting rent from cultivators.

This, however, was a particular form of landlordism. In general, khots were not indigenous farmers thrown up as rentiers and entrepreneurs by a buoyant agriculture and their position never rested on any direct day-to-day supervision of cultivation activities. Instead, khots owed their power to their total control of village administrative office and the sources of local information. Since most khoti villages had no independent accountant to parallel the Deccan Kulkarni,[70] the khots held all village documentation: 'the only records are their own accounts, which they never produce to Government, and can make up as they like'.[71] In addition, they monopolised even district office, for 'most of the hereditary district officers, deshmooks and deshpandeys are also khots'.[72] So government, by the onset of British rule, had been simply excluded from direct contact with village society in the khoti areas. Of necessity, the first British revenue settlement in the Konkan in 1818 was predominantly a farming

67 The historian of the Bombay revenue system claimed that 'they were introduced in the revenue settlement made by Yusuf Adil Shah of Bijapur'. See A. Rogers, *The Land Revenue of Bombay: a History of its Administration, Rise and Progress* (London, 1892), Vol. 2, p. 2.
68 *SBG*, NS, No. 134, Selections, with Notes, from the Records of Government Regarding the Khoti Tenure by E. T. Candy (hereafter 'Khoti Tenure Selections'), p. 1.
69 *BG, Vol. 10, Ratnagiri and Savantvadi*, p. 138.
70 Ballhatchet, *Social Policy and Social Change in Western India 1817–1830*, p. 152.
71 Wingate Papers, Box 293/3, Diary, 26 January 1850.
72 Revenue Report for 1821-2 by J. A. Dunlop, Collector of Ratnagiri, quoted in *SBG*, No. 2, Report by Capt. Wingate on Introducing a Survey and Revision of Assessment in the Ratnagiri Collectorate, 30 January 1851 (hereafter 'Wingate's Ratnagiri Report'), para. 23.

arrangement with the khots and 'Government did not know what the tenant paid or whether the khot gained or lost by the farm.'[73]

So the khots were left free to enforce their claims to proprietorship by squeezing what they could from the cultivators. In practice, though, their power on the ground was not without limitations: lack of contact with the mechanics of farming and of everyday village life ensured this. Within the khoti village, the special rights of particular cultivators had always to be recognised. The dharekaris, for instance, the descendants of Baden-Powell's original proprietors, were in practice mostly still occupancy tenants and they maintained their traditional prerogatives over their holdings, such as those of sale and mortgage.[74] Most cultivators, it was true, were by 1850 formally 'badhekaris', tenants at will of the khot, but even these could, by length of service, acquire de facto occupancy tenure rights, with which it would not be politic for a khot to interfere.[75] Economically, too, the khot was not automatically supreme. One defendant of the khoti system claimed in 1874 that 'there are now dhara holdings larger and more valuable than many khoti villages',[76] and certainly some evidence suggests that assertive cultivators could still deny the khot control over the best land.[77] In some villages, in fact, the existence of the khoti system was attributed to a desire by the leading cultivators to protect themselves against possible exactions by revenue farmers, for 'in some sanads conferring a village as a khoti watan, it is actually stated that the cultivators had petitioned for a khot. He acted as a buffer ...'[78] In such cases, effective control over the village might still reside with the leading dharekaris.

In addition, just as the Deccan saw no simple ryotwari uniformity, so there was no khoti monopoly in Ratnagiri and Kolaba Districts. Large groups of villages resisted khots' incursions, maintained a roughly ryotwari tenure system and continued to settle for their land revenue directly with government. Inamdars other than khots owned lands in the Konkan as throughout Bombay Presidency. Thus in 1880 in Ratnagiri District

---

73 *BG, Vol. 10, Ratnagiri and Savantvadi*, p. 224.
74 See ibid., pp. 203–13.
75 See Lt. Dowell's notes to his Survey of Ratnagiri taluk, 1826–9, quoted in Khoti Tenure Selections, p. 27.
76 N. V. Mandlik, ed., *Writings and Speeches of the Late Hon. Rao Saheb Vishvanath Narayan Mandlik* (Bombay, 1896), p. 91, 'Brief History of the Watandar Khotes', 1 February 1874.
77 For example, Lt. Dowell wrote of one part of Ratnagiri taluk in the late 1820s: 'nearly the whole of the rice land in every village in this Mahal is cultivated by the Kunbis, and is their watan land; scarcely any is the khot's land, though no doubt the khots would get it if they could...' Lt. Dowell's notes to his Survey of Ratnagiri taluk, 1826–9, quoted in Khoti Tenure Selections, pp. 23–4.
78 Ibid., p. 19.

there were 607 khoti villages and 397 'khichadi', a mixed system where the khot simply acted as the revenue collecting agent and received a percentage in return, as against 210 ryotwari villages and over 100 held in inam.[79] In Kolaba at the same time the number of ryotwari villages slightly exceeded that of the purely khoti.[80]

For all this, in the early nineteenth century the khoti system must be seen as a powerful, expansive force. On the ground, 'the khoti system is gradually extending itself' as, in the 'mixed' villages, khots assumed direct control over land which fell out of cultivation: subsequently any new peasant cultivators who took up such land did so on strictly defined tenancies.[81] At the same time, the elite was not entirely closed. 'It is rare to find a village in which some share of the khotiship had not been alienated either by sale or mortgage',[82] so that a prosperous dharekari might hope to buy his way in. This process probably acted as an effective safety valve for the system. Further, particularly at times of maverick government revenue demand, cultivators' interests did not automatically lie in resisting khoti incursions. Some clearly preferred to take their chance on settling with the khot rather than deal with the anonymous and unpredictable forces of government officialdom.

In these circumstances the accession of a foreign government gave the khots an admirable opportunity to extend and confirm their position. The barriers to local information, strongly erected against the Mahratta authorities, could now be made fully impregnable against the bewildered newcomers. Added to this was early British officialdom's own ideological confusion over the khoti system. Were the khots worthy improving landlords or deceitful parasites? Elphinstone, on tour in Ratnagiri District in 1823, recommended the conversion of the khoti villages to a ryotwari system[83] and the prejudices of most of his contemporaries were certainly hostile to the pretensions of the khots. Yet they still had their advocates. W. H. Chaplin, Commissioner of the Deccan, was one prominent early official sympathetic to their proprietary claims[84] and J. H. Pelly, the first Collector of Ratnagiri, felt that, although many khots could rightfully be

79 *BG, Vol. 10, Ratnagiri and Savantvadi,* pp. 201–3.
80 *BG, Vol. 11, Kolaba and Janjira* (Bombay, 1883), p. 162.
81 Khoti Tenure Selections, p. 18, Revenue Report for 1827–8 by Mr Reid, 26 August 1828, para. 70.
82 *SBG,* NS, No. 254, Papers on the Settlement of Khed Petha, Old Suvarndurg Taluk of Ratnagiri District, A. K. Nairne, Assistant Collector, to the Collector of Ratnagiri, 27 November 1868, para. 3.
83 *BG, Vol. 10, Ratnagiri and Savantvadi,* p. 230.
84 He had argued that the cultivator in khoti villages had no right to the land 'except on the terms which he can agree upon with the khots'. See Wingate's Ratnagiri Report, para. 21.

displaced by government at will, others, the 'wuttundar' khots, should be regarded as hereditary proprietors.[85] These sentiments were widespread enough to prevent any immediate, determined attempt to undermine the khots' position. And such inactivity soon looked like acquiescence in khoti claims. As early as 1825 the Bombay government was conceding that 'whatever may have been the origin of their title, they appear now to have acquired a right with which it would be neither just nor politic to interfere'.[86]

In the end, also, financial realities placed the khots in a strong position in the early years of British rule. The British inherited a situation whereby the khots provided much of the revenue supply from the Konkan and, in the khoti village, controlled the records and mechanism for its collection. Hence any threatening moves against them could be met with dangerously dwindling revenue returns. In fact experimental land revenue surveys were attempted by Dunlop in 1820–4 and Dowell, in just Ratnagiri taluk, in 1827–30[87] but both foundered on the rock of non-co-operation from the village and endangered the regularity of revenue supply. The subsequent abandonment of any attempt at revenue survey throughout the 1830s and early 1840s meant that government could not acquire the documentation it would need even to attempt to eliminate intermediaries. By 1850, although most tenurial forms could be found in the Konkan, the khot landlords seemed to have confirmed their proprietary claims over a significant slice of the region and their social supremacy over most of it.

Gujarat

The Konkan's cultivating tradition was, however, still at root ryotwari since khoti overlordship rested on relatively recently acquired control over administrative office, not long-established agricultural supremacy. In Gujarat, the social and tenurial pattern was more complex, more mixed, considerably influenced by the region's turbulent political history. The Marathas were foreign conquerors here. Their eighteenth-century invasions had created intense administrative pressure on some Gujarati localities, with, for example, overwhelming revenue demands, but Rajput and Muslim influences invariably remained strong. Reacting to political events, the diverse groups had all established claims and rights in the land of the region. Hence, whilst according to Baden-Powell 'the bulk of the villages' in Gujarat 'exhibit the usual raiyatwari features' common to

---

85 See ibid., para. 20.
86 Khoti Tenure Selections, p. 15, Report by Bombay Government, 10 January 1825.
87 See *BG, Vol. 10, Ratnagiri and Savantvadi*, pp. 203–13.

western India,[88] the most striking characteristic of the tenurial system in 1850 was the massive extent of land alienated, even greater in proportion than that in the Southern Maratha Country. An 1855 estimate was that nearly Rs. 30 lakhs of revenues were alienated in Gujarat's four districts of Ahmedabad, Kaira, Broach and Surat,[89] well over a third of Goldsmid's 1850 figure for the whole Presidency.[90] Much of these alienations were, as in the Deccan, individual inams not vast estates and their existence suggested a similar pattern of a large, diverse, cross-caste social elite of cultivators. In the Daskroi taluk of Ahmedabad, for example, an early report found sizeable portions of alienated lands owned by every major local group; by Brahmins, Rajputs, Kunbis and the low-caste Kolis.[91]

Nevertheless, one substantial departure from western Indian norms marked Gujarat's tenurial structure and awakened comparison with north Indian experience: the existence of the co-sharing joint village where, on the North-Western Provinces pattidari principle, landlord rights and the liability for land revenue were shared among a number of leading cultivators. Throughout Gujarat, there were 347 such villages in the mid-nineteenth century, of which the 'bhagdari' villages of Broach District numbered 244 and the 'narwa' villages in Kaira District 90.[92] British authorities differed over the origins and significance of these systems. Baden-Powell, anxious to stress a comparison with northern pattidari villages, saw the narwadars and bhagdars as descendants, somewhat reduced socially, of 'an old territorial nobility' or of the grantees of large revenue farms.[93] In contrast, W. G. Pedder, a Gujarat settlement officer and enthusiastic supporter of the systems, argued that they were originally schemes of joint responsibility for government demand, evolved as a protection against the harsh and uneven exactions of Maratha revenue farmers.[94] Partly Pedder's aim was to ensure approval

88 Baden-Powell, *The Land Systems of British India*, Vol. 3, p. 259.
89 *SBG*, NS, No. 29, Correspondence Regarding the Concealment by the Hereditary Officers and Others of the Revenue Records of the Former Government and the Remedial Measures in Progress (hereafter 'Correspondence on Concealment of Records'), T. A. Cowper, Inam Commissioner, Northern Division, to H. L. Anderson, Secretary to Bombay Government, 7 August 1855, para. 50.
90 See the earlier discussion on inams.
91 *SBG*, No. 10, Report on the Portion of the Daskroi Purgunna Situated in Ahmedabad Collectorate by Capt. J. Cruikshank, 1825, para. 52.
92 *SBG*, NS, No. 114, Correspondence Relating to the Introduction of the Revenue Survey Assessment in the Kaira Collectorate (hereafter 'Kaira Revenue Papers'), W. G. Pedder, Survey Settlement Officer, to C. J. Prescott, Superintendent Revenue Survey, 21 March 1862, para. 29.
93 Baden-Powell, *The Land Systems of British India*, Vol. 3, pp. 260-1.
94 See Kaira Revenue Papers, W. G. Pedder to C. J. Prescott, 21 March 1862.

36 *The village in 1850: land tenure, social structure and revenue policy*

for the systems and hence their continuance in the face of the Bombay administration's ryotwari sympathies but he made a powerful case, supported by documentary evidence, that 'the nurwa and bhagdars are merely the old proprietary cultivators'.[95]

Initially, the narwa and bhagdari systems seemed in a state of recent evolution. Pedder discovered one narwa village only organised as such as late as 1771–2[96] and in the early nineteenth century the first British officials could still describe others apparently falling out of co-sharing into ryotwari arrangements.[97] During early British rule, however, the structures solidified. By the mid-1850s the Collector of Kaira was writing that 'no case had for years occurred of a sharehold community breaking',[98] whilst there were no signs of any new narwa or bhagdari villages in the nineteenth century. Within the co-sharing villages proprietary arrangements could vary considerably and numbers of bhagdars or narwadars owning each village fluctuated from four up to over two hundred.[99] Nevertheless, one striking common factor existed: the co-sharing principle seemed mainly the prerogative of the more economically successful cultivators, typically in the more prosperous parts of Gujarat. 'Many of the narwadars have always been the richest and most influential men of the district,' Pedder commented,[100] and the geographical distribution of the system bore this out. Within Kaira District differences of climate and soil were marked, conditions steadily improving as one moved south and east, deteriorating progressively to the north and west. The south grew cotton and tobacco whilst the north and west stood at risk from scarcity in poor seasons. The narwa system's home was overwhelmingly in the south. Borsad, the most south-easterly and prosperous taluk, boasted 54 narwa villages, over half of the total within the system and nearly half of all the villages in Borsad.[101]

Most important, the narwa and bhagdari systems were closely linked to the social aspirations of their participants. Most of the narwadars were

---

95 Ibid., para. 28.
96 See Kaira Revenue Papers, W. G. Pedder to C. J. Prescott, Report on Koobudthal Village, 1861.
97 For example, Cruikshank wrote of Daskroi taluk in 1825: 'there are only two villages in which the nurwas, with an assessment apportioned to each, exist entire, and in about 16 they have been relinquished within the last ten years'. *SBG*, No. 10, Report on the Portion of the Daskroi Purgunna Situated in Ahmedabad Collectorate by Capt. J. Cruikshank, 1825, para. 41.
98 Revenue Report by A. Elphinston, Collector of Kaira, for 1854–5, quoted in *BG, Vol. 3, Kaira and Panch Mahals* (Bombay, 1879), p. 105.
99 See Kaira Revenue Papers, C. J. Prescott to the Acting Collector of Kaira, 17 October 1867, para. 29.
100 Kaira Revenue Papers, W. G. Pedder to C. J. Prescott, 21 March 1862, Appendix 1.
101 Kaira Revenue Papers, C. J. Prescott to Collector of Kaira, 13 October 1867, para. 3.

Kunbis, the same broad caste group as the Maratha peasantry, but of the assertive Leva sub-caste. Increasingly the term 'Patidar' came to be adopted by and for the sub-caste and the designation acquired crucial implications as not only 'a mark of improved status' but also 'both a name and an ideal'.[102] The word 'Patidar' was finally recognised officially in the 1931 Census.[103] 'Patidars' were not just the co-sharers within the narwa system – the caste name had far wider application – but the caste, with their acute social consciousness, recognised many divisions within their ranks and ownership of a genuine 'pati' was the most distinct source of honour. Pedder claimed that 'the pattidar Koonbees of some villages now refuse not only to intermarry with but even to eat with non-proprietary Koonbees of the same sub-division of the caste as themselves'.[104]

By the British period, then, the bhagdari and narwa systems predominantly amounted to a device to confirm, preserve and enhance the social authority of the leading cultivators in these, by western Indian standards, prosperous villages. This was genuine economically based landlordism in the sense that the elite had sprung from the soil, but the systems posed no challenge to central government's authority. Initially, on the British takeover, the villages had been 'taken under the direct management of the Collector and administered ryotwar'.[105] The subsequent restoration of the systems deprived the revenue administration of direct contact with non-proprietary cultivators who were left to fix their own terms as tenants of the narwadars and bhagdars, but, as Pedder insisted, the authorities lost not a rupee by the arrangement.[106] Indeed they often gained. Any land where cultivation had temporarily lapsed yielded no revenue under ordinary ryotwari arrangements, this being one of the much vaunted advantages of the system to the cultivator. Co-sharers, however, were liable for a fixed proportion of revenue due on all cultivable village lands, so that land waste produced revenue too. For this reason, the introduction in 1861–2 of a new narwa settlement in one village, Koobudthal, produced results which were 'all that Government could desire financially'.[107] Sometimes, then, the co-sharers were paying more revenue to maintain their special arrangements. The gain was the right to sub-let entirely on their own terms and the more formally evidenced control over the land, guaranteed by complete exclusion of government revenue officers.

102 Pocock, *Kanbi and Patidar. A Study of the Patidar Community of Gujarat*, p. 1.
103 Ibid., p. 61.
104 Kaira Revenue Papers, W. G. Pedder to C. J. Prescott, 21 March 1862, para. 16.
105 Ibid., para. 26.
106 Ibid., para. 28.
107 Kaira Revenue Papers, H. E. Jacomb, Under-Secretary to Bombay Government, to Revenue Commissioner, Northern Division, 30 June 1864, para. 5.

The co-sharing systems, though, were by no means the only deviations from ryotwari norms in Gujarat. Throughout the region Rajput and some Mughal chieftains had survived as owners and rulers of estates; in Broach where the 'Thakurs' owned estates in the north and, most conspicuously, in Ahmedabad District where 'talukdars' held nearly half the land.[108] The talukdars included men 'of all sorts and kinds, ranging from petty yeomen with shares in a few fields apiece to wealthy Thakers with numerous villages and wanta estates',[109] but they were all at root political rulers, 'rajahs' rather than 'landlords'. Hence their position had depended not on the economic determinants of landlordism but on political resistance to central authority. By 1850, consequently, talukdari estates survived mainly in the outlying parts of the region, notably in the four western taluks of Ahmedabad, 'the border land between Guzerat proper and the peninsula of Katheewar'.[110] Such talukdari estates had had little opportunity to develop an efficient, improving agriculture, for the western taluks of Ahmedabad could 'nowhere approach to the high cultivation of Duskroee or Nuriad', the areas to their east.[111] To this natural economic weakness was added a long tradition of revenue over-assessment. Maratha revenue farmers squeezed the talukdars hard and early British agreements then assumed these former demands were a minimum level to be built upon. Inefficient and corrupt estate administration whereby talukdars were 'the victims of their own agents as well as their creditors'[112] completed the cycle of poverty and deep indebtedness: by the 1860s 'they have during the last half century sunk deeper and deeper'.[113] A settlement officer reporting on Gogha taluk on the Gulf of Cambay in 1865 could condemn most local talukdars as 'indistinguishable from the Kolie rayat, with scarcely a vestige of the Rajput pride lingering about them'.[114]

However, the talukdari system had survived because a wide range of

108 The talukdari estates covered 1,922 out of 4,414 square miles in Ahmedabad District in 1866. *SBG*, NS, No. 106, An Account of the Talukdars in the Ahmedabad Zillah by J. B. Peile, Talukdari Settlement Officer, 6 June 1866 (hereafter 'Peile's Report'), p. 1.
109 I(ndia) O(ffice) (Records and) L(ibrary, London), Government of India Revenue Progs., Vol. 3670, September 1890, A. No. 6, Memorandum by H. E. M. James, Collector of Ahmedabad, 27 February 1890, para. 2.
110 Peile's Report, p. 4.
111 Ibid., p. 2.
112 M(aharashtra State) A(rchives, Bombay) R(evenue) D(epartment Papers), Vol. 149 of 1862–4, No. 841, Part 1 of 1862, A. D. Robertson, Chief Secretary to Bombay Government, to E. C. Bayley, Secretary to Government of India, No. 2232, 6 June 1862, para. 5.
113 Ibid.
114 *SBG*, NS, No. 279, Settlement Report on the Talukdari Villages of Gogha Taluk, Ahmedabad District by N. B. Beyts, 30 October 1865, para. 12.

interests benefited from its continuance. Besides the talukdar himself, all the junior branches of his family, the 'Bhayad' or brotherhood, enjoyed their share of the proceeds and privileges: in Gogha 'the junior branches are in a manner pensioned off by the enjoyment of land which they and their descendants hold rent-free'.[115] Often such sub-sharers were among the most successful beneficiaries of the talukdari system. This was apparently so in Vatva village of Daskroi taluk, Ahmedabad, owned by the 'Barra Meea', a member of 'one of the oldest and noblest Mussalman families of Gujarat'.[116] Besides the Barra Meea, 73 sub-sharers, or 150 individuals including their sub-sharers, held some small stake in the proprietary rights to the village, shares which totalled together came to a substantial amount. In the year Samvat 1944 (A.D. 1887–8), for example, Rs. 3,650 revenue receipts were available for distribution, following payment of the government demand and village expenses, among the talukdari sharers.[117] Of this, the sub-sharers received Rs. 2,569 and from the remainder the Barra Meea had to pay heavy personal expenses like the upkeep of the family tomb. The Collector of Ahmedabad strongly criticised this division: 'the sub-sharers get more than their shares; while the Barra Meea has only his own bit of land rent-free, is responsible for the jama and had to pay into Court on account of debts'.[118]

Even where talukdars tried to renege on traditional responsibilities – in contrast, the Barra Meea and his father had accepted their liabilities and 'considered themselves always as fathers to their humble relatives'[119] – this was clearly far from a regime of oppression or of exploitation. Cultivators in such a ramshackle system could normally protect their own interests successfully enough. Regulated by tradition, it was as difficult to force up demands on cultivators as to reduce obligations to sub-sharers, for 'the exact proportions in which the crop has to be distributed are fixed in each village by immemorial custom, they are well known to every cultivator'.[120] The social prescriptions which regulated the

---

115 *SBG*, NS, No. 279, Papers on the Settlement of the Talukdari Villages of Gogha Taluk, Ahmedabad District, A. Rogers, Revenue Commissioner, Northern Division, to the Chief Secretary to Bombay Government, 8 January 1866, para. 7.
116 *SBG*, NS, No. 365, Papers on the Settlement of Vatva Village, Daskroi Taluk, Ahmedabad District, in 1896, C. J. Prescott quoted in M. C. Gibb, Acting Collector of Ahmedabad, to the Survey Commissioner, 4 February 1897, para. 4.
117 Supplement to *SBG*, NS, No. 365, Papers on the Settlement of Vatva Village, Daskroi Taluk, Ahmedabad District, in 1889–91, H. E. M. James, Collector of Ahmedabad, to W. H. Propert, Commissioner Northern Division, 11 November 1889, para. 7.
118 Ibid. The jama was the government revenue assessment on the village.
119 Ibid., para. 6.
120 IOL, Government of India Revenue Progs., Vol. 3669, June 1890, A. No. 50, A. D. Younghusband, Talukdari Settlement Officer, to the Chief Secretary to Bombay Government, 27 November 1889, para. 5.

relationship between talukdar and cultivators were also jealously guarded. In Vatva, for example, the Barra Meea was traditionally forbidden from entering the village except that part on the public road.[121]

Yet the talukdari system was, by any standards, an inefficient way of distributing the income from a relatively poor agriculture. Many could gain some financial reward from the system but only on the basis of small-scale handouts. The system contrasted with the co-sharing arrangements designed to provide social distinction for the pioneers of a far more productive agriculture. Hence Gujarat's social and tenurial structure, even more so than those of the other regions of western India, was in 1850 extremely varied. The narwadars and bhagdars then provided the most precisely defined and assertive cultivating elite in the Bombay Presidency. Elsewhere, in the talukdari and ryotwari villages, power was widely shared among inamdars, talukdari co-sharers and leading cultivators.

British revenue policy and the village

Our review suggests that land tenure and social structure in Bombay Presidency in 1850 were genuinely marked by the prevalence and dominance of the peasant proprietor. The extent of land alienated throughout the Presidency and the character of the alienations showed this for the inam was usually the perk of the small man. Political and economic factors had underpinned this: the high and uneven revenue assessments of late Mahratta and early British administrations linked with a probable downswing in aggregate agricultural performance[122] to undercut potential landlords' positions, whilst the relatively limited extent of land under cultivation encouraged peasant ownership by removing pressure on potential cultivators to accept tenancy. Even so, local variations were considerable, particularly in the degree of power and independence of village office. The khot built up a landlord's position from it; in the Deccan its rewards were far less. The Gujarati co-sharers, in turn, provided a peasant elite with an exclusive group identity, linking direct economic control over cultivation with an acknowledged source of social status.

But what was the impact of British revenue policy, as by 1850 it became clarified and sharpened, in shaping agrarian society? There is, of course, a widespread view about the character and role of this shaping

---

121 *SBG*, NS, No. 365, Papers on the Settlement of Vatva Village, Daskroi Taluk, Ahmedabad District, in 1896, C. J. Prescott to T. C. Hope, Collector of Ahmedabad, 27 August 1863, para. 30.
122 For a discussion of this issue see Chapter 3.

process. The argument runs that, throughout the south and west of India, British revenue policy determinedly aimed to break down overlord pretensions and any superior tenurial or social claims in favour of the small peasant. The driving force behind this policy, the interpretation goes, was ideological: the early nineteenth-century reaction in political philosophy replacing the Whiggism, which had influenced the Bengal zamindari settlement, with the utilitarian principles of Bentham and Mill. Under the latters' strictures, the superior holder was converted in zealots' minds from a symbol of social stability and agricultural improvement into a parasitical imposition on the cultivator's labours; hence the development of the ryotwari ideal. In Stokes' central study, the Bombay revenue system emerged as the most prominent product of the change in official ideology, at once 'a system closest to the Utilitarian ideal' and 'an aggressive instrument of social change'[123] and Ravinder Kumar's major examination of western India inherited this viewpoint.[124] Yet an alternative interpretation is possible. British policy can be seen as basically empirical, worked out to accord with local reality. This would make its impact less direct and therefore less significant in influence on rural society. Klein has argued that the mid-nineteenth century saw the emergence of a more 'pragmatic method of assessment' at Bombay and that 'the pragmatic method partly signalled the crumbling of English economic theories in a foreign social environment'.[125]

It cannot be denied that Ricardian economics and utilitarian philosophy always had considerable influence on members of the Bombay administration, particularly the revenue department. Elphinstone read Bentham, though without committing himself wholeheartedly to all the implications of utilitarianism.[126] R. K. Pringle, however, the moving spirit behind the land settlements of the 1820s and 1830s, was an enthusiastic protagonist of the new philosophies. He was 'one of Malthus's best pupils at Haileybury' and the winner of a prize medal in political economy.[127] It is tempting, then, to regard his settlement work as inaugurating the most ideologically motivated of the British Indian land revenue systems, and to see its attitudes continuing to influence the evolution of the agrarian structure of western India. Certainly, right down into the twentieth century, Bombay revenue officials could be found discoursing in Benthamite and Ricardian vein on the rent theory of

---

123 Stokes, *The English Utilitarians and India*, p. 126.
124 See Kumar, *Western India in the Nineteenth Century*.
125 Klein, 'Utilitarianism and Agrarian Progress in Western India', p. 585.
126 Ballhatchet, *Social Policy and Social Change in Western India 1817–1830*, pp. 35–6.
127 Stokes, *The English Utilitarians and India*, p. 99.

land revenue or the glories of small peasant proprietorship. As late as 1929 F. G. H. Anderson, Bombay's Settlement Commissioner, produced in his *Facts and Fallacies about the Bombay Land Revenue System Critically Expounded* a straight catechism of Ricardian economics, insisting on land revenue simply as rent and rent as the unearned surplus from agriculture.[128]

Nevertheless, the impression of a consistent ideology dictating the pattern of Bombay land policy from Pringle to Anderson is something of an illusion. Ideology could be paid lip service. It could even win passionate commitment without always directing day-to-day actions and policies. A Bombay revenue official of the mid-nineteenth century would have simply accepted, say, Ricardian concepts of rent, but they did not tell him everything about the complexities of local circumstance and may have been lightly assumed in practice.[129] Pringle's revenue report on the Deccan of 1828 had undoubtedly a strong ideological stamp: the abrupt ending of Elphinstonian sympathy for the established rights of landlords and village headmen revealed this. However, although utilitarianism remained official orthodoxy, the degree of prominence of ideological presumptions varied greatly over time. Empiricism staged a firm revival after Pringle and, arguably, then determined the vital decisions about the long-term shaping of revenue policy.

The change came because Pringle's work, and hence the overt use of ideology, was largely discredited in the eyes of subsequent Bombay officials by its practical failure. Not only did the over-assessments of the 1820s and 1830s limit cultivation and hence revenue receipts. In addition, the practice of the settlements was in such dramatic contrast to the scientific precision which Pringle avowed as his aim. Throughout the Pringle period arrangements on the ground were simply cobbled up as well or as badly as individual local officers could manage. Land revenue was realised through ad hoc compromise demands on assessees, but 'even this assessment was usually higher than it was found possible to collect, so that large remissions had frequently to be made, and considerable balances were left unrecovered'.[130]

---

128 See F. G. H. Anderson, *Facts and Fallacies about the Bombay Land Revenue System Critically Expounded* (Poona, 1929).
129 In any case, as Klein well points out, there is a distinction between the narrow use of the Ricardian rent concept and acceptance of the wider doctrines of utilitarianism: 'English administrators in India naturally cited English economic ideas to explain or justify, . . . but this would only bespeak the clear dominance of Utilitarian concepts if, in addition to Ricardo's rent doctrine, the essence of Mill's programme, or of a strictly Benthamite programme was carried into being.' Klein, 'Utilitarianism and Agrarian Progress in Western India', p. 582, n. 1.
130 H. Green, *The Deccan Ryots and their Land Tenure* (Bombay, 1852), p. 6.

Pringle does not quite deserve his bad press. He was not the only early British revenue official to pitch assessments too high and overassessment, in turn, was merely one element in the contraction of cultivation in the Deccan. Nevertheless, the failure of the first twenty years of British revenue administration in the Bombay Presidency was self-evident. The Bombay revenue tradition was, therefore, to rest on a different foundation. In 1835 George Wingate had come together with H. E. Goldsmid, a civil servant, to conduct an experimental settlement in Indapur taluk of Poona District. Both men were certainly well versed in the fashionable ideologies: Wingate, in particular, knew his Mill.[131] However, their major concern in Indapur was the immediate, practical one of balancing revenue demand correctly and extending cultivation and revenue receipts. Thereafter Wingate rarely escaped from the burden of such aggressively practical matters. His diaries, covering sections of the years 1838–53[132] reveal vividly the physically demanding, nomadic existence of the busy field-working settlement officer. There is little impression here of leisure to reflect on the niceties of political philosophy. Instead the day-to-day solution of the individually minor but in sum crucial local problems of village revenue and tenure predominate, as he tours from village to village.

Throughout the 1840s and early 1850s Wingate, often in partnership with Goldsmid, travelled throughout the Presidency revising the assessments of the Pringle years and elsewhere planning the first effective surveys. Always the object was the same: to try to make local land revenue administration more efficient, based on more accurate land surveys, and to moderate and equalise the burden of assessment. What marked out these settlements was their conspicuous success judged by the standards by which Pringle had failed. From about 1840 a substantial expansion of the cultivated acreage, particularly in the Deccan, began. By July 1852 alone, Wingate was claiming that, thanks to his new surveys, well over a million acres of land in just the four districts of Poona, Ahmednagar, Sholapur and Dharwar had been newly brought under the plough.[133] The more land which was tilled, the more the revenue coffers swelled. The long-term result was the conversion of the Bombay revenue system into one of the most lucrative and regular money-raisers in all India. Total land revenue receipts for the whole Presidency increased by 37 per cent between 1856–7 and 1870–1 and by a further 17.5 per cent between the

---

131 'When he came to refute the proposal for a permanent redemption of the land revenue in 1862, he quoted extensively from the Political Economy.' Stokes, *The English Utilitarians and India*, p. 127.
132 For these, see Wingate Papers.
133 Ibid., Box 293/8, G. Wingate to H. Green, 31 July 1852.

latter date and 1890.[134] At both stages the percentage rises were larger than any of those in the United Provinces, the Punjab, Bengal, Madras or the Central Provinces. By 1890 the Government of India could justly award Bombay its hard-earned approval: 'the growth of land revenue has been more satisfactory in Bombay than in any Province'.[135]

Ironically, this triumph was partly accidental since agriculture, in any case, was emerging from depression, but, whether deserved or not, it made Wingate and his colleagues' reputations. Subsequently, Bombay revenue officials chose to regard the post-1840 successes as the basis of the tradition on which they worked, contrasting its achievements with earlier falterings. The settlement report on Indapur taluk of Poona, revising in 1867 the first of the Wingate–Goldsmid settlements, elaborated a theme common to many administrative accounts of the 1860s: formerly 'there were hundreds of acres lying waste and unprofitable which no man cared to cultivate, owing to the oppressive nature of the assessment and our defective system of revenue administration. All these broad acres have not merely been brought under tillage, and thereby rendered profitable to the State, but their occupancy right has become a valuable private property.'[136]

The tone of the Wingate and subsequent policies was struck by the Joint Report of 1847, a series of guidelines for settlement officers compiled by Wingate, Goldsmid and a colleague, D. Davidson. Utilitarian philosophies – the assumption that the small peasant should be the unit of revenue administration coupled with the belief in detailed administrative organisation – and Ricardian rent theory certainly underlay the report, but such ideas were taken for granted, never explicitly stated. The object was to provide a practical handbook for settlement officers in the field and the pragmatic approach predominated. For example, the report, acknowledging the necessity to vary assessment according to local conditions, did not define what proportion of a ryot's produce land revenue should take. Wingate felt that, in practice, his settlements took for the state between 50 and 75 per cent of the surplus production after deducting all the expenses of cultivation, but considered that 'it is a matter of impossibility to ascertain what is a fair rent in the Political Economist's sense of the word either in this or in any other country'.[137] Reacting against Pringle, there was now to be no attempt to estimate the actual

---

134 IOL, Government of India Revenue Progs., Vol. 3669, May 1890, A. No. 15, Financial Statement for 1888–9, para. 37.
135 Ibid., para. 47.
136 *SBG*, NS, No. 151, Settlement Report on Indapur Taluk, Poona District, by J. Francis, 12 February 1867, para. 98.
137 Wingate Papers, Box 293/8, G. Wingate to H. E. Goldsmid, 12 July 1852.

market value of a holding. Only the investigation of soil and ground conditions would be used to give it a classification. The new overall rates fixed on all or part of a taluk at a settlement – which would then determine the holding's assessment when applied to its classification – were to be decided entirely from relative data like price rises or improvements in transport facilities since the previous settlement. The idea was clearly that the local settlement officer should know by 'feel' at what level to fix land revenue rates. The great object and virtue of the Joint Report scheme was security: to the government in the reliability of its revenue supply and to the cultivator in its assurance of his proprietorship at reasonable rates and the flexibility which permitted the taking up and lapsing of land on an annual basis. 'I believe', wrote Wingate in 1848, 'that in no country of the world are the cultivators so thoroughly independent as they will be in our surveyed districts, when the rules of the joint report shall be fully enforced.'[138] This, he would have said, was an empiricist achievement.

Can we, then, conclude that revenue policy had little original impact on the character of rural society in Bombay; that it always simply moulded itself to fit local circumstance? Interpretations stressing the pragmatic nature of policy often proceed to argue this. Yet the dangers of this approach to British revenue policy are at least as great as seeing ideologically motivated official decisions transforming a pliant Indian society. Pragmatic policies develop their own momentum. Even if a policy was completely based on empiricism, in India it could still generate in time a commitment to its methodology among its administrators, which might become 'ideological' in its strength. In turn, this sort of commitment could produce an intense impact and reaction within local Indian society, where the methodology was rigidly enforced. This was what happened in Bombay. By the 1860s the new generation of revenue officials, notably J. T. Francis, W. Waddington and W. C. Anderson, had succeeded Wingate and their enthusiasm for the Joint Report arrangements was such as to philosophise them into a 'system' with a harder intellectual cutting-edge. It was Francis who wrote the Indapur report of 1867 which so lauded the expansion of cultivation since the Joint Report. Another aspect of the mid-nineteenth-century situation influenced this process. Until Baden-Powell assured his readers in the 1890s that the most suitable land system for one province of India would not necessarily work elsewhere, the quest was on for the most perfect land revenue system.

---

138 PP, 1852–3, Vol. 75, Official Correspondence on the System of Revenue Survey and Assessment in the Bombay Presidency, G. Wingate to E. H. Townsend, 23 December 1848, para. 6.

Officials of the different provinces propounded the virtues of their arrangements with competitive zeal. The Wingate settlements soon came under attack from north India men like Thornton, who denounced the Bombay government for its obstinate failure to discover village communities with which to settle.[139] In reaction, the Bombay defenders inevitably grouped to present their arrangements as a viable, fully-fledged 'system', the true solution to the problems of the Indian land.

The central feature of this system was an intensely detailed investigation of local conditions. The Joint Report laid down that surveys should be made of fields no larger 'than may be cultivated by ryots of limited means'.[140] Such fields were to be demarcated by ridges or mounds of earth[141] and the class of their soil, according to a fixed table, discovered by taking the average of several samples from different parts of the field.[142] The classifier had to take into account the depth and quality of the soil, plus any faults like a sloping surface, and diligently record the facts in his notebook. The details of field and village boundaries were to be recorded in accurate maps.[143] They would be used every thirty years when the revenue assessment was revised.

The creation of this massive Domesday Book was a relentless process which generated its own momentum. It was this determination to survey the land comprehensively, not any special ideological distaste for landlordism, which was to bring revenue policy into conflict with those, like the khots, who claimed superior privileges or any right-holder who distrusted official intrusion. The system declared in 1847, by its precise methodology, made government a more prominent actor on the village stage and hence involved fiercer demands on the agrarian society it encountered. Mahratta administration had never attempted so intrusive a state presence in village life. It had always acknowledged the need for a vast collaborating elite: hence the development of the intricate functions of district and village officer and the elaborate system of perks and land alienations which rewarded them. However, the policies evolved by Wingate and Goldsmid aimed anew to by-pass many of their traditional functions; for example, the government revenue officials would now control the allocation of waste lands so that the Patel would no longer take his cut from introducing new cultivators. In the Deccan such arrangements

139 The North India attack on the Joint Report arrangements was certainly swift and aggressive. See *PP*, 1852–3, Vol. 75, Memorandum by J. Thornton on the Bombay Plan of Survey and Assessment, 11 July 1850.
140 Joint Report, 1847, para. 13.
141 Ibid., para. 20.
142 Ibid., paras. 47–8.
143 Ibid., para. 34.

would significantly reduce the incomes of watandars, already far from large and secure. As, therefore, the Wingate settlements spread, clashes with those who held the range of vested interests they endangered seemed likely. By 1892 W. W. Hunter saw what had happened in Bombay as follows: 'in no other province has the impact of the British land-system been so close and in no province has it met with a Hindu race so capable of resisting it'.[144] The struggle between the 'close impact' and the resistance it called forth was waged intently for twenty-five years after the Joint Report.

The ryotwari system and its enemies

The efficient drive of the Wingate settlements to cut costs and establish detailed contact with cultivators had its earliest impact on the Deccan. The first victim was the faltering institution of district officer, the old Deshmukh. Wingate and Hart's recommendation in 1846 that 'there ought not to be in our opinion any link between our Talook and village establishments'[145] destroyed any hope of the revival of the Deshmukh's administrative function. Subsequent settlements usually gave district officers some compensation for loss of former huks but, typically, it was small and, in cases in the Southern Maratha Country, non-existent.[146]

This, though, was merely the final accomplishment of a long decline. The existence of the Patel could not be threatened in the same way. He was still a vital cog in the administrative machine and any government would require the support of the village authority which only he could provide. In legal terms, the ryotwari system meant the Patel's 'degradation to the level of other cultivators'.[147] Where the state aimed to settle with each individual peasant, then the Patel was simply just another landowner: this was why perks like controlling the allocation of waste lands or selling the holdings of extinct families were now abolished. However, the Patel's functions as an administrator within the village remained, indeed were sometimes increased. In return the Patel was to be regularly paid by the state. The old huks and allowances were normally replaced by some fixed percentage of local land revenue collections: 3 per

---

144 W. W. Hunter, *Bombay 1885–90: a Study in Indian Administration* (London, 1892), p. 12.
145 Wingate Papers, Box 118, W. Hart and G. Wingate to A. N. Shaw, Collector of Sholapur, No. 462, 30 July 1853, para. 3.
146 As in Badami and Bagalkot taluks of Belgaum District. See Wingate's Belgaum Report, para. 36.
147 Sykes, Report on Land Tenures, p. 29.

cent of the revenue of Sholapur District, for example, was in the early 1850s formally reserved for its Patels.[148]

Again, then, a pre-British trend was brought to its logical conclusion. The Patel became a paid government employee. This neutralised the effects of the old struggle between Patel and villagers over payment of huks: if the powerful Patel lost out, then those in areas like Belgaum District, where cultivators had resisted full payment, were now assured of regular remuneration. What the Joint Report system did was to emphasise and highlight the Patel's vested interest in the efficient rendering of land revenue to the central authorities. The real key to the impact of the new arrangements, though, was, clearly, how much the Patels were actually paid. The Wingate settlements were taking a calculated gamble in seeking to increase total revenue receipts by reorganising and mitigating local assessments to encourage cultivation. There was, initially, little spare cash available. Hence the formal levels of Patel payment now introduced were a considerable reduction on old claims. In Badami taluk of Belgaum District Wingate's settlement of 1852 gave the village officers a total of Rs. 1,218 whereas the village documents recorded that the old huks had totalled just over Rs. 2,530.[149] In neighbouring Bagalkot taluk the new award amounted to just over 57 per cent of the official valuation of the huks.[150] How this worked out for the individual can be seen in Dharwar District where the new settlement, involving similar formal reductions, gave an average of around Rs. 43 per year to each village office.[151] And, typically, there were a large number of sharers in these offices.

The Wingate settlements therefore confirmed that under British rule in the Deccan and Southern Maratha Country a Patelship would not provide for an independent power base. Holding village office was still a source of some prestige but its financial rewards appeared to have declined significantly. Why, then, was there not firm resistance from office-holders to British revenue policy? Its absence perhaps shows how much the impact of British rule was merely completing a process already in full motion. Most cultivators stood to gain from the new surveys which

---

148 Wingate Papers, Box 118, G. Wingate to T. Loughran, Collector of Sholapur, No. 462, 30 July 1853, para. 3.
149 Wingate's Belgaum Report, paras. 35–6. Of course, as Wingate pointed out, the old official valuation was a great exaggeration of the actual situation in 1852, for this was an area where huks had been reduced to a minimum. Possibly, then, this settlement was a good deal for the village officers of Badami. Yet its reduction of formal claims here by around a half was typical of settlements of village officers' remunerations everywhere.
150 Wingate's Belgaum Report, paras. 62–3.
151 Wingate Papers, Box 118, G. Wingate to I. S. Law, Collector of Dharwar, No. 272, 11 August 1852, para. 31.

were making revenue assessment less burdensome and inequitable. Over most of the Deccan their interests sufficiently predominated over the office-holders' to ensure that agrarian society acquiesced.

Only in one outlying part of the region was that acquiescence momentarily disturbed. Early in 1852 Wingate went to Khandesh District in the north to prepare the way for the introduction of a new survey and settlement there. Khandesh was then very much a frontier district, largely overrun with bush-jungle and sparsely populated and cultivated.[152] The potential for expansion, however, was considerable. Already the district's central position on the cotton trade route from Berar over the Ghats ensured a lucrative carrying business and Wingate thought that the Khandesh peasants were 'in much more easy circumstances than those of the Deccan',[153] particularly those living on the fertile plain of the river Tapti in the east. This seemed the obvious place to begin the surveys, an area with 'vast capabilities for the production of cotton' but 'now greatly hindered by the high rates of assessment'.[154] So in late 1852 when the settlement officer A. F. Davidson arrived in eastern Khandesh to commence operations in the Sowda, Yawul and Chopra taluks his mood was one of optimistic confidence. 'The appearance of the country around here', he told Wingate, 'quite delights me. Under a more favourable assessment I confidently predict that there will not be a single yard of this beautiful soil waste.'[155]

Clearly many of the local population saw matters differently. When Davidson reached Yawul in mid-November 1852 he soon encountered opposition. On the morning of 15 November a crowd of 'between two and three thousand people' assembled at his tents, in a threatening manner and insisting on 'no survey'.[156] Davidson, believing that any action by him would lead to violence, suspended the revenue survey. For the next month the opponents of the settlement held 'riotous assemblies' – around 12,000 peasants were reported to have gathered at Yawul[157] – and suc-

---

152 In 1851 its population density amounted to 65 persons per square mile, much lower than the Bombay average. Only 14 per cent of the estimated cultivable area was actually under the plough. See *SBG*, No. 1, Report by Capt. Wingate on the Survey and Assessment of Khandesh District, 29 March 1852, paras. 15, 18.
153 Ibid., para. 55.
154 Ibid., para. 45.
155 Wingate Papers, Box 117, A. F. Davidson to G. Wingate, 13 November 1852. (This is not the Davidson of the Joint Report.)
156 Ibid., Box 117, A. F. Davidson to G. Wingate, 17 November 1852. There is a detailed discussion of the events of November and December 1852 in Khandesh in J. F. M. Jhirad, 'The Khandesh Survey Riots of 1852: Government Policy and Rural Society in Western India', *Journal of the Royal Asiatic Society*, 1968, Parts 3 and 4, pp. 151–65.
157 Wingate Papers, Box 117, A. F. Davidson to G. Wingate, 17 November 1852.

ceeded in preventing the revenue officers from crossing to the east of the Tapti. Then the movement was suddenly broken. At dawn on 16 December a contingent of troops surrounded and seized the leading protesters, assembled at Faizpur and at Sowda, and by the following day Wingate could report the whole area as 'quite pacific'.[158]

The Khandesh Survey Riots were hardly the most violent episode in the history of the British Raj but they did show, briefly, how much land revenue settlements depended for their operation on local acquiescence. Why, then, was this acquiescence withdrawn here? A number of complex minor factors were possibly involved. Conceivably the peasants took exception to the survey's administrative demands, particularly the order that landholders should provide stones to set up the boundary marks of their fields.[159] Elphinston, the Collector of Khandesh, was felt by Davidson to have abetted the opposition by his lukewarm attitude to the new survey. He certainly avoided going to mediate between the protesters and the survey team,[160] instead writing to Davidson: 'I think the government will not renew the survey in the talookas where the ryots will not aid it.'[161]

Elphinston made a convenient scapegoat for the troubles but it is hard to assign administrative mistakes more than a contributory role. Perhaps the most tempting interpretation is to see the Khandesh riots as, in 1852, an advanced agrarian movement of an economically progressive area, with the 'rich peasant' flexing his muscles behind the scene. Wingate argued that the peasants of Khandesh, for all its undeveloped nature, were much more commercially orientated than their central Deccan counterparts because the district lay astride the major trade routes from the north: 'the great Agra and Berar roads act as teachers and civilisers and are fast rousing the minds of the people and changing entirely their views and habits of thinking'.[162] Davidson's eventual settlement report on Sowda and Yawul taluks spoke of an elite of peasant traders: they 'secure both the grain and the monopoly of the market' and 'many of them have accumulated vast fortunes'.[163] This elite, the four wealthiest

---

158 Ibid., Box 293/4, Diary for 17 December 1852.
159 This, together with the fear of rising taxation, is the major reason given by S. B. Chaudhuri for the Khandesh troubles. See S. B. Chaudhuri, *Civil Disturbances during British Rule in India, 1765–1857* (Calcutta, 1955), p. 171. Davidson himself, however, thought this of very minor importance: on this issue, he claimed, he was easily able to pacify complaints. See Wingate Papers, Box 117, A. F. Davidson to G. Wingate, 17 November 1852.
160 For Davidson's grumbles on this, see Wingate Papers, Box 117, A. F. Davidson to G. Wingate, 17 November 1852.
161 Ibid., Box 117, A. Elphinston to A. F. Davidson, 23 November 1852.
162 Ibid., Box 293/8, G. Wingate to A. Mackay, 27 February 1852.
163 *SBG*, NS, No. 43, Settlement Report on Sowda and Yawul Taluks, Khandesh District, by A. F. Davidson, No. 206, 23 December 1854, para. 23.

families of which Davidson described as 'millionaires',[164] might seem the natural leadership for the no-survey movement and certainly economic factors have been prominent in most modern interpretations of the riots.[165]

Yet why should rich peasants oppose the new survey for economic reasons alone? The 'millionaires' of east Khandesh were the very sort of peasant entrepreneur whom the Wingate settlements aimed to encourage. This was what so baffled the survey team: the eastern taluks had been especially chosen because they were 'the most oppressed by the present rates'.[166] In fact the Khandesh disturbances only make sense when viewed as a protest (the only significant one in the Deccan) by village officialdom, fighting the threat of the survey to its autonomy. What is striking is that the protest was opposition in principle to the whole idea of any new survey. The redress of specific economic grievances, such as fears about the new rates, seemed unimportant. Davidson realised this when he met a group of peasant representatives on 15 November. They did ask some detailed questions such as whether waste land would be taxed but in general 'they wished for no survey. No compromise on my part would be of any avail.'[167] Elphinston, who had at first argued for the boundary stone regulations as the reason for the disturbances, conceded by late November that 'the ryots' objection to the survey is objection in toto'.[168]

Such objection would have served the interests of a small but dominant group within Khandesh society. The Wingate settlements generally reduced revenue levels but they were notably harsh on the financial and social pretensions of district and village officers. In east Khandesh, significantly, office holding seems to have produced more substantial rewards than further south. Thus in a region where the bulk of fertile, cultivable land still lay waste, the Patel's right to supervise and take his cut from the allocation of new land was a lucrative source of income. Even high revenue rates hardly harmed office-holders who were powerful enough to keep a significant share. The difference between the wealth and influence of the office-holder in Khandesh and in the bulk of the Deccan was revealed by the virility of the Khandesh district officer: in contrast to the dying breed further south who faced administrative extinction under

164 Ibid.
165 This is one of the main planks in Miss Jhirad's thesis. She points to the economic hierarchy among the Khandesh cultivators and concludes that 'an expansion of the economic functions and personal fortunes of the traditional peasant capitalists' had taken place, making them prepared to fight the survey. Jhirad, 'The Khandesh Survey Riots of 1852', p. 161.
166 Wingate Papers, Box 117, A. F. Davidson to G. Wingate, 20 October 1852.
167 Ibid., Box 117, A. F. Davidson to G. Wingate, 17 November 1852.
168 Ibid., Box 117, A. Elphinston to A. F. Davidson, 23 November 1852.

the new settlements, the Collector of Khandesh could speak of 'the hereditary revenue officers, whom the ryots obey as the sheep the shepherd'.[169] These men, Elphinston eventually told Davidson, provided the real momentum behind the movement. They 'have secretly instigated resistance to the survey partly because the reductions of revenue reduce their percentages, but chiefly because you order the allowances of these hereditary officers to be mostly stopped'. Hence 'this is their fight for their pockets'.[170] Patels, too, were clearly among the leadership. Wingate suspected from the start that 'the village officers ... are at the bottom of the affairs',[171] and later events confirmed their role. When, for example, the chief protester, Heeramen Alkuree, escaped from the rounding-up operation at Faizpur, it was with the connivance of the Patel of Faizpur, Bursu Chowdree.[172]

Attribution of the Khandesh riots to the local district and village officers later became established orthodoxy within the Bombay revenue department,[173] and this view does seem substantially correct. Davidson's belief that the disturbances were originally created by a few 'influential persons' rather than being based on spontaneous mass protest makes sense given, particularly, the speed with which the movement collapsed in December. The Khandesh disturbances, then, need to be located firmly within their context. They were not a forward-looking protest about the economic grievances of any rising peasant groups but a reaction, a desperate fling by the threatened official to protect his privileges.[174] Their failure confirmed the impotence of the old Deccan village and district officer before the settlements which set the seal on their decline. In east Khandesh the officer class was unusually powerful and the revenue authorities' lines of communication most stretched. Even here, though, the limited demonstrations quickly collapsed, once the ringleaders were rounded up.

Nevertheless, we have been talking about the decline of an *institution*, particularly in the Deccan. The office-holder may have flourished wearing a different hat, for even in the Deccan, as we saw earlier, the vast extent of

169 Ibid., Box 117, A. Elphinston to A. F. Davidson, 15 November 1852.
170 Ibid.
171 Ibid., Box 117, G. Wingate to H. Goldsmid, 22 November 1852.
172 Ibid., Box 293/4, Diary, 17 December 1852.
173 See Rogers, *The Land Revenue of Bombay*, p. 290.
174 Miss Jhirad discusses this aspect of the riots, describing them as 'a mobile monster petition organized with traditional methods of social control by those with an interest in maintaining the existing agrarian arrangements'. See Jhirad, 'The Khandesh Survey Riots of 1852', p. 164. However, she somewhat spoils her case by pressing a comparison also with the Deccan Riots of 1875 and the origins of the twentieth-century Ryots Associations. The argument for contrast between the protests against the revenue survey and these later events will become clear as this study continues.

land alienated still gave the opportunity for a variety of groups to buttress their economic and social position. How, then, did the inamdar fare in the era of the Wingate settlements? The extent of alienations had been the product of widespread official ignorance and inevitably formed a challenge to the mood of the Bombay government in the 1840s. The sums of revenue involved, also, were massive and during the late 1830s and early 1840s, as Bentinck's budget surpluses were spent and, particularly, the Afghan campaign opened a new era of wars, the income of Indian governments came severely under pressure.[175] The response was the appointment of an 'Inam Commission' in June 1843, at first to investigate titles to alienated land in Dharwar District.[176] The following year W. Hart was appointed the first 'Inam Commissioner' and he began full-scale operations testing rights to alienations throughout the Southern Maratha Country. Financially, the work showed steady results and by the end of 1847 some Rs. 40,000 of unauthorised annual inam revenue, recoverable by government, had been brought to light by the Commission.[177] In 1852 legislation gave the Bombay government the power to extend the Commission outside the Southern Maratha Country to any part of the Presidency.[178]

Here, it might seem, was the inexorable pressure of government action again bearing down on local privilege. Yet the inamdar had plenty of weapons with which to retaliate, a powerful armoury of silence and concealment. The legal validity of most land alienations could only be tested by documentary evidence, but many of the relevant records were destroyed or hidden from the British authorities. The means of disposal were varied and original. Some valuable documents, for example, were discovered in 1852 being used as waste-paper by grocers and firework makers in Poona City.[179] The find was part of an enquiry in the early 1850s which unearthed many records whose very existence had been emphatically denied by local officials.[180] One Belgaum district officer who in 1846 had produced 29 accounts as the sum total of all his public records was found in 1852 to possess over 1,800 'most valuable' docu-

---

175 On the rapid growth of the Indian Public Debt in the early 1840s, see Romesh Dutt, *The Economic History of India in the Victorian Age* (London, 1903), p. 217.
176 See Inam Commission Papers.
177 Ibid., W. Hart to D. A. Blane, Revenue Commissioner, No. 597, 1 July 1848, Postscript, Tables 2 and 3.
178 See ibid., Act No. XI of 1852.
179 *SBG*, NS, No. 29, Correspondence Regarding the Concealment by the Hereditary Officers and Others of the Revenue Records of the Former Government, T. A. Cowper, Inam Commissioner, Northern Division, to H. L. Anderson, Secretary to Bombay Government, No. 876, 7 August 1855, para. 41.
180 See ibid.

ments.[181] As a result of the enquiry the Bombay government prepared legislation in 1856 to outlaw and provide penalties for the concealment of public revenue documents but the effectiveness of such enactments remains unknown.

Again, even if obstructionism at local level could be frustrated and information obtained, government still faced the insidious threat of corruption within its own departments set up to examine claims. The potential for bribery and pressure was clearly immense where lesser, poorly paid officials with their own contacts and position in local society were called upon to adjudicate rights claimed by the wealthy and influential. In 1852 an enquiry into the holding of grants of surinjam in the Deccan revealed massive corruption within the Agent for Sirdars Department, which had judged these claims.[182] The rules for testing claims to surinjams had been strictly established: those granted before A.D. 1751 were to be fully recognised as hereditary rights, surinjams awarded between 1751 and 1796 were to be allowed to the holder and his successor, but the third generation would receive just half the former proceeds and that would end the entitlement. Rights granted after 1796 (i.e. by the last Peshwa) were allowed in full to the holder only, with a pension of half the proceeds just to his successor.[183] Supposedly on this basis, lists of the valid claims admitted by government were made by the Agent for Sirdars Department in both 1844 and 1847. Yet the Inam Commission's enquiry in 1852 found that surinjams worth nearly Rs. 90,000 a year were then being held improperly.[184] One notorious case was that of the Kuddum Banday family whose three villages in Khandesh had been recognised as a hereditary surinjam, whereas even the Peshwa had confiscated the grant.[185] The Agent for Sirdars Department, the Governor of Bombay was moved to comment, had proved 'an organisation for the deliberate robbery of Government'.[186]

This story, of the deceiving of the imperial establishment by local elites, of course, makes familiar reading for students of the Raj. The other fundamental problem created by the enquiries into alienations also typified an intrinsic weakness of imperialism: its continual need for local acquiescence to protect its security. The long, painstaking investigation of the Inam Commission was a tightrope-like operation. The revenue it

181 Ibid., para. 21.
182 See *SBG*, NS, No. 31, Correspondence Exhibiting the Results of the Scrutiny by the Inam Commission of the Lists of Deccan Surinjams.
183 Ibid., T. A. Cowper to H. L. Anderson, No. 1108, 29 October 1855, para. 6.
184 I.e. they were placed in a higher one of the three categories than they should have been. Ibid., para. 83.
185 Ibid., Appendix A.
186 Ibid., Minute by Lord Elphinston, 8 July 1856, para. 4.

returned to government was invaluable but the process also risked alienating many influential local interests. From the start the Inam Commission's judgements erred on the side of generosity, often allowing completely unauthorised alienations to incumbents.[187] The Mutiny of 1857-8 increased the nervousness of the Bombay government, for the parallel between the Oudh talukdars and the Bombay inamdars was too close for comfort. The dangers were enhanced by the uncertainty which the slow working of the Inam Commission's operations created: 'until they are brought to a close, no man, whose title to exemption or pensionary provision has not been adjudicated, feels himself secure, and this feeling of insecurity must unsettle men's minds, and predispose them to discontent and sedition'.[188] To end this insecurity, the Bombay Governor felt in the aftermath of the Mutiny, was 'worth the sacrifice of a large part of the reversionary interest of government in all alienated revenue'.[189]

So the Bombay government, during the early 1860s, abandoned the ideal of bringing all alienated lands within its full scrutiny. Two Acts of 1863 instead set up a 'summary settlement' mechanism for alienated lands. The inamdar was to be offered full proprietary, hereditary rights over his estate, provided he accepted a small summary assessment made on it by the Collector. If the inamdar refused to pay any such revenue, then an Inam Commission type enquiry would be held which might declare the lands absolutely rent-free but might alternatively rule the estate liable to full levels of assessment.[190] The resultant compromise produced new revenue, quite apart from the saving on the administrative costs of the Inam Commission. The annual amount of alienated land revenue was reduced from 132 lakhs of rupees to just over 80 lakhs by the late 1880s.[191] Yet, as these figures suggest, the bulk of inam privilege remained, a survival which had seemed unlikely when the Inam Commission had begun operations. The inamdar's power-base, an estate not liable to the full land revenue assessment, survived until Independence, when some 12–13 per cent of both the revenue and occupied area of

---

187 For example, of the Rs. 40,000 of unauthorised inam revenue discovered by the end of 1847, only Rs. 23,334 worth had been ordered for immediate resumption by government. See Inam Commission Papers, W. Hart to D. A. Blane, No. 597, 1 July 1848, Postscript, Tables 2 and 3.
188 Wingate Papers, Box 120/7, Papers Relating to a Summary Settlement of Alienated Revenues in the Bombay Presidency, Minute by Lord Elphinston, 4 April 1859, para. 1.
189 Ibid., para. 70.
190 The arrangements are fully discussed in Baden-Powell, *The Land Systems of British India*, Vol. 3, p. 304.
191 Ibid.

Maharashtra was still alienated.[192] Only a comprehensive programme of legislation in the late 1940s and early 1950s finally destroyed the system.[193]

In the Konkan region, another major source of conflict with superior right-holders existed. British policy down to 1850 had largely avoided the challenge posed by the khoti system or at least temporarily accepted defeat and here, therefore, the change represented by the more active mood of the Wingate settlements was to be most evident and to produce the most intense struggle. For Wingate had strong views about the khots, opinions which he presented forcefully in his Report on Ratnagiri District of 1851.[194] The report, more so than any other Wingate wrote, showed a firm utilitarian distaste for landlordism. He argued that the rights of the khots 'fall very far short of an absolute proprietorship of the soil'[195] and, further, that the khoti system was notably less productive of agricultural improvement than the ryotwari.[196] Yet if this sounds like simple political ideology, one suspects that the cardinal sin of the khoti system, in Wingate's eyes, was that it resisted a revenue survey, remained beyond the pale of the system of measurement and assessment delineated in the Joint Report. This was certainly seen as the first task: Wingate's report recommended that a full revenue survey be introduced as soon as possible in Ratnagiri District.

Even so, there was no speedy start. Waddington began operations during the mid-1850s but the Mutiny outbreak of 1857 led to their suspension, the product of the same fears within Lord Elphinston's government which had motivated re-examination of the Inam Commission's work. A full survey in Ratnagiri, directed by J. T. Francis, was resumed in 1859. Francis stressed that the object was survey-measurement and compilation of records in the Wingate tradition. The aim was not to challenge the khots' overall right to settle with government for the khoti villages, but to delineate holdings and to establish rents and occupancy rights of the peasant cultivators. While the work was confined to survey, it was accepted. However, from the mid-1860s, settlements – the fixing of revenue assessments – based on the new surveys began and now proceeded swiftly: in Bankot taluk in 1866, and in Khed and Saitavda, in Dapoli, in Ratnagiri and in Chiplun in successive years. A storm of pro-

192 The figures for Maharashtra in 1947–8 were 3.75 million acres of alienated lands out of a total occupied area of just over 30 million acres and nearly 32 lakhs of rupees alienated out of a total revenue of almost 234 lakhs. See G. D. Patel, *The Land Problem of Reorganised Bombay State* (Bombay, 1957), Appendix H, p. 445.
193 See ibid.
194 See Wingate's Ratnagiri Report.
195 Ibid., para. 35.
196 Ibid., para. 18.

The ryotwari system and its enemies

tests resulted. The khots complained bitterly that the new rates were too high and that the survey records on which they were based were proving defective.[197] More likely, the implications of the operations were now becoming clear to the khots. With a full survey and settlement established, the authorities could always protect the position of any khoti tenant and might even, at some future date, arbitrarily abolish khoti privilege, all without undermining the security of their own revenue supply. In 1868 Francis warned the government that the opposition of the khots was one of principle and that therefore a fierce battle was likely to impose the settlement: 'some of them [the Khots] have openly declared from the first that they won't have the survey in their villages, and, however fair the assessment may be, we may still expect to see them opposing it'.[198]

As the settlements continued, Francis' prophecy was fulfilled. Indeed in the early 1870s opposition hardened into total non-co-operation. In 1873 the Collector of Ratnagiri reported that of 682 khoti villages in the newly settled areas the khots of 417 were refusing to accept the assessment.[199] Their tactics now even included legal action against the revenue authorities.[200] From the government's point of view, this situation was unsatisfactory in a number of ways. Firstly, many khots seemed to be reacting to the settlements by clamping down on cultivators' rights and even ejecting tenants to prevent them being granted full occupancy rights. One official claimed in 1873 that 'hundreds of suits have been brought by the khots against their ryots merely on account of the survey'.[201] To increase tenant insecurity was clearly the exact opposite of what was intended. Most important, however, the khots' campaign against the settlements considerably reduced revenue returns from Ratnagiri, so that in the early 1870s large balances of land revenue remained uncollected.[202] Here was the nub: if the new settlements were not increasing revenue receipts what was their point?

These difficulties led to reconsiderations of the policy. The whiff of

---

197 See *SBG*, NS, No. 377, Settlement Report on Ratnagiri Taluk, Ratnagiri District, by J. L. Lushington, No. 148, 30 March 1897, Appendix R, J. T. Francis, Survey and Settlement Commissioner, to the Acting Commissioner, Southern Division, No. 193, 29 February 1868.
198 Ibid., para. 24.
199 MARD, Vol. 61 of 1875, No. 526, Collector of Ratnagiri to the Revenue Commissioner, Southern Division, No. A.421, 10 March 1873, para. 8.
200 In 1868 and 1869 over one hundred suits were filed by khots against the Collector as a legal challenge to the settlement's validity. Ibid., para. 7.
201 MARD, Vol. 61 of 1875, No. 526, E. T. Candy to J. Elphinston, Collector of Ratnagiri, 9 January 1873, para. 12.
202 See ibid., Collector of Ratnagiri to the Revenue Commissioner, Southern Division, No. A.421, 10 March 1873, para. 10.

utilitarian ideology, in the air with Wingate's report, had long since been replaced in the revenue department by a more customary pragmatism in policy towards the khots. If it made practical sense now to recognise their position, then advocates would be found to justify the change. Prominent among these by the early 1870s was E. T. Candy, the editor of an official publication about the khoti system,[203] who argued that 'the watandar khots have important rights' whilst some of the cultivators in khoti villages were 'purely tenants-at-will'.[204] In 1874 the Bombay government recognised the reality that the new settlements were not working by appointing a Khot Commission to reconsider the situation in the Konkan. Thereafter a compromise was hammered out in the late 1870s. The khots' full rights over their villages were conceded, but most of the cultivators became customary tenants with hereditable but not transferable rights entered in the village register and only around 5 per cent of cultivators were officially recorded as tenants-at-will.[205] This was legalised by a Khot Act in 1880.

The settlement was formally a compromise but in reality the khot had won. In the khoti regions the principles of the Joint Report – effective government knowledge of and contact with the cultivating situation – were abandoned. Officially, rules regulated relations between khots and tenants but how could the authorities know or influence what actually happened in the khoti village? Here the machinery of the revenue department would only be used to collect dues for khots from obstreperous tenants, a function on which the khots soon seemed ready to draw.[206] If anything, the failure of the push against the khoti system had exacerbated the position of some cultivators by causing khots to sharpen their control over the tenurial situation.

Yet it is hard to see how the khots, granted their complete determination to oppose the settlements, could have been broken. As we saw earlier, the interest in the khoti system was quite broadly based within Konkan society and some officials argued that cultivators occasionally fought the settlements as bitterly as the khots.[207] In addition, the sales of

203 See *Khoti Tenure Selections*.
204 MARD, Vol. 61 of 1875, No. 526, E. T. Candy to J. Elphinston, 9 January 1873, para. 29.
205 *BG, Vol. 10, Ratnagiri and Savantvadi*, p. 257.
206 In 1880 13,433 notices, 1,085 attachment warrants and 45 sales were authorised or issued by government to aid khots in recovering their rents. MARD, Vol. 85 of 1880, No. 1060, Revenue Department Resolution No. 6026, 15 November 1880, para. 5.
207 For example, A. K. Nairne, an Assistant Collector, suggested that in his sub-division 'the rayats are not merely the tools of the khots, but are really fighting for themselves'. *SBG*, NS, No. 254, Papers on the Settlement of Khed Petha, Old Suvarndurg Taluk, Ratnagiri District, A. K. Nairne to the Collector of Ratnagiri, No. 77, 27 November 1868, para. 2.

khoti office or of shares in it had brought in groups who proved the fiercest opponents of the settlement: they had 'laid out their money, in very many cases quite recently, in the trust that things would always remain as they were'.[208] At the same time, potential sellers could only, in future, raise money on their khotships if khoti rights and privileges remained unaffected. Finally, the khots' claims were publicised by an effective propaganda network, which efficiently sowed doubts in officials' minds about the rights of the issue. Unlike Deccan Patels, Brahmin khots had close links with literate, urban Indian opinion. V. N. Mandlik, for example, the influential lawyer, was an articulate spokesman on their behalf,[209] and most of the press highlighted the issue, supporting them wholeheartedly: even fresh official enquiries about khoti rights in 1890 brought a storm of protest from newspapers.[210] This combination, underpinned by the crucial denial of full land revenue supply, was of the type that imperial rule could not easily resist.

Against such political realities, any comment on the legality of the khots' claims sounds like mere academic semantics. Nevertheless, investigations after 1880 did suggest that the khots' proprietary rights scarcely differed from those of other village officials. In the late 1880s the first detailed enquiry took place into the wording of the khoti sanads, the documents which had originally prescribed the form of the grants to the khots. The report revealed that not a single Ratnagiri khot could produce 'an express recognition by the former government of his having a right to the soil of his village'.[211] Instead, it was argued, the Peshwas had treated the khoti watan as an ordinary service watan: the justification for the khots' existence had been 'the promotion of cultivation' and 'the payment of government assessment' and, for instance, lapsed holdings in khoti villages had reverted not to the khots but to government. Hence the report concluded that the khot was legally a very similar official to the Deccan Patel: 'a khot does not possess proprietary rights any more than a gaokar or a patel above the Ghats does'.[212] This revelation came far too late for any further reversal of policy. The authorities could only accept that, as a Collector of Ratnagiri put it, events had converted the khots

---

208 Ibid, para. 3.
209 See 'Brief History of the Watandar Khotes', in Mandlik, ed., *Writings and Speeches of the Late Honourable Rao Saheb Vishvanath Narayan Mandlik*, p. 91.
210 See Report on the Vernacular Press in the Bombay Presidency, week ending 6 September 1890, paras. 19–21.
211 *SBG*, NS, No. 446, Papers Regarding the Proprietary Rights of Khots in the Ratnagiri District, Achyut Bhaskar Desaie, Special Mamlatdar, to J. R. Naylor, Remembrancer of Legal Affairs, No. A.105, 31 May 1886, para. 12.
212 Ibid.

'from the position of heavily assessed farmers of the land revenue to that of landlords'.[213]

Yet, to stress again the local variations, the khoti system far from covered the whole Konkan. In the rest of the region, throughout Thana District for example, the post-Joint Report settlements proceeded efficiently and against a background of local acquiescence. Only one serious challenge emerged, the exact Konkani equivalent of the Khandesh Survey Riots. This occurred in North Kanara, the most southerly of the Konkan districts which was only transferred to the Bombay Presidency from Madras in 1862. A new revenue survey and assessment in the district then began and in 1870 the operations of W. C. Anderson and his settlement team reached Karwar taluk. Here they found a more strictly stratified social system than in most of the non-khoti areas of the Konkan. A Brahmin elite owned and sub-let most of the land and, Anderson claimed, had the reputation 'of being most hard and exacting landlords'.[214] Most of the peasant cultivators held land from them on annual leases, but an important and influential group were superior occupancy tenants called mulgenidars, enjoying permanently fixed rent levels.[215] Like east Khandesh, Karwar taluk was potentially a wealthy area. Its coconut and cashew-nut trees provided cash earnings and its rice lands were capable of producing sugarcane every three years or the occasional second crop of grain. Land values and the profits of cultivation were, Anderson argued, rising fast in 1870.

The survey team successfully carried out their measurements in Karwar, but thereafter trouble began. At the announcement of the new rates of assessment for 18 villages 'a large proportion of the most influential landholders absented themselves, under the idea that they might thus plead ignorance of the survey rates for a time'.[216] The commitment of the opposition in these villages, however, only became clear when the new land revenue assessments first fell due, part in December 1870 and the remainder in February 1871. By March only just over Rs. 5,000 of the total Rs. 20,394 due from the 18 villages had been paid[217] and, like the khots, many of the landholders concerned began taking to legal action against the revenue authorities. This had now become a concerted cam-

213 MARD, Vol. 61 of 1875, No. 526, Collector of Ratnagiri to the Revenue Commissioner, Southern Division, No. A.421, 10 March 1873, para. 13.
214 SBG, NS, No. 158, Settlement Report on Karwar Taluk, Kanara District, by W. C. Anderson, No. 168, 21 February 1871, para. 22.
215 Ibid., para. 31.
216 Ibid., para. 30.
217 SBG, NS, No. 158, Papers on the Settlement of Karwar Taluk, Kanara District, Collector of Kanara to the Revenue Commissioner, Southern Division, No. 703, 3 March 1871, para. 14.

paign and two officials, sent out in March 1871 to investigate matters, reported back that 'there is a very general combination amongst the entire body of cultivators to resist the payment of the revised rates of assessment'.[218]

Why should this occur in just these 18 villages? Rumours and administrative complexities created by the transfer from the Madras to the Bombay revenue system may have exacerbated grievances for the protesters alleged that their old assessment had been permanently fixed and that therefore any acknowledgement of a thirty-year settlement would infringe their rights.[219] In addition, the obvious financial motive, denied potential opposition to the settlements of 1840–60, now existed. With the increase in crop prices associated with the American Civil War cotton boom, Wingate's successors were, from the 1860s, looking to raise revenue rates again, and in Karwar the new settlement did involve a moderate, overall rise. Nevertheless, the same paradox as in east Khandesh occurred: many individual assessments were lowered and yet landowners concerned often still opposed the settlement.[220] In fact, the whole affair seems highly reminiscent of the Khandesh Survey Riots. Here was a relatively prosperous area with an elite who combined social and administrative position with the power to raise or coerce the peasant masses: the Karwar Brahmins were both village officers and dominant landowners. This position had provided the opportunity for the creation of curious inaccuracies which were now discovered in the Karwar revenue records. Some cultivated land, for example, had apparently remained unassessed and the overall increase in land revenue created by the new settlement, it was later argued, 'arose not from high rates but from the taxation of this hitherto concealed cultivation'.[221]

So the Karwar elite had a strong vested interest in what was again, as in east Khandesh, a struggle against the principle of a new survey and settlement. Brahmins, or at least occupancy tenants, seem to have led the movement and in the eleven northernmost villages 'not a single Brahmin is among the number of those who have paid the assessment due in full'.[222] Most of the peasant cultivators were drawn into the campaign, but the official attitude, as with the Khandesh Survey Riots, was that they had been bludgeoned into opposing a settlement fundamentally in their

---

218 Ibid., H. Ingle, Deputy Collector, to the Collector of Kanara, No. 92, 17 March 1871, para. 18.
219 Ibid., para. 2.
220 Ibid., R. E. Candy, Assistant Collector, to Collector of Kanara, No. 124, 15 March 1871, para. 6.
221 Rogers, *The Land Revenue of Bombay*, Vol. 2, p. 412.
222 *SBG*, NS, No. 158, Papers on the Settlement of Karwar Taluk, Kanara District, R. E. Candy to Collector of Kanara, No. 124, 15 March 1871, para. 18.

interests: most, an Assistant Collector claimed, 'when questioned by me, can give no intelligent or connected account of their reasons for their refusal to pay'.[223] Certainly, in so far as most land and the obligation to render revenue was owned by Brahmins, the grievances of most tenants towards any settlement would be entirely indirect.

However, in just these 18 villages of one taluk, opposition could hardly be as widespread and menacing as that organised by the khots. Officials, too, had no qualms here over the legality of their actions and the cases against the authorities led to a firm decision by the High Court in 1875 in favour of the government's right to raise the land revenue in Karwar. Yet valuable concessions, especially designed to appeal to the landowning elite, had been necessary to end the resistance. Firstly, owners of land uncultivated at the time of the settlement were allowed their right of occupancy at only one-eighth of the full assessment for five years, a significant aid to large landowners. The other major concessions also helped them: where assessment of over Rs. 25 had been raised more than 50 per cent, the increase was only to be levied fully after three years.[224]

Our review so far of the evolution of British revenue policy and practice begins to suggest different themes for the different regions. In the Deccan there was the efficient, mainly successful imposition of the ryotwari survey and settlement, by-passing many old watandar functions. In the Konkan the theme was more of struggle with administrative and landlord elites and of a more equivocal outcome. Moving on to Gujarat, the mood, in contrast, was of compromise with the more numerous but less threatening tenurial variations. It was epitomised by the summary settlement of claims to inams, which affected Gujarat with its high proportion of alienated land more than any other region. Gujarat, also, illustrates another important theme of the evolution of revenue policy and its effects. The pattern of interaction – of conflict and compromise – between adminstrative action and the agrarian social system was never consistent on either side. Just as the response from the village varied, from the wholehearted obstructionism of the khot through to the impotence of the Deccan watandar, so too policy constantly fluctuated in its force and sentiment. Thus, while ryotwari principles remained unchallenged as the mainstream of Bombay revenue policy, the enthusiasm and universality with which they were applied wavered. Sympathisers with landlord aspirations proved at times an influential minority, particularly during the 1860s and 1870s when the warning provided by the Oudh talukdars

---

223 Ibid., H. Ingle to the Collector of Kanara, No. 92, 17 March 1871, para. 19.
224 See Rogers, *The Land Revenue of Bombay*, Vol. 2, p. 413.

was freshest in official minds. They included Raymond West, the High Court judge[225] and Richard Temple, who as Governor of Bombay between 1877 and 1880 played an important role in framing the compromise Khot Act of 1880.[226] Admittedly, such men rarely included the revenue officers on the ground: they – the Wingates, Francises and Waddingtons – fiercely resented deviations from peasant proprietary norms for their threat to the efficient progression of survey and settlement. Yet where it could be shown that tenurial variations did not challenge that efficiency or oppress already low peasant living standards (the claim that was made against the khots) then even settlement officers could tolerate them. In Gujarat these conditions existed.

The bhagdari and narwa systems, for example, fitted them. When the post-Joint Report surveys and settlements began in Gujarat, the complete abolition of the systems might have been expected. However, as we saw earlier, the advocacy of W. G. Pedder, combined with their reliability of revenue supply, led to the British revenue authorities accepting their existence. Perhaps, though, the key factor was that the principle of survey was not here under challenge. The bhagdars' and narwadars' rights and responsibilities remained unchanged and the surveys which were carried out simply gave the co-sharer the opportunity at any time to end the arrangement and opt for a straight individual assessment on his holding. By the 1880s Baden-Powell thought this was frequently happening[227] but later evidence – that, for example, provided by the pattern of Gandhi's Kaira satyagraha – confirmed that the systems had survived intact into the twentieth century. The bhagdars and narwadars remained a group apart, aided by the consistent generosity of settlement policy towards them.[228] In Broach, for example, although 'a very high demand for agricultural lands' existed by the end of the nineteenth century, still bhagdari land was 'seldom brought under the hammer'.[229]

---

225 West earned prominence with his *The Law and the Land in India* which warned about the dangers of Deccan peasants losing their land to alien moneylenders. Yet the same work also argued strongly for bolstering traditional landlord groups where they existed, commending the 'certain kindliness in the relations between the old proprietors and those under them'. Raymond West, *The Law and the Land in India* (Bombay, 1872), para. 36.
226 Temple's attitude to the khots was illustrated by his remark in 1877 about the settlement attempts of the 1860s: 'a status was being conferred on large numbers of the tenantry beyond what could be established by custom or prescription'. IOL, Papers of Sir Richard Temple, Vol. 193, Minute by Richard Temple, 4 June 1877, para. 2.
227 See Baden-Powell, *The Land Systems of British India*, Vol. 3, p. 268.
228 Crispin N. Bates, 'The Nature of Social Change in Rural Gujarat: the Kheda District, 1818–1918', *Modern Asian Studies*, 15, 4, October 1981, pp. 780–3.
229 *SBG*, NS, No. 407, Settlement Report on Broach Taluk, Broach District, by P. R. Mehta, No. 593, 11 August 1899, para. 43.

64 *The village in 1850: land tenure, social structure and revenue policy*

The real test of the flexibility of British revenue policy in Gujarat came in the early 1860s. By then the affairs of many of the Ahmedabad talukdars had reached crisis point: talukdari villages began to be sold off by the courts to satisfy creditors' claims, often for very low prices. In September 1862, for example, three-quarter shares in three villages, involving a land area of over 6,000 acres, were attached and sold for only Rs. 231 and the next month, four villages, 5,030 acres in extent, fetched just Rs. 256.[230] J. B. Peile, a local official sympathetic to the talukdars, presented the problem to government starkly: 'the state must either help these men or complete their ruin'.[231] In terms of the ideological commitments usually seen as ruling in Bombay, the latter may have seemed the obvious, realistic policy and yet sympathisers of the talukdars were soon able to construct a reasonable case to support them. Firstly, Peile played cleverly on the security aspect: 'it certainly is a question of some political gravity, whether the state can safely strip of all their property a class of men of warlike race and character and scatter them penniless along the British frontier'.[232] Then there was the guilt of official complicity in the talukdars' downfall: the effect of early settlements had been 'to annihilate the Talookdars' ancient proprietary rights, and convert them into mere leaseholders'.[233] And yet, as Peile stressed, 'their loyalty has not been impaired'.[234]

These arguments helped to ensure the enactment, late in 1862, of the Ahmedabad Talukdars' Relief Act. The Act enabled hopelessly indebted talukdari estates to be taken over by a Talukdari Settlement Officer who would administer them and eventually hope to pay off the debts. Its operation soon showed the massive extent of talukdari debt. During 1863 the Act was applied to 55 talukdari villages, against which claims totalled over 12 lakhs of rupees, and in the following year 116½ villages with debts of nearly 10 lakhs were brought under management.[235] In this way, over a third of talukdari villages were brought under the Act's provisions in its first two years. All this, however, made manifest sense in revenue terms. Bringing poverty-stricken estates under direct official administration offered the only opportunity of reorganising their finances and eventually making them pay a reasonable assessment and total revenue receipts

---

230 Peile's Report, p. 16.
231 MARD, Vol. 149 of 1862–4, No. 841, Part 1 of 1862, Minute by J. B. Peile, 30 September 1862.
232 Ibid.
233 MARD, Vol. 149 of 1862–4, No. 841, Part 1 of 1862, A. D. Robertson, Acting Chief Secretary to Bombay Government, to E. C. Bayley, Secretary to Government of India, No. 2232, 6 June 1862, para. 7.
234 Ibid., Minute by J. B. Peile, 30 September 1862.
235 Peile's Report, p. 18.

from the talukdars actually increased by over 16 per cent between 1863–4 and 1866–7.[236] This successful precedent led in time to legislative protection for the other debt-ridden estate holders of Gujarat. The Thakurs in Broach and Kaira Districts, who had already required periodic grants to keep their heads above water, were provided with the same mechanism as the talukdars by an Encumbered Estates Act in 1877.[237]

Such legislation was a holding operation. It preserved intact the social structure of the areas concerned rather than rehabilitating landlord wealth and position. Heavily indebted estates continued to be brought under the legislation's mechanism and by 1890 the Talukdari Settlement Officer was managing a grand total of 275 estates.[238] Settlement reports of this era remained necessarily pessimistic about most talukdars' position and prospects. In Gogha taluk, for example, by 1893 'most of the talukdars are probably worse off now' than at the previous settlement.[239] Talukdari society during the late nineteenth century continued to function but not to flourish.

Revenue policy and social structure in the Bombay village

Let us now attempt to draw together the different themes and arguments of this chapter and provide it with an answer to the central issue: what type of social structure had emerged and how in the Bombay village by the 1860s? Our argument has tried to stress the diversity of influence. Ecological and climatic conditions combined with historical experience to shape local social and tenurial patterns, but they might be subtly changed again by the impact of British policy. The role of the latter once more needs reassertion in case enthusiastic empiricists are drawn by the story we have described to derogate its importance. Certainly on the surface the Wingate settlements seemed to be shaped substantially by local circumstance. They undermined the position of the Deccan village officer whose wealth and authority were in any case insubstantially based and conceded victory to the khots who closely controlled local revenue supply and sources of information. Yet the thrust of policy, however modified by conditions in the village, still produced effects. The khot in many ways emerged stronger from his battle against Francis' settle-

236 Ibid., Appendix 10.
237 See MARD, Vol. 9 of 1881, No. 317.
238 IOL, Government of India Revenue Progs., Vol. 3670, September 1890, A. No. 6, A. D. Younghusband, Talukdari Settlement Officer, to the Commissioner, Northern Division, No. 103, 19 February 1890.
239 SBG, NS, No. 279, Papers on the Settlement of Gogha Taluk, Ahmedabad District, H. O. Quin, Talukdari Settlement Officer to the Collector of Ahmedabad, No. 937, 25 August 1893, para. 11.

ments, for it had led him, typically, to define more closely the rights of tenants. The Deccan watandar lost, through the Wingate settlements, much in simple monetary terms which the more forceful might have hoped to retain. And the power and sanctions the Raj could, in the end, control should be remembered. This study, like so many, stresses the security fears which influenced policy but the authorities themselves decided how much they weighted these. If they chose to ignore them then the constraints on action, in the mid-nineteenth century, were as limitless as on any other 'Oriental Despotism'. Hence a fanatically ideologically motivated government could, for instance, have crushed the khots, for the Khandesh Survey Riots were eventually ended by military action. In practice, there was compromise with the khots because policy was shaped by pragmatism and also by the khots' propaganda and a minor resurgence of landlord sympathies during the 1860s. This, though, is to recognise British policy formation's active role: if different decisions had been taken, then a different type of society might have emerged in the khoti regions.

Policy itself, as we have seen, was never consistent but, in the Bombay revenue department, although sympathies towards landlords were by no means as absent as is often assumed, a dominant but distinctive ryotwari philosophy can be discerned. Within this, utilitarianism was undoubtedly the root ideological justification, but its practical lessons on the need for legal and social efficiency were much more noticeably influential than its political judgements about the ownership of local power. To repeat our thesis, the Bombay 'system' after Wingate was about the effective and economical garnering of information at the grass roots to ensure a reliable revenue supply. This, in passing, challenged the privileges of many local elites but the major struggles tended to focus on one specific issue: opposition to the principle of revenue survey. Clashes developed where groups, like the khots, felt that any full revenue survey would threaten their basic rights and position in local society. Yet where elites – the classic examples are the bhagdars and narwadars – could come to terms with the principle of survey, then official attitudes might suddenly reveal a tolerance for tenurial diversity. This was why the bolstering of the Ahmedabad talukdars was a not inconsistent action. Here were traditional overlords benevolently regarded because they posed no conceivable obstacle to officialdom's attempts to establish relations with the grass roots: indeed, their financial disorganisation positively aided the process. To launch a set-piece struggle over the surveys, local elites had to be sure of their village power base. The strength to coerce the mass of cultivators and, preferably too, the existence of influential peasant collaborators with a vested interest in the system – like the Karwar mulgenidars or the khoti regions' dharekaris – were needed for effective resistance. The absence of these conditions inhibited the Deccan watandar from any major show of opposition, except in the special case of east Khandesh. Even where they existed, the capacity to withhold large absolute amounts of land revenue, as the khots could do, was an essential for any long-term success.

How then did the survey and settlements of 1840–70 affect and change agrarian society? Ryotwari tenure remained the norm and was now grounded on deeper official knowledge of local rights and circumstances. However, a diversity of overlord and privileged tenure remained, strengthened, if anything, by battles with the revenue authorities or their acquiescence or support. In 1880 the simple statistics concerning the major tenurial forms in Bombay Presidency were as follows. There were 299 talukdari villages in Ahmedabad and 47,017 acres of talukdari estates in Broach. Khots owned 607 villages in Ratnagiri and 166,181 acres in Kolaba; 70,750 acres were held on the narwa arrangement.[240] Inamdars and jagirdars owned villages covering more than 4 million acres throughout the Presidency.[241] Overall in Bombay Presidency in 1880 over 8 million acres' worth of villages were held on some overlord or revenue-free tenure, against just over 28 million acres' worth of villages, ryotwari in character.[242] These statistics, however, considerably underestimate tenurial diversity because they classify villages according to their general character and large inam, talukdari and khoti holdings remained in villages predominantly ryotwari. In sum, at least a third of the occupied area of the Bombay Presidency was, in 1880, non-ryotwari.

In the Deccan, and to a lesser extent the Southern Maratha Country, ryotwari tenure was most widespread. The egalitarian stereotype of ryotwari systems was nearest to reality in the subsistence agriculture, much governed central Deccan. Here, too, the weakness of the watandar's position by the late nineteenth century was most evident. In Poona District there was comment, during the famines of the late 1870s and the late 1890s, that Patels were experiencing real hardship. In November 1896, for example, the Collector of Poona requested permission to include the Patel of Vangali village, Indapur taluk on the gratuitous relief register. His cash allowance came to only just over 10 rupees and at famine times

---

240 *PP*, 1881, Vol. 71, Part 2, Report of the Indian Famine Commission, 1880, Appendix 3, Ch. 1, Question 10, Statement by J. Peile.
241 Baden-Powell, *The Land Systems of British India*, Vol. 3, p. 251.
242 Ibid.

this Patel was 'very badly off'; 'the same is probably the case in other villages here and there'.[243] The poverty of the one group who might have provided an easily identifiable social elite did genuinely create something of a power vacuum in the villages of the region.

This had wide consequences. There were, for example, fewer established elite networks within the village to provide credit or organise trade and this partly explains why immigrant professional moneylenders and traders, typically Marwaris and Gujarati Vanias, came to play a much more substantial role in the village economy of the Deccan than elsewhere in the Presidency. In 1875 these moneylenders became the leading victims of the Deccan Riots, an outburst which, as we shall see, could be partially attributed to the social structure of the region.[244] The central Deccan, one might argue, would remain potentially volatile until elites founded on new sources of wealth and authority could carve out their position.

In the regions where established elites survived, then, in the short run, greater political and social stability was guaranteed. Noticeably, immigrant commercial interests were less conspicuous and powerful. Traditional elites could link economic activities with social power over the village, advancing loans, paying land revenue for the cultivators and trading their crops. In Ratnagiri District, for example, 'when his circumstances allow, the khot secures the monopoly of the village moneylending and grain-dealing business. His position gives him a great advantage over professional usurers such as Marvadis.'[245] However, one could argue that these characteristics threatened to create closed economic systems. To take, for example, the talukdari areas of Ahmedabad. As we saw earlier, a wide range of interests was here supported from the uncertain agricultural output of these poorer parts of Gujarat and official comment in the late nineteenth century liked to emphasise the stability of arrangements within the talukdari system. In 1889, for example, the Talukdari settlement officer replied reassuringly to a Government of India enquiry about the possible rack-renting of tenants: 'the exact proportions in which the crop has to be distributed are fixed in each village by immemorial custom, they are well known to every cultivator'.[246] Yet, if that was so, where were the mechanism and drive for

243 IOL, Bombay Revenue Progs., Famine, Vol. 5087, December 1896, No. 10126, Report by the Collector of Poona, No. 10141, 27 November 1896, para. 3.
244 These themes will be developed in Chapter 4. See also my 'The Myth of the Deccan Riots of 1875', *Modern Asian Studies*, 6, 4, October 1972, pp. 401–21.
245 *BG, Vol. 10, Ratnagiri and Savantvadi*, p. 138.
246 IOL, Government of India Revenue Progs., Vol. 3669, June 1890, A. No. 50, A. D. Younghusband to the Chief Secretary to Bombay Government, No. 776, 27 November 1889, para. 5.

the economic change necessary to rehabilitate talukdari finances or improve peasant agriculture? On this line of argument, the Bombay revenue system was not sharply ideological enough to disrupt and stimulate anew the traditional societies supervised by overlords.

The khoti system, also, could be accused of limiting peasant motivation and opportunity for agricultural expansion in a region where, with high population density and limited resources of land, substantial raising of productivity was necessary to create more wealth. The potential for change seemed limited when the khot could continue to enforce collection of a standard proportion of the crop and the tenant had no necessary involvement in the cash economy. Even contemporary officials noted that the survival of such systems, alongside ryotwari arrangements, might be insulating a more static, less commercialised system. W. G. Pedder, hardly a ryotwari ideologue, noted, in analysing the licence tax receipts for 1880, 'that on the whole, under the khoti tenure, as compared with the very poorest ryotwari or peasant proprietary districts, the trading, manufacturing, and artisan proportion of the population is smaller; that it is generally poorer, and that it comprises fewer well-to-do people'.[247] Yet this is to anticipate too much. These brief comparisons will have served their point if they demonstrate that tenurial patterns could influence economic performance as well as vice versa. The realities of the economics of agriculture in the mid-nineteenth century must be our next concern.

---

247 *PP*, 1881, Vol. 71, Part 2, Report of the Indian Famine Commission, 1880, Appendix 3, Ch. 1, Question 10, Statement by W. G. Pedder.

3
The village in 1850: land and agriculture

The traditional beginning for any examination of agriculture in mid-nineteenth-century India is to emphasise its technical backwardness and the relative absence of stimulants to decisive change. For Bombay Presidency, this might seem particularly appropriate. The province, as we have seen, was predominantly millet-growing and in 1873–4 out of a total of 19.1 million acres under crops in the whole Presidency, the two great staples of jowar and bajra accounted for over 10 million acres.[1] Millets were hardy and adaptable to climatic and ecological variations, but their capacity for high output and rapid yield improvement was limited. They were, for example, incapable of anything like that substantial leap in production sometimes created by the initial expansion of high-yielding rice cultivation – what Braudel calls 'the miracle of the paddy fields'.[2] At the same time, millet cultivation in South Asia made large demands on land and inhibited diversification into areas like livestock farming. The physical basis of Bombay agricultural production also seemed to suggest those restrictive features traditionally presented as archetypally 'Asiatic'. Villages, particularly in the Deccan and the Southern Maratha Country, were large, distinct and, superficially, self-contained. Hence the idea of a primitive, subsistence isolationism gained wide currency among mid-nineteenth-century Bombay officialdom. Each village, one commentator noted, formed 'a kind of separate republic, with its own peculiar interests, and entirely independent of those around it'.[3] Within it, another wrote, existed an 'individual equality of poverty' so that 'there is wanting that example of prosperity, which is the surest guarantee for inciting emulation'.[4]

The notion of a primitive, subsistence agriculture, however, is too simple. Alongside the foodgrains, crops for sale on the market were widely

---

[1] *PP*, 1881, Vol. 71, Part 2, Report of the Indian Famine Commission, 1880, Appendix 3, Ch. 1, Question 3, Statement by J. Peile.
[2] See Fernand Braudel, *Capitalism and Material Life, 1400–1800* (Miriam Kochan trans., London, 1973), p. 100.
[3] *PP*, 1852–3, Vol. 75, Papers on the Settlement of Bunkapur Taluk, Dharwar District, D. Young to G. Wingate, No. 2, 29 June 1846, para. 21.
[4] MARD, Vol. 116 of 1860, No. 497, H. E. Jacomb, Assistant Collector, to C. Fraser Tytler, Collector of Ahmednagar, No. 13, 6 December 1858, para. 7.

The village in 1850: land and agriculture 71

cultivated. In particular, in 1850 the Bombay Presidency, of all British possessions, seemed to offer large opportunities for British cotton manufacturers, anxious about their sources of raw material supply. Over the whole Bombay Presidency plus the neighbouring princely states, Amalendu Guha estimates that as much as 2.6 million acres of land were growing cotton in the mid-nineteenth century.[5] Certainly trade statistics indicate that in 1849–50 Bombay contributed over 150 million pounds weight out of India's total cotton exports of 165.66 million pounds.[6] Ninety-four million pounds of this were produced in Gujarat, from where a visiting representative of the British cotton industry described cotton exports in 1850 as 'fully seventy per cent of the gross exports of all the products of the province'.[7] Clearly, so large a business required a widespread distributory and marketing network: the crop, grown in the village, had to be prepared for sale and conveyed, often long distances, to the port of export. In these cases, at any rate, commercial interchange between the 'village republics' could hardly have been unknown.

Cash-crop production was underpinned by diversity in ownership of resources, access to credit and levels of overall income, for most peasants needed, first, to ensure their foodgrain supply before producing for the market on any scale. Ownership of, say, inam land offered opportunities for agricultural diversification, but there were also substantial inequities in the holding of ryotwari land. Distribution of ryotwari landownership was pyramidal: above the mass of smallholders came a decreasing number of increasingly substantial holders.[8] Hence, in theory, even those excluded from the massive privilege system of the non-ryotwari tenures could differentiate themselves from any subsistence stereotype. Wingate met one ordinary peasant in Dharwar District in 1844 who owned 50 acres of good soil out of a total cultivated acreage in his village of 355 acres.[9] Even in the poorest areas of the Deccan 'a few cultivators hold farms of upwards of 200 acres, and have 20 or 30 bullocks'.[10] Perhaps the classic description of a 'rich peasant' in the Bombay official literature comes from a settlement report of 1854 on a group of villages in eastern Ahmednagar 'decidedly inferior to any district hitherto assessed':

---

5 Amalendu Guha, 'Raw Cotton of Western India: 1750–1850', *The Indian Economic and Social History Review*, 9, 1, March 1972, pp. 1–42.
6 *PP*, 1852–3, Vol. 69, Statistical Papers Relating to India, Section on Cotton, p. 52.
7 Alexander MacKay, *Western India. Reports Addressed to the Chambers of Commerce of Manchester, Liverpool, Blackburn and Glasgow* (London, 1853), p. 43.
8 For statistical demonstration, see the information in the District Gazetteers on size of holding.
9 Wingate Papers, Box 293/2, Diary, 27 December 1844.
10 *SBG, NS*, No. 123, Settlement Report on Six Taluks of Ahmednagar District by G. S. A. Anderson, 31 January 1854, para. 20.

in Hingri... a ryot of Indapur was found to be cultivating 82 acres of land. He told me that all the land in his own village having been taken up for cultivation, he could only, on becoming more prosperous, enlarge his farming operations by crossing the river and taking up land in Korti. He now travels on horseback from one village to the other, and I dare say considers himself quite a gentleman farmer.[11]

This makes the point not only that agriculture was far from solely subsistence orientated but also that its values and ambitions were hardly constrained by some primitive village egalitarianism. Commercialisation in Indian agriculture was not introduced as a novelty by a rampant British capitalism: the Bombay village, before the advent of the railway and before any supposed effects of British legal and revenue arrangements had time to bite, was already commercially orientated in values and patterns of production. Technically too, agriculture in mid-nineteenth-century Bombay has to be seen as a relatively sophisticated system, adapted over long years to local conditions. Absolute levels of output, relative to other countries, may not have been high but technology was neither as backward nor as static as often argued. This is evident from an extensive review in 1846 by the revenue official Young of the agricultural implements then in use in Dharwar District.[12] They included two main ploughs of different weight – the 'neguloo' and the 'rentee' – and a variety of scarifiers and hoeing implements. The materials used in construction were certainly basic and on the most common form of scarifier 'the only part iron is the blade in front, the rest being wholly of wood'.[13] Yet the implements themselves seemed well devised to perform a wide range of functions with considerable labour-saving effect. For example, the 'kooreegee', a seed sowing machine, could sow four rows simultaneously by means of separate conductors or pipes. Agricultural technology, too, was capable of some evolution and adaptation. The bullock carts in use in the early nineteenth-century Deccan were widely criticised by British officials as heavy and clumsy and Young described the typical carrying cart of Dharwar District as 'a huge unwieldy machine'.[14] In practice, however, this was largely necessitated by the then state of tracks and roads which, in any case, limited speed and demanded solidity of construction. As some improvements in communications facilities were made in the mid-nineteenth century, lighter

---

11 Ibid., para. 195. This area, where the easternmost boundaries of Ahmednagar and Poona Districts adjoined Sholapur District, was one of the least favoured of the whole Presidency, with a low annual rainfall and high famine risk.
12 See *PP*, 1852–3, Vol. 75, Papers on the Settlement of Bunkapur Taluk, Dharwar District, D. Young to G. Wingate, No. 2, 29 June 1846.
13 Ibid., para. 15.
14 Ibid., para. 16.

carts became more common and, in particular, a prototype, developed by Gaisford of the revenue department in 1836, became extensively copied and manufactured.[15]

In these ways, then, Bombay agriculture of 1850 can hardly be seen as simply 'backward' or 'primitive'. Nevertheless, having attacked the easy stereotype, extending the argument with more substantive comment on agricultural performance proves difficult. Ideally, we should now proceed to some firm judgement about levels of per capita agricultural output and the productivity of different crops. This, however, with no full census before 1872 and few even superficially reliable aggregate agricultural statistics before the late nineteenth century, is impossible. What can be outlined are the secular trends as they seemed to be developing in the middle years of the nineteenth century: like the Wingate revenue settlements, we have to base our conclusions about the state of agricultural development on empiricist observation of comparative drift, rather than absolute statement concerning one moment in time.

This, as Wingate argued on revenue practice, is not necessarily inferior technique, particularly where secular trends were strong. And this did seem to apply to the mid-nineteenth century. After the long, severe agricultural depression, stretching in most localities from the 1810s down into the 1840s, the period was one of notable upturn, marked by rapid growth in population and, simultaneously, substantial increase in cultivated acreage. As we have seen, the Bombay revenue department claimed that the credit for the latter lay with the Wingate settlements. The moderation of assessment may certainly have played a contributory role, but in general the process was part of a wider, all-Indian cycle of expansion, caused by a recovery in prices and trade.[16] In the Deccan, perhaps, the extension of cultivation could be especially swift after about 1840, because of the degree of the previous contraction. In some areas in the 1840s land availability was so considerable as to encourage a form of 'slash and burn' agriculture, exploiting the windfall gains from newly cultivated land and then moving on. 'The peasant', Wingate noted, 'finds it more profitable to work his land until it becomes exhausted, and then throw it up and take fresh land into cultivation, than to cultivate continuously the same land.'[17]

In the more intensively populated regions of the Konkan and Gujarat, such techniques were not, of course, possible. Yet the expansion in cultivated acreage between the 1840s and the 1860s was undoubtedly a Presidency-wide phenomenon, even if the Deccan was always its most

---

15 BG, Vol. 18, Poona (Bombay, 1885), Part 2, p. 10.
16 See Farmer, Agricultural Colonization in India since Independence, pp. 10–11.
17 Wingate Papers, Box 293/8, G. Wingate to H. Green, 5 January 1852.

spectacular site. In Gujarat, settlement reports suggested increases in land use of 17 and 34 per cent respectively in the Kaira taluks of Mahunda and Thasra between 1842 and 1862.[18] In Poona District the official figures for the total cultivated area rose from 1.37 million acres to 1.74 million acres in the twelve years before 1866[19] and even the former figure represented a substantial expansion on the levels of the 1830s. Similar, or even greater, rates of increase were achieved in the north Deccan districts of Nasik and Khandesh where the jungle frontier was being steadily eroded. Wingate, speaking of the Deccan phenomenon, claimed that 'so remarkable and rapid a development of agricultural industry is almost unexampled in any part of the world, without aid from immigration'.[20] The consequences of this expansion, however, were not clear-cut in terms of per capita production and living standards. Population, too, was rising rapidly. Settlement reports on taluks of Poona District in the early 1870s then revealed that population had grown by about 40 per cent in the previous thirty years: from 20,401 to 28,467 in Bhimthari taluk[21] and from 37,695 to 53,829 in Haveli taluk,[22] for example. The increase in numbers seemed, therefore, often at least as large as the expansion of the cultivated acreage. Indeed, statistics later collected by W. G. Pedder on 18 taluks throughout the Presidency showed that in 11 the rate of increase in population had exceeded that in cultivation in the years between the 1840s and the 1870s.[23] This evidence suggests, then, that the whole process was merely at best maintaining an existing equilibrium in per capita land use and production, creating a greater volume of economic activity without any qualitative changes within agriculture.

Indeed, the situation may have been even less favourable. One obvious possibility is that land availability and productivity became a serious problem. The new land brought under the plough may have been increasingly marginal, capable only of low yields and clearly, if this were so, per capita agricultural output might have been declining perceptibly

18 IOL, Home Miscellaneous Series, No. 796, Indian Papers of Sir James Caird, Enclosed Return in W. G. Pedder to Caird, 2 August 1879. These and similar statistics of the same period, it goes without saying, can only be regarded as a general indication of trends.
19 Rogers, *The Land Revenue of Bombay*, Vo. 2, p. 133.
20 Wingate Papers, Box 120/4/1, Memorandum Regarding Proposals by the Government of India for the Sale of Waste Lands and Redemption of the Land Revenue, 2 May 1862, para. 12.
21 *SBG, NS*, No. 151, Settlement Report on Bhimthari Taluk, Poona District, by W. Waddington, 12 July 1871, para. 6.
22 Ibid., Settlement Report on Haveli Taluk, Poona District, by W. Waddington, 30 November 1872, para. 17.
23 See Indian Papers of Sir James Caird, Enclosed Return in W. G. Pedder to Caird, 2 August 1879.

by the 1860s and 1870s. There is some evidence that, by the latter period, good quality, uncultivated land was becoming hard to find in some areas: in Satara District by the mid-1860s, commented one official, 'land is not allowed to lie waste without a good reason'.[24] Cultivated area, even in the Deccan, was not now greatly in arrears of what officialdom considered the total 'cultivable' acreage and some would argue that these estimates were often over-optimistic.[25] The statistics submitted to the 1880 Famine Commission quoted over 21½ million acres of land as in occupation in the Bombay Presidency out of 24 million acres of estimated cultivable government land.[26] Against this, expansion in land use, as we shall see, was able to continue steadily in the years after 1880, so that no absolute limitations on access to land were in sight during the mid-nineteenth century.

The story, nonetheless, might suggest that Malthusian constraints were exerting a powerful influence on the agricultural economy of western India. Since the period of expansion would end amidst serious famine in Bombay Presidency in the late 1870s, the possibility exists that population increase was outrunning the growth of the means of subsistence from the 1850s. The notion, however, of Malthusian overpopulation eventually rectified by an inevitable high mortality seems not to fit with actual historical cases, for a number of responses to rising population – changes in fertility and marriage patterns, more intensive use of land – are usually possible.[27] Given the shortcomings of our data, it is impossible to test accurately for these responses in mid-nineteenth-century western India. Yet it is highly unlikely that population pressure, either in a Malthusian way or as a stimulus to agricultural innovation in the manner suggested by Esther Boserup,[28] had a dominating impact on the agrarian economy at this stage. For we need to remember that the long-term trends, like those in population and extent of cultivation, however important, existed within the context of short-term peasant decisions. These were determined in the mid-nineteenth century predominantly by reaction to prevailing annual conditions and the uncertainty and variability of harvest performance.

In short, agriculture in Bombay Presidency, as much as any Indian

---

24 Wingate Papers, Box 119/1, Remarks by A. Wingate, Assistant Collector of Satara, 1874.
25 See, e.g., Farmer, *Agricultural Colonization in India since Independence*, pp. 29–35.
26 *PP*, 1881, Vol. 71, Part 2, Report of the Indian Famine Commission 1880, Appendix 3, Ch. 1, Question 3, Statement by J. Peile.
27 D. B. Grigg, *Population Growth and Agrarian Change. An Historical Perspective* (Cambridge, 1980).
28 Esther Boserup, *The Conditions of Agricultural Growth: the Economics of Agrarian Change under Population Pressure* (London, 1965).

province, was a gamble on the weather. In the absence, before 1850, of new irrigation facilities or even the rehabilitation of traditional tank systems,[29] reliable rainfall was of overriding importance, especially to Deccan agriculture. Rainfall, however, was everywhere subject to intense fluctuation: there was no such thing as the 'average year'. During the generally good seasons of the 1860s, for example, annual rainfall varied from 17 to 47 inches in Poona City;[30] from 38 to 57 inches even in a more climatically stable place like Belgaum.[31] This helped to create massive variations in output, according to the luck of local conditions and climate, from holding to holding and from year to year on each holding. The phenomenon was striking even on the government-owned farms which undertook special crop experiments under conditions far more favourable than those available to the average peasant. In a year, 1873–4, when bajra production on the government farms averaged 331 pounds per acre, some plots on one Deccan holding produced as little as 112 pounds per acre and others as much as 790 pounds.[32] Annual variation was illustrated by the case of jowar cultivation on the same government farm: 'the favourable season of 1872–73 gives about 1,000 pounds to the acre; but next year the failure is almost complete'.[33]

Climatic influences probably enjoyed a yet wider role. From the late 1840s through to 1870 was overall a long period of relatively favourable conditions. Rainfall, despite the large annual variations, never failed seriously in a major region of the Presidency. This comparative climatic stability itself shaped patterns of farming and the judgements required about them. Poor quality land newly brought under the plough, for example, was not necessarily 'marginal' in the context of a series of good seasons. More extensive cultivation and the greater labour input possible with large families might bear satisfactory dividends with consistent rainfall. Climatic factors, in short, were probably of the essence of mid-nineteenth-century agricultural trends. Bombay agriculture in 1850 was enjoying an expansionary phase based on favourable conditions and the prior recovery of trade and prices and this had no Malthusian inevitability about its progress.

It was not, then, necessarily the population–land ratio any more than simple 'backwardness' which inhibited long-term agricultural develop-

29 V. D. Divekar, 'Regional Economy 1757–1857. 3. Western India', in Kumar, ed., *The Cambridge Economic History of India, Vol. 2*, p. 335.
30 *BG, Vol. 18, Poona*, Part 1, p. 16.
31 *BG, Vol. 21, Belgaum* (Bombay, 1884), p. 43.
32 *PP*, 1881, Vol. 71, Part 1, Report of the Indian Famine Commission 1880, Appendix 1, No. 2, Part 2, Note on Crop Experiments in the Bombay Presidency, 1 October 1878.
33 Ibid.

The village in 1850: land and agriculture

ment and transformation in 1850. The constraints need to be viewed from a different angle. Although Bombay agriculture was not unsophisticated technically, there still remained the question of allocation and ownership of resources and of the technology which had evolved. The range of implements described earlier as in use in Dharwar District is impressive, but how widespread was ownership of them? More than any single asset, in the dry-crop regions of Bombay, the use of agricultural bullocks was vital to efficient farming operations. Bullocks were needed to pull carts and ploughs. The large, heavy Deccan plough was pulled by six or eight bullocks and even the scarifier which followed the plough, levelling and breaking up clods, needed four.[34] Carts were yet more demanding of cattle labour. In Dharwar, the large carrying cart was 'drawn when loaded by at least four yoke of bullocks, and often by as many as five or six': even a manure cart for use in the fields required two or three yoke.[35] Bullocks, too, were of potentially great importance for more than motive power, for their manure was the richest, cheapest and most extensive source of fertiliser available to the village economy.

Agricultural bullocks, however, needed much care and attention for their breeding and survival. In addition, scarcity and disease were always a threat to them. This raises, then, the crucial question of the numbers and availability of these considerable but highly vulnerable capital assets. Our evidence for the mid-nineteenth century is extremely patchy but what we have suggests that bullock numbers were not sufficient for optimal farming techniques. In Rahuri taluk of Ahmednagar, for example, a survey of 1849 revealed a total 8,475 working bullocks in the taluk dealing with a cultivated area of some 60,000–80,000 acres, which represented an average of only just over one bullock per holding.[36] Wingate had remarked on a bullock shortage problem when surveying in the east Deccan district of Sholapur in 1839. Here he examined bullock ownership in five villages and discovered that of a total of 373 cultivators, only 57 owned eight bullocks and more, 126 possessed between four and seven and 190 had three or less bullocks.[37] This was a heavy soil area where eight bullocks were really needed to pull the required plough. Hence under a sixth of the peasants could have worked a plough with their own cattle and 'nearly a half of the whole body of cultivators have actually not the

---

34 *PP*, 1852–3, Vol. 75, Papers on the Settlement of Bunkapur Taluk, Dharwar District, D. Young to G. Wingate, No. 2, 29 June 1846, para. 15.
35 Ibid., para. 16.
36 *PP*, 1852–3, Vol. 75, Official Correspondence on the System of Revenue Survey and Assessment in the Bombay Presidency, G. S. A. Anderson to E. H. Townsend, No. 17, 21 February 1849, para. 4.
37 Wingate Papers, Box 118, G. Wingate to I. J. Law, Assistant Collector of Sholapur, No. 86, 15 June 1839, para. 19.

means, without receiving assistance from the others, of tilling their lands at all'.[38] The problem was complicated by unequal distribution not only between individual cultivators but between different villages and areas. In one village, 33 of the 40 cultivators owned three or less bullocks and any form of effective ploughing of village lands must have proved difficult even with sophisticated pooling arrangements. Such shortages were not confined to the central Deccan. In Amalner taluk of Khandesh, for example, there were just 21,755 working bullocks according to statistics collected at the settlement of the 1850s;[39] in Bagalkot taluk of Bijapur District only 15,352 in 1850–1.[40] In both these taluks few cultivators could have pulled a plough with their own cattle. Bullock numbers, of course, may have fallen during the long agricultural depression between the 1820s and the 1840s. On the other hand, the contracted extent of cultivation offered favourable opportunities for a rapid increase in the cattle population, since grazing was so readily available. Given these circumstances, it seems likely that limited bullock availability was an endemic difficulty.

Similar problems, in many cases, applied to agricultural implements. One visitor to mid-century Gujarat, Alexander MacKay, was told by the mamlatdar of Nadiad, Kaira District, that 'the mass of the cultivators are so poor that they have to hire not only the bullocks, but most of the implements with which they work'.[41] This, the mamlatdar estimated, cut the profits of their cultivation by half. It was often, therefore, limited ownership of resources rather than shortcomings of technique or in the implements used themselves which created the appearance of slovenly cultivation condemned by British officialdom. Limited bullock numbers and plough ownership explained, fundamentally, why in the central Deccan 'land is not ploughed more than once in three or four years'[42] and why MacKay could comment that 'it is not often that Indian husbandry soars as high as the use of manure'.[43]

In a sense, stressing the problem of limited ownership of resources is merely to state the obvious; that the typical Bombay peasant was poor. Yet the handicap of bullock shortage, for example, can be defined more clearly. Cattle were expensive to feed and if the peasant farm was threatened by any crisis or by scarcity, they would have to be the first to

---

38 Ibid., para. 21.
39 *SBG*, NS, No. 229, Settlement Report on Amalner Taluk, Khandesh District, by W. Turnbull, 13 February 1889, para. 20.
40 *SBG*, NS, No. 229, Settlement Report on Bagalkot Taluk, Bijapur District, by W. M. Fletcher, 30 June 1883, para. 14.
41 MacKay, *Western India*, p. 120.
42 *SBG*, NS, No. 151, Settlement Report on Supa Petha, Poona District, by W. Waddington, 5 September 1873, para. 4.
43 MacKay, *Western India*, p. 301.

The village in 1850: land and agriculture 79

suffer. Thus even if the cultivator built up a small herd of bullocks, a year of bad scarcity or disease might wipe them out. This returns us, of course, to the central influence of climate. Foodgrain production, outside the years of especial calamity or a series of poor seasons, could and did adapt to these variables, for the classic technique of Moses was well established in India too: 'to save the surplus of a good season to meet the deficiencies of bad years'.[44] More difficult to gauge but possibly more important was the influence of unevenness of annual production and uncertainty about resource availability on the process of agricultural commercialisation. The vagaries of performance may have dictated a necessarily cautious attitude to diversification, limiting the impact of market opportunities and inhibiting the further accumulation of capital and resources. Official comment agreed that in the mid-nineteenth century, cash-crop cultivation was typically only pursued when a reliable food supply had first been assured. Even in the best cotton-growing areas of Dharwar District, Wingate believed that 'jowarree is the first necessary with the cultivator ... His first care is, therefore, to provide himself with a sufficiency of these, and it is only when this is done that he can turn his attention to cotton.'[45] Cotton cultivation, in any case, required considerable care, prior investment and planning. Additional wage labour, for example, might have to be employed and ideally a complex rotational system was necessary, growing cotton on the same ground only one in three years.[46] Under these circumstances, extent of cotton cultivation in the localities tended to fluctuate considerably. In Poona District, for example, it varied from over 4,600 acres in 1850–1 down to merely 602 acres in 1855–6.[47]

The ability to produce extensively for the market in mid-nineteenth-century western India was also circumscribed by wider restrictions. Problems of communications were an obvious factor. In 1850 good roads were 'few and inadequate to the wants of the country'.[48] Alexander MacKay found only 24 miles of made road in the whole of Gujarat and this mostly existed for military purposes or for Europeans' use. Elsewhere there was only track, some of it barely passable: MacKay had to experience an unpleasant seven-hour bullock cart journey to cover merely the 12 miles between Jambusar and the port of Tankaria Bunder.[49] The Deccan and the south were, perhaps, better served with

44 *PP*, 1878, Vol. 58, Report of the Deccan Riots Commission, 1875–8 (hereafter 'Deccan Riots Report'), para. 57.
45 *PP*, 1852–3, Vol. 75, Settlement Report on Bunkapur Taluk, Dharwar District, by G. Wingate, 29 September 1846, para. 10.
46 *PP*, 1852–3, Vol. 69, Statistical Papers Relating to India, Section on Cotton, p. 53.
47 *BG*, Vol. 18, Poona, Part 2, p. 48.
48 *PP*, 1852–3, Vol. 75, Settlement Report on Bunkapur Taluk, Dharwar District, by G. Wingate, 29 September 1846, para. 65.
49 MacKay, *Western India*, p. 211.

road systems centring on Poona and on Belgaum but, even so, north–south links were limited. Throughout the Presidency the onset of the monsoon always caused a substantial interruption to communications. Tracks became impassable and in the Deccan even main routes were cut, for the many streams which became a torrent in the rainy season were rarely bridged. MacKay estimated that of the 700 miles of made road 'only 140 miles are available for traffic throughout the year'.[50]

Communications difficulties particularly afflicted export of agricultural produce. Even in Gujarat, as MacKay's painful experience showed, roads to the ports were not good. Further south, however, the produce of the Deccan and the Southern Maratha Country had to overcome the great natural barrier of the Ghat spurs before it could reach the coast. Most Dharwar cotton sent for export went to the Konkan port of Coompta. In 1850 the difficult journey by bullock took around a fortnight[51] and the conditions encountered adversely affected the quality and hence the value of the cotton. One Collector of Kanara remarked of the cotton received at Coompta: 'it is exposed throughout the journey to the sun by day, and the dew by night, to the sweat of the cattle, and the dust... and on crossing the Tuddri and other rivers, it is liable to be saturated with water'.[52] The Coompta journey, though, was relatively easy. Immediately to the north, 'it is only at two points... along a line of fully 500 miles' that wheeled vehicles 'can ascend with anything like ease, or descend with anything like safety'.[53]

All these problems, it might be argued, imposed insurmountable constraints on the extent of agricultural production for the market. Many mid-nineteenth-century officials claimed, for example, that the limits of Indian cotton cultivation had been reached. Their logic seemed impeccable: cotton should be grown on a given piece of land only once every three years and its output, in Western India, totalled no more than 100 pounds per acre. Hence to maintain export levels at the 160 million to 170 million pounds of the early 1850s seemed to require around 5 million acres of land; a figure which, considering the predominance of subsistence needs and the other restrictions outlined, did not appear capable of further expansion.[54] And yet, as we shall see, both acreage under cotton and export levels were to increase substantially with the American Civil War cotton boom of the early 1860s. This suggests considerable facility

50 Ibid, p. 398.
51 Wingate Papers, Box 293/2, Diary, 31 May 1844.
52 T. L. Blane, Collector of Kanara, quoted in W. R. Cassels, *Cotton: an Account of its Culture in the Bombay Presidency* (Bombay, 1862), p. 299.
53 MacKay, *Western India*, p. 369.
54 See *PP*, 1852–3, Vol. 69, Statistical Papers Relating to India, Section on Cotton.

The village in 1850: land and agriculture 81

among the peasantry to circumvent the difficulties we have discussed when it became clearly profitable to do so. Yet, in turn, this raises another problem for commercial agriculture in the period before 1860; that of demand. It might be suggested that the major limitation on cotton cultivation in 1850 was that demand seemed strictly finite. British cotton manufacturers were always prepared to voice great hopes of the Indian supply and to damn East India Company rule for alleged failure to stimulate commercial agriculture, but they rarely backed their words with hard cash. American cotton, with its longer staple, was always preferred. Indeed in the 1840s its steady expansion of supply seemed, Wingate thought, to threaten the Bombay cotton trade with 'extinction or great contraction at no distant date'.[55] The result was that demand from Britain (overwhelmingly the main export market) varied wildly according to short-term vacillations in the American supply: during the 1850s, cotton exports to Britain fluctuated from year to year by as much as 60 or 70 per cent.[56] In short, British industry wanted Indian raw cotton as a sort of permanent twelfth man, always ready in the pavilion but only occasionally brought on to the field of play. This role hardly produced the consistency of demand necessary to promote a more extensive commercial agriculture.

It can be seen, then, that a variety of restrictions hemmed in Bombay agriculture of the mid-nineteenth century. They were enough to ensure that the expansionary trend then in train remained essentially a short-term phase, not a force for qualitative transformation. They prevented any break-out. They also influenced the patterns of behaviour which often puzzled Western contemporaries. For example, the apparently 'extravagant' expenditures of many peasants, frequently condemned by officialdom as the product of an irrational social value system, are largely explicable granted the unevenness of agricultural performance and the imperfections of market and fluctuations of demand. British administrators bewailed that, in the Deccan, 'the manifestations of prosperity are to be looked for in the greater expense of the wedding ceremonies and other social entertainments'.[57] Yet the certain return in social prestige from a lavish wedding made that, often, a more attractive and 'rational' investment than sinking funds in agricultural innovation where output levels and market demand were subject to such vacillation.

The types of problem we have outlined are familiar ones, traditional

---

55 *PP*, 1852–3, Vol. 75, Settlement Report on Bunkapur Taluk, Dharwar District by G. Wingate, 29 September 1846, para. 64.
56 *PP*, 1852–3, Vol. 69, Statistical Papers Relating to India, Section on Cotton, p. 53.
57 *Census of India, 1881, Bombay Presidency*, Vol. 1, Ch. 1, p. 13.

elements in economists' 'low level trap' model. Nevertheless, it is important to try to distinguish where lay the most fundamental obstacles to development. The purely physical difficulties – land supply, limitations of communications – might be surmounted: despite them, steadily more cotton was grown and exported from Bombay Presidency. In 1850, too, the railway offered hope of future substantial improvement in transport and consequent cuts in costs for those trading produce over distance. Even the climatic problem was not as insoluble as it first seems. Irrigation facilities were extremely limited in mid-century Bombay, but any wider provision of reliable water supply could revolutionise conditions in 'famine belt' agriculture.

Even more so, agricultural technology and the capacity for adaptation per se were not automatically major problems in the mid-nineteenth century. The last example illustrates this: ameliorating problems of water deficiency did not necessarily require any sophisticated technical development, merely the much wider adoption of the ancient device of the tank. It was, however, limited ownership and application of the resources and technology already existing which formed the most fundamental constraint. For most peasants, new implements and facilities were beyond their means: it cost over Rs. 500 to build a well to irrigate 7 acres in the Deccan in the 1870s.[58] If a modern development economist had been set loose on the agriculture of western India in 1850, he should have diagnosed future requirements as, not any particular concern with technical improvement, but the greater, wider and cheaper provision of existing resources and facilities.

Credit and rural trade

Yet in many ways, it might be argued, the village economy had recognised the nature of the problem and had tried, so far as was possible, to adapt to it. In a poor economy where most individuals gravely lacked capital and resources, it was vital that surplus funds, where they existed, should be made available as expeditiously as possible to potential users. This was the developmental aspect. Equally, there was the problem of sheer survival: without periodic access to the surplus of others, many peasants would face ruin and starvation in years of crop failure. In these ways, the existence of a vast and complex system of credit and debt is often seen as underpinning the whole operation of the nineteenth-century Indian rural economy.

[58] *PP*, 1881, Vol. 71, Part 2, Report of the Indian Famine Commission, 1880, Appendix 3, Ch. 1, Question 11, Statement by W. G. Pedder.

However, in the official literature of Bombay as of other provinces, the issue is hardly presented in such strictly functional terms. By all the economic standards which made sense to the Victorian mind, the apparently massive volume and irredeemable nature of many peasants' debts suggested widespread existing destitution and the constant threat of further deterioration in peasant living standards and status. In Poona District, for example, 'at least 90 per cent of the cultivators are not merely in debt but hopelessly involved in it. Men paying government Rs. 10 to 20 owe the sowkars Rs. 1,000 to 2,000.'[59] In Shevgaon taluk of Ahmednagar in 1851 around a quarter of all cultivators seemed to risk 'daily likelihood of having their effects attached and sold' for debt.[60] What especially perplexed officialdom was the evident existence of widespread debt in the more buoyant and agriculturally secure areas as well as in the famine belt region. The Khandesh peasants, who rallied against the introduction of the survey in 1852, 'are just as deeply enthralled in the bonds of the moneylenders'.[61] The phenomenon in western India was universal. Submissions to the 1880 Famine Commission estimated that some 75 per cent of all Bombay peasants were in debt, half of them 'hopelessly'.[62]

Nobody now would see the problem exclusively in such light. However, the terms of amendment still require emphasis. Firstly, British officialdom believed it was dealing with a novel situation of rapidly extending indebtedness. Yet the existence of the credit and debt nexus was undoubtedly long-standing. Wingate, for example, was told by a Deshmukh of Ahmednagar even that 'creditors had far more power over their debtors under the Mahratta Government than they have now',[63] and the French traveller, Victor Jacquemont, was writing of a traditional situation when he described, in 1832, how the typical western Indian peasant 'nait, vit et meurt en état de dettes'.[64] Nearly a century later, the *Report of the Bombay Provincial Banking Enquiry Committee* was able to make the identical comment.[65] Even in the 1970s assessments have been produced of apparently horrendous levels of individual peasant debt: an

---

59 MARD, Vol. 7 of 1875, No. 960, C. G. W. Macpherson, Assistant Collector, to G. Norman, Collector of Poona, No. 72, 17 July 1875, para. 11.
60 Wingate Papers, Box 293/3, Diary, 20 December 1851.
61 SBG, NS, No. 93, Settlement Report on Sowda and Yawul Taluks, Khandesh District, by A. F. Davidson, 23 December 1854, para. 23.
62 PP, 1881, Vol. 71, Part 2, Report of the Indian Famine Commission, 1880, Appendix 3, Ch. 1, Question 9, Statement by Mahadeo Wasudeo Barve.
63 Wingate Papers, Box 293/4, Diary, 13 January 1852.
64 Victor Jacquemont, *Voyage dans l'Inde pendant les années 1828 à 1832* (Paris, 1841), Vol. 3, p. 559.
65 In 1929, still, 'a very large proportion of the agricultural population is born in debt, lives in debt and dies in debt'. *Report of the Bombay Provincial Banking Enquiry Committee, 1929–30* (Bombay, 1930), para. 45.

enquiry then estimated average amount of debt liability per household in Maharashtra at more than Rs. 320 for smallholding cultivators and over Rs. 1,200 for large holders.[66] In practice, though, the pervasiveness of credit and debt seems a consistent feature of many peasant societies. Its existence in pre-industrial Western economies has been long ignored simply because their historians were blinded with the model of the 'credit economy' as fundamentally a new feature of burgeoning capitalism and yet, it is now clear, a complex and massive network of agrarian credit was, for example, 'routine in English rural life' before the industrial revolution.[67] The worries of British Indian officialdom about the alleged 'dangers' of indebtedness were matched by those of commentators on nineteenth-century Sicily or seventeenth-century Germany.[68] In viewing Indian rural credit, it must be remembered, we are examining simply one variation on a central theme of the peasant economy world-wide. It is ironic that so fundamental a feature of most traditional societies should be regarded as the leading source of dramatic social change in India.

Nevertheless, credit systems clearly varied greatly in kind or even simply in degree. Were the sturdy yeomen of early modern England 'in debt' to anything like the extent of the peasant in British India? A major enquiry in the late 1920s estimated that the Bombay Presidency's total agrarian debt then stood at fifteen times the value of the province's annual land revenue.[69] Such statistics, however, probably meant little in terms of actual day-to-day reality. No peasant was going to 'pay back' formally a sum equivalent to fifteen times what he paid in land revenue and such debts, therefore, were unlikely to be realised in any monetary sense. Nor, in developmental terms, did they represent, in any way, disposable capital resources: it might have been a very good sign if they had. Similarly, formal interest rates, widely condemned by officialdom as usurious, hardly gave any accurate indication of actual returns on lending. The actual working capital of the village credit system was, in sum, much more limited.[70] All this is not to say that the formal degree and conditions of the debt may not have been important in determining the balance of the

---

66 See Government of Maharashtra, Agriculture and Co-operation Dept., *Report of the Committee on Relief from Rural and Urban Indebtedness* (Bombay, 1975).
67 B. A. Holderness, 'Credit in English Rural Society before the Nineteenth Century, with Special Reference to the Period 1605–1720', *The Agricultural History Review*, 24, 1976, p. 98. For the role of credit in the medieval period, see M. M. Postan, *Medieval Trade and Finance* (Cambridge, 1973), Ch. 1.
68 Holderness, 'Credit in English Rural Society', p. 97.
69 *Report of the Bombay Provincial Banking Enquiry Committee, 1929–30*, para. 49.
70 This point has been well made, for north India, in P. J. Musgrave, 'Rural Credit and Rural Society in the United Provinces, 1860–1920', in Dewey and Hopkins, eds., *The Imperial Impact*, pp. 216–32.

creditor–debtor relationship. In economic terms, however, the aggregate figures conjured up for official use tell us little of value.

Again, officialdom's assumption that indebtedness was a sign of poverty or even destitution is now widely amended. Volume of individual debt was never a simple index of wealth. Thus the Provincial Banking Enquiry discovered, to their surprise, that 'the agriculturist with a small holding is not more heavily involved than the agriculturist with a larger holding'.[71] Indeed, the rich often borrowed more extensively. The common official explanation for this type of phenomenon was some type of moral degeneration caused by wealth: 'the extravagance of the cultivator, born and bred in him by the fruitfulness of his land, that launches him into debt'.[72] However, for the peasant who owned valuable resources, there was much to be said for exploiting his consequently greater credit to the full. Certainly expansions in cash-crop cultivation frequently rested, as might be expected in view of their commonly heavy capital requirements, on credit. An investigation into the activities of cotton growers in Khandesh and Gujarat in the 1920s revealed a relatively high volume of debt with at least 70 per cent of the cultivators owing an average of well over Rs. 100 each.[73]

Such remarks about the Indian credit system are hardly now very original. Their spirit reflects, in part perhaps, the personally more sanguine view of borrowing and debt required of the heavily mortgaged late twentieth-century historian compared to the nineteenth-century civil servant. Even so, the justification for a strictly functionalist view of credit and debt in India is clear. The fundamental problem faced by agriculture in western India, as we saw earlier, was limited ownership of resources and the highly uneven nature of the surplus. Borrowing offered the only easily available facility to mitigate these circumstances, particularly in years of total crop failure. Even in the best season, the peasant's receipts of income came irregularly at particular times, especially after the harvest. These often ill-matched the pattern of expenditure demands, such as those on land revenue. Wingate visited, in 1840, one Poona village where the bulk of the crop was only ready for market in February. The land revenue, however, was due in two instalments in December and January 'which obliges the cultivators to resort to the sowkars for their payment'.[74] Here, of course, the accusation that British

---

71 *Report of the Bombay Provincial Banking Enquiry Committee, 1929–30*, para. 53.
72 *SBG*, NS, No. 412, Papers on the Settlement of Jambusar Taluk, Broach District, Report by G. Gopalji, Mamlatdar of Broach, No. 430, 23 May 1902, para. 1.
73 The Indian Central Cotton Committee, *General Report on Eight Enquiries into the Finance and Marketing of Cultivators' Cotton, 1925–28* (Bombay, 1929), Table 7.
74 Wingate Papers, Box 293/1, Diary, 19 May 1840.

innovation compounded indebtedness seems more capable of sticking, for now land revenue was required in cash at a fixed time. It was also an inflexible assessment on the holding, not its output so that, in one official's criticism, 'Rs. 20 are levied whether the crop is worth Rs. 960, Rs. 320 or Rs. 80.'[75] Yet British land revenue practice made merely one addition to the many traditional expenditure demands which necessitated borrowing. Many peasants, for example, needed financial or practical aid before starting the season's operations. In Khandesh even in the mid-1920s 54 per cent of one sample of cotton growers relied on moneylenders for their seed and for them loans were most important 'at or just before the commencement of cultivation operations'.[76]

In these ways, it can be argued, the necessities of agriculture demanded the development of a sophisticated credit system. Credit as, in part, a village social security system was an argument whose force did not escape some commentators. Pedder pointed out pertinently how, without moneylenders, 'the land revenue would never be realised fully or punctually'.[77] William Wedderburn, the district judge, was another nineteenth-century official to promote a strongly functionalist view: 'the existence of the money-lender in the village polity', he argued, 'is "as essential as that of the ploughman" '.[78] On such interpretations, even interest rates taken at face value were not necessarily usurious. Their apparently high levels might well be justified in classic economic terms in view of the very bad nature of the risk being taken. As the Collector of West Khandesh commented in 1927 of many local cultivators: 'their holdings are uneconomical, their reputations in the village bad, and beyond the sowkar who will often... advance money on very frail security, they have no hope of getting credit'.[79]

Equally, a functionalist view can be taken of the lender's role and motives. In western, as in northern, India lending money was, in the nineteenth century, 'virtually the only way of investing relatively small

---

75 William Wedderburn, *A Permanent Settlement for the Deccan* (Bombay, 1880), p. 11.
76 The Indian Central Cotton Committee, *General Report on Eight Enquiries into the Finance and Marketing of Cultivators' Cotton, 1925–28*, p. 10.
77 IOL, Political and Secret Department Library, Papers of Sir William Lee-Warner, Deccan Ryot Series, No. 2, W. G. Pedder, Note on the Indebtedness of the Indian Agricultural Classes, its Causes, and Remedies, 29 July 1874, p. 5.
78 Wedderburn, *A Permanent Settlement for the Deccan*, p. 7. For a study of official attitudes to indebtedness, see Clive Dewey, 'The Official Mind and the Problem of Agricultural Indebtedness in India, 1870–1914' (University of Cambridge, unpublished Ph.D. thesis, 1973).
79 *Report of the Royal Commission on Agriculture in India, 1926–28* (London, 1928), Vol. 2, Part 1, p. 287, Evidence of H. F. Knight, Collector of West Khandesh.

sums profitably'.[80] If there was more risk than in simply hoarding money or spending it conspicuously, the returns involved social advantages of patronage and prestige. As a result, the creditor's role was widely performed. In western India, the typical lender in villages was 'a man of very small means' and 'frequently unites with the transactions of a capitalist the business of a cultivator, or of an artisan'.[81] In practice, most with any surplus funds must have lent, just as those with any needs borrowed. Indeed, often the same individual was lending in one market and borrowing in another. In turn, this line of interpretation has significant implications for development strategy. If the credit and debt system performed a central function within the village economy and most participated in it to their essential advantage, then the system possibly needed extension and wider facilities, not the reform which those who condemned moneylender 'exploitation' demanded. The way forward might be, as William Wedderburn argued, to pump more funds into the credit system; to encourage cultivators to lend and borrow more. This, at least, forms a consistent prescription when linked with our earlier advocacy of the need for more extensive provision of basic resources.

A strictly functionalist view of credit, therefore, works; and it has enjoyed the additional advantage, in the modern literature, of elegant revisionism in damning both imperial officialdom and the Marxist tradition as simplistic. Nevertheless, common assumptions within so unholy an alliance must command respect. Even in non-Marxist terms an economic system entirely functionalist in roots may acquire far wider social implications as it develops and evolves. The functionalist approach might be correct to stress that the credit system arose from the mutual economic interest of both creditor and debtor, but their relationship involved far wider personal and social implications which might sometimes subordinate economic rationale. For the Marwari moneylender to 'attach and sell his debtor's cooking and drinking vessels even when the family are in the midst of a meal'[82] seems to have been against even his own long-term interest: pauperised clients are of little value to any businessman. Yet, clearly, deteriorations in personal relationships or social considerations did occasionally justify, in creditors' eyes, such violent action.

The most fundamental question, however, concerns how economic change in the nineteenth century may have affected the credit system. In

---

80 Musgrave, 'Rural Credit and Rural Society in the United Provinces, 1860–1920', in Dewey and Hopkins, eds., *The Imperial Impact*, p. 217.
81 MacKay, *Western India*, p.51.
82 *BG, Vol. 18, Poona*, Part 2, p. 106.

so far as commercialisation was intruding or was likely to intrude into peasant agriculture, this may have altered the values of resources and commodities used as security and so given enhanced powers and leverage to those who lent money. Not only might the creditor stand between the bulk of peasants and the market securing any rewards from its expansion predominantly for himself, but also market opportunities might lead to the imposition of onerous new burdens on debtors. Moneylending, it can be argued, was, in the nineteenth-century world, substantially about social power and commercial control. Hence one official's epigram: 'of the wealth of the province the "savkar" has the oyster, and Government and the rayat each a shell'.[83]

It is certainly true that, in mid-nineteenth-century Bombay, a stake in the debtor's crops seems often to have formed the price of regular provision of credit. Pedder noted, of the heavily indebted, that 'when one of these peasants has reaped his crop, he keeps in his house what grain he wants for the food of his family and hands the rest over to his banker'.[84] Further, many lenders consciously looked to establish this situation of commercial control. In Poona District, according to Wingate, it was 'a common practice for the Marwarees to enter in their bonds that if the debt is not paid within a few months, it is to be converted into grain'[85] and, in cases, they might then be able to 'take away the entire crop the moment it is ripe'.[86] Such activities could offer obvious economic advantages. There was the simple matter of timing within the agricultural year. The value of crops taken by the moneylender would, in theory, be credited against the peasant's debts, but clearly at the point of the season – harvest time – when prices were lowest. Simply to hold the produce for a few months and then sell might present significant gain.

Fundamentally, though, the degree of economic advantage obtained from regularly acquiring debtors' crops depended on secular trends in prices. At one level, it might seem correct to borrow at times of actual or threatened inflation, but where the obligation was met in a significant share of the debtor's harvest this tendency was countermanded. In such circumstances, the moneylender would appear to gain most in inflationary times, advancing in the case of, say, cash against land revenue an asset deteriorating in real terms whilst receiving produce appreciating in value.

---

83 SBG, NS, No.165, Settlement Report on Jhalod Taluk, Panch Mahals District, by N. B. Beyts, 16 February 1881, para. 45.
84 IOL, Lee-Warner Papers, Deccan Ryot Series, No. 2, W. G. Pedder, Note on the Indebtedness of the Indian Agricultural Classes, 29 July 1874, p. 2.
85 Wingate Papers, Box 118, G. Wingate to the Registrar of the Court of Suddur Dewanee Adawlut, Bombay, No. 319, 24 September 1852, para. 15.
86 Wingate Papers, Box 293/4, Diary, 28 March 1852.

In addition, inflation might prompt those with resources more consciously to seek such terms. Remembering the very limited opportunities open to rural investors in nineteenth-century India, inflation represented a serious threat to the holder of cash: to seek to acquire a greater stake in the appreciating assets of agriculture was an obvious response. What, then, of price trends in mid-nineteenth-century Bombay? Before the American Civil War cotton boom – which certainly produced substantial rises in cash-crop prices – data is, as ever, sketchy and unreliable. However, it seems most likely that a steady if unspectacular price rise accompanied from around 1840 the process of population growth and expanding use of land, itself indicative of rising demand for agricultural produce. As we shall see more fully later, the long period between 1840 and 1920 was, with the probable exception of the 1870s, predominantly a time of rising prices. This itself might suggest a long-term shifting balance of power towards creditors, for, as one official put it, 'as regards the peasant who has hypothecated his crop in advance to the Bania, and who is in debt to him, a rise in price is first of all in favour of the Bania, with only a poor contingent remaining to himself'.[87]

The implications of control, however, are wider than this. If production for the market was increasing and where the degree of the debtor's dependency was extreme, the moneylender might be able to enforce choice of crops grown, according to what was of most use to him on the market. Wingate quoted the case of one moneylender in Dharwar District who died in 1852 leaving Rs. 900,000, mostly made by contracting with peasants, who owed him substantial obligations, to grow cotton for him.[88] Such a situation, in turn, might endanger the peasant's long-term security by undermining his foodgrain production. Further, control over the village marketing mechanism by moneylenders would raise larger issues concerning Indian rural trade. India as the 'paradise of the middleman trader' was a frequent image presented by contemporary Western commercial interests. Within the British cotton industry, for example, it was often alleged that the existence of a mass of intermediaries between the peasant producer and the foreign buyer raised costs and undermined the competitiveness of Indian raw cotton.[89] Village power for the moneylender entrepreneur, therefore, might even impose structural disadvantages on Indian agriculture's power to compete in international markets.

87 MARD, Vol. 29 of 1883, No. 1934, A. D. Fforde, Assistant Superintendent, Revenue Survey, to T. H. Stewart, Survey and Settlement Commissioner, No. 40, 13 November 1883, para. 3.
88 Wingate Papers, Box 293/8, G. Wingate to H. Green, 19 July 1852.
89 See, e.g., Cassels, *Cotton: an Account of its Culture in the Bombay Presidency*.

At root the issue becomes the validity of the different stereotypes. Was the normal implication of the credit system in mid-nineteenth-century Bombay the useful provision of social security or considerable control by creditors over debtors' produce and choice of crops, and hence a 'trade monopoly of the Banias'?[90] If we investigate the problem from the other end – the characteristics of dealers, marketers of produce and moneylenders – a highly complex and diverse situation immediately becomes evident. In the Dharwar cotton trade, for example, as early as the 1840s agents of businessmen from the coastal towns, notably Coompta, competed for cotton directly in the village with 'a great number of merchants or dealers who buy it from the ryots... on their own account'.[91] Moneylenders, too, in even the poorest village, were not members of a tiny, socially close knit group, making gentlemen's agreements not to compete or encroach on rival interests. Most people with any capital resources, to reiterate a fundamental point, lent money. Hence there was everywhere a vast range of potential creditors, from the 'professional' shopkeeper moneylenders – often, in the Deccan, Marwaris or Gujarati Vanias – through to the smallest agriculturist with spare cash. All overlord groups too – talukdar, khot and inamdar – lent money to their dependants and clients. Thus, when Western observers comment of Bombay rural trade that 'the dealer is almost always a moneylender',[92] the implication is not necessarily that local commerce was the monopoly of a few.

The diversity and number of creditors, in turn, suggests that lenders could hardly impose monopolist terms on their debtors. Musgrave neatly turns traditional approaches to the social implications of credit and debt on their head, by pointing out that moneylenders, too, whatever their desires and ambitions, were confined by social restrictions.[93] In mid-nineteenth-century western India, genuine competition for lending business must have existed, in many villages, between, say, professional shopkeepers and those with traditional social or tenurial status. In some Ratnagiri villages, dominated by khoti privilege, and in some Deccan areas, lacking an allodial elite, monopoly may well have existed. Elsewhere, peasant debtors should have had some freedom to negotiate terms. In practice, too, control over marketing was by no means always a

---

90 This, at any rate, was one official summary of the situation in the 1880s. MARD, Vol. 29 of 1883, No. 1934, A. D. Fforde to T. H. Stewart, No. 40, 13 November 1883, para. 2.
91 *PP*, 1852–3, Vol. 75, Papers on the Settlement of Bunkapur Taluk, Dharwar District, J. T. Francis, Assistant Superintendent, Revenue Survey, to G. Wingate, 18 June 1846, para. 13.
92 MacKay, *Western India*, p. 50.
93 See Musgrave, 'Rural Credit and Rural Society in the United Provinces, 1860–1920', in Dewey and Hopkins, eds., *The Imperial Impact*, pp. 216–32.

condition of the regular supply of credit. Some cultivators always had the freedom of action, born, it seems, of ownership of resources or social status, to borrow heavily with very few strings attached. By the 1920s cotton growers in Gujarat 'as a whole... were not under any obligation to sell to or through their lenders'.[94]

The latter, of course, may be substantially a later development created, among commercial producers, by widening market opportunities. It was certainly true that, whatever the degree of competition among lenders, in 1850 most peasants exchanged marketable produce in the village: communications limitations, even where there were no conditions imposed by debt, dictated this. Most exchanges, then, were inevitably with the groups who acted as lenders and petty traders and the condition of local exchanges, even if no individual could ever stand as a monopsonist, might well have given them substantial advantages. Systems of weights and measures were, for example, rudimentary in the village. Weighing procedure, involving, typically, use of stones or bricks in bags, was highly inaccurate, and also standards varied even within small localities. As late as the 1920s, 'in Khandesh each village seems to have its own weight system, and in some the maund has two different values'.[95] The opportunities for fraud, in such a situation, are self-evident and the skilful intermediary might gain a significant margin of benefit by consistently under-valuing peasant produce. The scale of such practices must be mere guesswork: we cannot even make accurate assessments about village prices and how they compared with those in the wider world in 1850. Suffice it to say that when such comparisons were later, occasionally, made, they tended to reveal surprisingly small disparities in prices in western India. A report on cotton cultivation and marketing in the 1920s considered that 'one of the most important findings of our investigation is that the rates in the villages compare very favourably with those prevailing in the markets'.[96] Again, this may be untypical and born of especially tight competition between traders in cotton. Nevertheless, the assumption that village prices left large margins to a middleman who exchanged produce over a distance cannot automatically be made.

To generalise about so complex a structure as the Bombay credit system, as it existed in the mid-nineteenth century, is to invite accusations of over-simplification. One missing factor is detailed evidence on how borrowed resources were utilised. Our assumption has been that they mainly met short-term deficiencies in production. Much borrowing was

---

[94] The Indian Central Cotton Committee, *General Report on Eight Enquiries into the Finance and Marketing of Cultivators' Cotton, 1925–28*, p.19.
[95] Ibid., p. 26.
[96] Ibid., p. 28.

also, undoubtedly, for consumption and 'social expenditure' on such as weddings and village ceremonials. However, some credit was always used for constructive investment in agriculture. To summarise, despite recent strictures,[97] the functionalist dimension has to be emphasised. Credit fundamentally existed in nineteenth-century western India because, as in most other peasant societies, the village economy and society could not operate effectively and reasonably humanely without it. This is not to say that there were not, often, social implications of control in the working of the system. Everything, however, depended on local conditions, particularly on the degree of competition between creditors and between debtors. One vital point stands: in most Bombay villages there was no usurious monopoly. This may do much to undermine simple assumptions about the degree of power exercised by a creditor in a still mainly uncommercialised situation. Yet the issue of change in the twenty or thirty years after 1850 remains. It is particularly important in western India in view of the existence of agrarian disturbances in 1875, the Deccan Riots, widely assumed to be evidence of breakdown in traditional relationships between creditor and debtor. The following chapter will examine these issues.

Before continuing, however, some firmer indication of the economic implications of the credit system, as it existed in 1850, might be attempted. Historians of India, arguably, have been too obsessed with the social implications of credit and debt. What did the existence and operation of the system suggest for India's economic development? Braudel poses the central question as to whether the Indian credit system is evidence of a burgeoning indigenous capitalism or a restrictive factor which locked development by confining markets to the village level.[98] This latter fear – the village credit and produce market as an inhibitive factor on urbanisation – matters, it seems to me, little for development. It is certainly true that exchange in western India remained substantially village based. Even of raw cotton sales by the 1920s, about 'two-thirds was sold in the village and one-third in the market [i.e. market town]'.[99] Braudel's implication, then, is correct in the sense that the existence of a vast, relatively sophisticated system of credit and exchange in the village

---

[97] For, inter alia, criticisms of a functionalist approach to the historical problem of credit in western India, see Frank Perlin, 'Of White Whale and Countrymen in the Eighteenth Century Maratha Deccan: Extended Class Relations, Rights and the Problem of Rural Autonomy under the Old Regime', *Journal of Peasant Studies*, 5, 2, January 1978, pp. 172–237.

[98] See Fernand Braudel, *Afterthoughts on Material Civilization and Capitalism* (Patricia M. Ranum trans., Baltimore and London, 1977), p. 31.

[99] The Indian Central Cotton Committee, *General Report on Eight Enquiries into the Finance and Marketing of Cultivators' Cotton, 1925–28*, p. 21.

may have obviated the need for the commercial producer to look wider afield. Nevertheless, commercial expansion could still be rural-centred: the world could come to the village. In cotton, for example, the 'purchasing organization has become extremely elaborate. The big purchasing or exporting firms have their agents and sub-agents all over the cotton growing tracts.'[100] Where demand for produce existed – and demand, as we emphasised earlier, was always one of the key variables for the rural economy's expansion – the predominance of village exchange was no handicap to economic development.

How 'capitalist', however, was the village credit system? The moneylender who insisted on a share of his debtor's marketable crops was obviously trying to increase his 'profit' but the relationship between most creditors and debtors involved other calculations too. Thinking in terms of any supposed transition from 'feudalism' towards 'capitalism',[101] the social impact of early British rule was, from what we have seen earlier, highly varied. In parts of the Deccan, perhaps, social and political conditions had been created for rural capitalism to emerge: the decline of the old office holding elite offered opportunities, in the buoyant conditions of mid-century, for thrusting agricultural entrepreneurs, like Indapur's 'gentleman farmer', to come to prominence. Yet such change, as we saw, was limited to particular areas. Elsewhere, established groups with power based on office holding, typically maintained or even strengthened their position; and their role involved, in turn, moneylending. Was the relationship between a Gujarat talukdar and a client-debtor in any sense capitalistic? Such overtones, of course, might still evolve. Some traditional elites – the Konkan khots, for example – may have been shifting 'from rajah to landlord',[102] changing in the process the nature of their relationships with others. Peasant and shopkeeper lenders, too, may themselves have been more 'capitalist' in attitudes and ambitions, but these, as we have argued, were each merely one among a range of lending groups. In the western India of 1850, competition among moneylenders had the additional important effect of inhibiting the spread of any truly 'capitalist' approach to lending. More fundamentally, too, polarisation of

100 Ibid., p. 28.
101 This ignores the issue of whether pre-British India was 'feudal', as well as the great debate over whether British rule can be said to have inaugurated 'capitalism'. However, most writers, including the Marxists, appear to have abandoned the 'Asiatic mode of production'. For an argument that Moghul India was, in fact, fundamentally feudal, see Irfan Habib, 'Potentialities of Capitalistic Development in the Economy of Mughal India', *Journal of Economic History*, 29, 1, March 1969, pp. 32–78.
102 The phrase is Metcalf's, describing the alleged change in the role of the Oudh talukdars. See Thomas R. Metcalf, 'From Raja to Landlord: the Oudh Talukdars, 1850–1870', in R. E. Frykenberg, ed., *Land Control and Social Structure in Indian History* (Madison, Milwaukee and London, 1969), pp. 123–41.

function and economic role in anything approximating to class terms had progressed very little.

Braudel's question, however, fundamentally concerns the long-term implications of the credit system and whether these were economically progressive. The international comparative dimension offers, perhaps, one means of judging this. In early modern England, for example, factors like the extensive diffusion of village borrowing and the complex range of creditors and debtors involved have recently been seen as creating a rural credit system which aided capital accumulation for development.[103] In terms of basic characteristics, it is hard to see any qualitative difference here from the village credit mechanism of mid-nineteenth-century Bombay. In western India, as in England, the credit system stands as a powerful means of utilising available funds and resources. Pedder's point, mentioned earlier, that the realisation of the land revenue depended on the moneylender illustrates this for the economy as a whole: land revenue was the major source of government funds and government's role as a consumer, in turn, might promote increased demand for a widening range of goods.

Nevertheless, as we noted earlier, the amount of individual debt typically seems so much greater in nineteenth-century Bombay than in Western economies on the verge of industrial revolution. What does this mean? Paradoxically, the large volume of many Indian peasants' debts was and is substantially a product of limitations of capital availability and of peasant sources of security to raise credit. In an economy where resources are widely owned and where capital and credit are extensive, loans can be steadily renegotiated and different sources of security used. In the Bombay countryside in 1850, many peasants could offer minimal security for loans. For the creditor, arithmetically 'high' interest rates and the consequent build-up of a large monetary volume of debt afforded the only means of protection for his investment. It was not, to reiterate an earlier point, that the recorded debt would normally be simply 'paid back', but at least it symbolised the scale of financial obligation of the peasant debtor. Massive indebtedness was a consequence of shortage of capital. In this area, too, the economic problems of the Bombay village in 1850 rested on basic limitations of resources.

103 See Holderness, 'Credit in English Rural Society before the Nineteenth Century', pp. 97–109.

4
Indebtedness and the Deccan Riots of 1875

Our concern, in the last chapter, was predominantly with the credit system as an economic structure. Viewed in this light, it can emerge as a relatively sophisticated mechanism evolved to utilise such resources as were available to the mid-nineteenth-century village economy. However, the social implications were possibly far wider than yet considered. The very pervasiveness of the credit system throughout western Indian society meant that there were many different types of relationship, each with greatly varying potential consequences. Sometimes, where one party's ability to negotiate continued loans collapsed, the results might be dramatic. As we saw earlier, it was indisputable that many talukdari estates in Ahmedabad were being sold up cheaply for debt in the late 1850s and early 1860s. Here was the credit system – although wider issues of economic performance were also involved – unquestionably acting as an agent of social change. The Bombay government declared during the course of the rescue operations of 1862 that it wished to revive the talukdars' 'ancient proprietary rights' but 'it is manifest that the object of Government would be entirely defeated... so long as the Talookdars are in debt'.[1]

Social changes, like the decline of the Ahmedabad talukdars, were no novelty in western India and we have learnt to be much more sceptical about assuming that the beneficiaries in such cases were always newcomers to landownership and the status it conferred. Yet the fundamental issue concerns the frequency of such events by the middle of the nineteenth century. The charge – made by even the more sanguine official writers on credit and debt, like Wedderburn and Darling – is that the innovations of British rule were altering the implications of indebtedness radically, with substantial social consequences. One engine of change is seen as commercialisation in the countryside making land, as well as its produce, an appreciating asset, which creditors might seek to obtain. Whatever the speed of this process in the Bombay locality, one might certainly expect extending population pressure by the 1860s to increase

---

1 MARD, Vol. 149 of 1862–4, No. 841, Part 1 of 1862, A. D. Robertson, Acting Chief Secretary to Bombay Government, to E. C. Bayley, Secretary to Government of India, No. 2232, 6 June 1862, para. 9.

demand for and the price of land and, in addition, one of the proudest boasts of the Wingate settlements was that their moderated assessment had given agricultural holdings a greatly enhanced market value. Thus even by 1858 the Collector of Ahmednagar was claiming that 'land has now become the most valuable real property in the country'.² If this were so, the nature of security demanded for debt and creditors' ambitions may also have changed.

Then, there were the range of administrative and legal reforms introduced by British rule. William Wedderburn spoke of a 'reckless subversion of old institutions' in the Bombay Presidency.³ His main target was the land revenue system which, he believed, oppressed the peasantry through its rigid cash demand, whatever the quality of the season.⁴ Raymond West, the Bombay High Court judge, argued that the revenue survey's major mistake was the introduction of laissez-faire arrangements for the ownership of land and that, to preserve social stability, the state should have retained customary rights over land's disposal.⁵ The land revenue system, it will be seen, was blamed both ways: for giving peasant proprietors dangerous new rights and responsibilities and for increasing debt through its rigidity. Oppression of land revenue, however, was a traditional charge against all Indian administration. More innovatory was the second element in Wedderburn's 'subversion': the introduction of Western legal forms. The British courts imposed a new formalisation on the agrarian credit system and the structure of legal rights as then determined by them may well have differed from traditional obligations and arrangements in the village, with socially disruptive consequences. Certainly the Bombay courts became widely used by creditors as a device for defining and strengthening their position. By the 1870s the civil mechanism was 'in too many cases a mere machine for collecting debts'.⁶ The vast majority of civil cases were uncontested,⁷ involving the automatic issue to moneylenders of writs against debtors. Again, the use of such writs may have decisively shifted the balance of the creditor–debtor relationship.

---

2 MARD, Vol. 167 of 1859, No. 322, C. Fraser Tytler, Collector of Ahmednagar, to J. D. Inverarity, Revenue Commissioner, Southern Division, No. 1397, 9 December 1858, para. 4.
3 William Wedderburn, *The Indian Raiyat as a Member of the Village Community*, London Institution Lecture, 10 December 1883 (London, 1884), p. 19.
4 For his main attack on this, advocating a permanent settlement taking a set proportion of the produce, see Wedderburn, *A Permanent Settlement for the Deccan*.
5 See West, *The Law and the Land in India*.
6 *Bombay Presidency Annual Administration Report, 1870–71*, p. 15.
7 Out of 135,083 suits disposed of in the Bombay district courts in 1870, 88,624 were dealt with ex parte (that is, without a contest). Ibid., p. 14.

Whether these developments really changed the Bombay village credit system at root remains the major issue to be considered, but it is indisputable that they could be deliberately used to promote particular economic and social interests. This is strikingly illustrated by another case, very different from that of the Ahmedabad talukdars. In the north-west of Khandesh District, most land was held by large landowners in farms of up to 1,000 acres and worked by employed labourers, mainly of the Bhil tribes.[8] By the late 1860s the traditionally tranquil relationship between the two parties seemed under stress. The Bhils refused to work on the farms and held protest meetings, whilst anxious officialdom wrote of 'the incubation of insurrection'[9] and warned of the dangers of another Santhal rebellion. The root of the problem was Bhil indebtedness to the landowning elite. Such debt was traditional and long-established, but in the conditions of the 1860s landowners seem to have been using it aggressively, backed by the formal rulings of the courts, to maintain their control in a potentially more fluid situation. The 1860s were a decade of inflation and in Khandesh, where the cultivated acreage was extending rapidly, 'the demand for Bheel labour has increased, and much higher wages are obtainable'.[10] Some landowners as a result, faced with rising wage costs, 'have made their unwary servants sign bonds of indebtedness to them, which they have only to go to the Civil Court to enforce'.[11] The financial opportunities this offered were revealed by the case of Maun Sing, a prominent figure within the Bhil community. In 1870 a court decree ordered him to work without pay for nine months to redeem his debt to his employer,[12] a novel legal formalisation of what was clearly a long-standing credit relationship.

The case of the Khandesh Bhils illustrates the potential implications of debt and also (since the concern, traditionally, is about its effects on landowners) how varied these could be. Yet even this example is not as simple as it first appears. Officialdom, with its constant security fears, may have exaggerated the extent and dangers of the situation, for by the mid-1870s, when lower price and wage levels may have altered the situation, there were apparently 'very few complaints' from the Bhils.[13] This shows that

8 See MARD, Vol. 30 of 1878, No. 231.
9 Ibid., L. R. Ashburner, Collector of Khandesh, to A. Rogers, Revenue Commissioner, Northern Division, No. 3500, 24 July 1870, para. 2.
10 Ibid., Memorandum by the Under-Secretary to Bombay Government, 14 July 1875.
11 Ibid.
12 Ibid., O. Probyn, Western Bhil Agent, to the District Magistrate, Khandesh, No. 390, 26 November 1870, para. 1. Maun Sing had been involved in the Mutiny uprising and Probyn was now frightened of the possibility of him going on the rampage.
13 Ibid., No. 85, O. Probyn, Principal Bhil Agent, to W. H. Propert, Collector of Khandesh, No. 1204, 24 December 1877, para. 16.

the social atmosphere of the credit relationship, no less than its economic rationale, was highly complex. In addition, in the tight-knit village community, so much depended on personal relationships. Clearly, even if the moneylender was seen as performing an acknowledged function and was a recognised member of the village community, personal contacts might be limited and sometimes strained. Peasants in societies like nineteenth-century Bombay are often seen as passive, accepting their lot. Quiescence is, to a large extent, forced on a peasantry because of its social condition and the demands of the economic struggle for survival, but most peasants also have a root recognition of how they would, ideally, like the world to be, that millenarian streak which often emerges overtly in the mass uprising. Amidst even the most constructive and tranquil relationship with a moneylender, one suspects, the nineteenth-century Indian peasant was dreaming of a world where such entanglements had no place.

The consequence, in mid-nineteenth-century western India, was sporadic and spasmodic sniping at moneylenders, particularly the 'professional' whose predominant role was as a creditor. Periodic violence was endemic. Wingate commented on two murders of moneylenders in 1852, one in Sholapur District 'in the midst of his village in broad day among a crowd of his neighbours' and another in Gujarat 'on the high road in open day by hired assassins'.[14] In 1874–5 of a total of 164 murder cases in the Presidency, seven at least were known to be motivated by enmity against moneylenders.[15] Dislike of moneylenders also prompted that typically peasant form of political action which Hobsbawm calls 'social banditry'.[16] The classic example occurred in the Deccan between 1873 and 1875 when the Koli dacoit, Honya Kenglia, established himself as Bombay's Robin Hood. Honya's gang, wearing as their Lincoln green a distinctive dress of red military coats, roamed the countryside, robbing moneylenders, firing their houses and often inflicting their own special brand of indignity: the cutting off of noses.[17]

Such outbursts of violence, it must be stressed, were not new to the 1870s and were, also, liable to occur throughout most areas of Bombay Presidency. The social bandits, for obvious technical reasons, fared best in jungle or hilly regions, like Honya Kenglia's west Deccan, but elsewhere the same forces which motivated them clearly existed. In Kaira

---

14 Wingate Papers, Box 118, G. Wingate to the Registrar of the Court of Suddur Dewanee Adawlut, Bombay, No. 319, 24 September 1852, para. 5.
15 *Bombay Presidency Annual Administration Report, 1874–75*, p. 68.
16 See E. J. Hobsbawm, *Bandits* (London, 1969). India is little mentioned here but the example which follows is, in every respect, an ideal example of Hobsbawm's phenomenon.

District of Gujarat, there were as many as nine murders of moneylenders between April 1871 and July 1875, far more than in an equivalent period in Poona District.[18] Yet we need to view the implications of this sceptically. Occasional violence against a small minority of personally unpopular moneylenders did not necessarily denote any general malaise in the credit system or novel developments in its operation. Indeed, amidst so massive and complex a system, periodic violence against individuals might be seen as a safety valve for the inevitable tensions. In the end, the village community could show through force that it was bigger and stronger than the errant moneylender who offended its rules: and, in turn, this realisation was a source of stability for the system.

In this way, the rampages of Honya Kenglia's gang or the angry demonstrations by the Khandesh Bhils far from prove that the Bombay credit system had become an instrument of intense social change. In the central Deccan, however, there is one famous manifestation of tension, which was apparently much more substantial and deep-seated. In 1875 full-scale rioting directed against moneylenders erupted in a number of villages in Poona and Ahmednagar Districts. Over 1,000 peasants were arrested and special police posts were quickly set up in a number of villages.[19] These Deccan Riots seemed entirely different, in scale and nature, from previous action against moneylenders. Whereas murders or social banditry were the work of small gangs or individuals, in Poona in 1875 'the whole country appears to be in league against the moneylenders'.[20] In some cases, the traditional village elite – Patel and even Deshmukh families – threw their weight behind the action: in the very first outbreak, for example, at Supa, Bhimthari taluk, 'the rioters... had the sympathy and countenance of some influential persons of their village'.[21] The Deccan Riots, therefore, may well be evidence of social breakdown, created by the operation of credit and debt in the region. The Bombay authorities implicitly supported this view, by quickly enacting legislation – the Deccan Agriculturists' Relief Act of 1879 – to regulate the working of the Deccan credit system. And the interpretation that the Deccan Riots reveal an agrarian social revolution in the region is still prominent in the modern historiography, advanced in particular by

---

17 For a description of their major escapades during 1875, see *Bombay Presidency Annual Police Report, 1875*, Southern Division, pp. 83–6.
18 Deccan Riots Report, para. 18.
19 Ibid., para. 10.
20 Ibid., Appendix C, p. 2, District Magistrate of Poona to the Police Commissioner, Southern Division, No. 736, 20 May 1875.
21 Ibid., para. 4.

Ravinder Kumar.[22] The forces for possible change are evident. In the quarter century before 1875 population and cultivated acreage had been extending rapidly in the central Deccan. Land had been in demand and had acquired a growing market value. At the same time, the institutional and legal reforms of British rule had borne most heavily on this region. If, for example, Wedderburn's view of the revenue system is taken – that it imposed a high, inflexible demand – the effects were probably most corrosive in Poona District, where the organisation was most refined and where substantial upward revisions of the Wingate settlements had already begun in 1867.[23] Consequently, Kumar argues that Deccan agrarian society had been 'transformed ... through the legal and administrative reforms carried out by the British Government'.[24]

Nevertheless, the precise mechanism of any supposed transformation requires much further definition. Even if the revenue system pauperised the peasantry and extending commercialisation gave fresh opportunities to the middleman, how did any new balance of economic power come to be formalised in the Deccan village? Peasant debt was certainly very heavy in the riots area. The official enquiry into the riots revealed that, in twelve typical riot villages, personal debt totalled over ten times the villages' land revenue.[25] As we argued earlier, however, arithmetical volume of debt per se meant little. Creditor control over debtors' crops was, again, extensive in the central Deccan, but here, too, was a traditional consequence of the system. Both of these characteristics suggested poverty and lack of capital within peasant agriculture rather than, necessarily, any progressive forces of change. The real agent of social revolution would be shifts in the ownership of land. This, unlike acquisition of a peasant debtor's produce, would affect manifestly and permanently the economic and social status of creditor and debtor. Such a process, too, would be particularly consequential where the moneylender traditionally owned little or no land. This, then, has to be the nub of the

22 See Ravinder Kumar, 'The Deccan Riots of 1875' *Journal of Asian Studies*, 24, 4, August 1965, pp. 613–35; *Western India in the Nineteenth Century*, Ch. 5. I. J. Catanach adopts a somewhat more sceptical view in his *Rural Credit in Western India. Rural Credit and the Co-operative Movement in the Bombay Presidency, 1875–1930* (Berkeley, Los Angeles and London, 1970) and in 'Agrarian Disturbances in Nineteenth Century India', *The Indian Economic and Social History Review*, 3, 1, March 1966, pp. 65–84. For my earlier revisionist tilt, see Charlesworth, 'The Myth of the Deccan Riots of 1875', pp. 401–21. L. Natarajan reasserts a more dramatic interpretation of the riots in 'Maratha Uprising: 1875', in A. R. Desai, ed., *Peasant Struggles in India* (New Delhi, 1979), pp. 159–69.
23 The first revision took place in Indapur taluk. In 1871 and 1872 revisions of assessment were also introduced in Bhimthari, Haveli and Pabal taluks as well as four taluks of Sholapur District.
24 Kumar, 'The Deccan Riots of 1875', p. 634.
25 Deccan Riots Report, para. 75.

issue. West claimed in 1872 that the Deccan peasants 'are rapidly losing their paternal acres without losing their attachment to them'.[26] Kumar argues, in the modern literature, that 'the dispossessed peasant was forced to live as a landless labourer, often on those very fields which he had formerly cultivated as an independent proprietor'.[27] If this is true, the case for British rule in the Deccan as an agent of agrarian social revolution stands.

To judge this, however, we need, in turn, to examine the mechanics of land transfer. Changes in ownership of peasant smallholdings could occur, fundamentally, in two ways: through sale and purchase on the market and through court decrees confiscating and selling land for debt. Free exchange on the market was almost certainly limited during the depression era between the 1820s and the 1840s. Wingate and Goldsmid commented in 1840 that 'sales of land have become as little known in the Deccan as in Madras',[28] and Alexander MacKay pointed to the depressed state of Gujarat under East India Company rule by asking 'who hears of a wealthy native in Guzerat investing his money in land?'[29] By the 1870s, as we have commented before, the root situation had radically changed and most land in Bombay Presidency did enjoy a genuine market value. Yet, official orthodoxy argued, this led most peasant proprietors to maintain their traditional landholdings 'with the utmost tenacity'.[30] Certainly, in the Deccan Riots region, the land market appears to have remained sluggish. The Riots Commission counted a total of 198 land sales in all the riot villages of Bhimthari taluk in the whole period between 1865 and 1875,[31] suggesting, then, the existence of just a few sales each year in the typical village. The evidence of individual cases seems to confirm this, for Kangaum, one riot village, had seen eleven land sales in the years 1865–74 and Ambi Khurd, another, as little as two sales in the same period. Supa, the large village where the 1875 disturbances began, boasted a more extensive land market with, on average, six or seven sales per year, but even here there had been no land transactions in the six months before the outbreak of the riots.[32] In this situation, such land exchanges as occurred were, very likely, mainly between village intimates. It would

---

26 West, *The Law and the Land in India*, para. 17.
27 Kumar, 'The Deccan Riots of 1875', p. 619.
28 *PP*, 1852–3, Vol. 75, Official Correspondence on the System of Revenue Survey and Assessment in the Bombay Presidency, H. E. Goldsmid and G. Wingate to John Vibart, Revenue Commissioner, Poona, 17 October 1840, para. 28.
29 MacKay, *Western India*, p. 134.
30 IOL, Papers of Sir Richard Temple, Vol. 193, Minute by Richard Temple, Governor of Bombay, 29 October 1878, para. 4.
31 Deccan Riots Report, para. 70.
32 For these statistics, see ibid., Appendix C, pp. 317–23.

have been highly difficult for individuals outside traditional, tight-knit groups of peasant proprietors to break into so limited and irregular a market. Thus the official who carried out the new revenue settlement of Bhimthari taluk in 1871 found that 'of the 13 instances of sale of land quoted by me, in 10 cases the buyers were patels or ryots'.[33] The land market, it seems, could hardly have acted as an agent of social revolution in the Deccan and nor was its operation the flashpoint in 1875.

The much stronger possibility is that rulings by the courts produced increasing transfer of land. Resort to the courts by Deccan creditors was undoubtedly growing rapidly in the years before the riots. The situation had been aggravated by the operation of the Limitation Act of 1859, which aimed to help peasant debtors by limiting the validity of court decrees to just three years. Instead, the Act produced more frequent court actions by moneylenders to protect their interests and was blamed by one member of the Deccan Riots Commission as the major root cause of the disturbances.[34] Certainly, after 1859, the number of civil cases throughout the Presidency had risen considerably – by over 50 per cent between 1860 and 1870[35] – and these were mostly petty cases, concerned with small debts: of all the suits disposed of in the Bombay subordinate courts in 1873, around 70 per cent were for sums not exceeding Rs. 20 in value.[36] It can also be demonstrated that, in contrast to the operation of the land market, litigation was increasing in the riot areas in the early 1870s. By 1873 the number of suits for debt in Poona District stood at double the level of the 1860s.[37]

Acquisition of court orders by creditors was, therefore, indisputably a fast growing habit by the 1870s, but the question of what moneylenders did with these orders remains. Officialdom often assumed that their issue would lead to the automatic dispossession of peasant debtors and the Deccan Riots Commission concluded its survey of the Bombay legal system: 'the description thus given of the powers of the decree-holder shows that they are practically unlimited in the fullest sense'.[38] Yet, in practice, the rights and rigours of the system seem often to have been moderated. Imprisonment for debts, for example, was a widely condemned provision, denounced by the Riots Commission as well as most reformers, but in the

---

33 *SBG*, NS, No. 151, Settlement Report on Bhimthari Taluk, Poona District, by W. Waddington, 12 July 1871, para. 12.
34 See Deccan Riots Report, Note on the Revision of Assessment by Shambhuprasad Lakshmilal.
35 See *Bombay Presidency Annual Administration Report, 1860–61*, Appendix A, p. 101; *Bombay Presidency Annual Administration Report, 1870–71*, p. 12.
36 Deccan Riots Report, para. 91.
37 Ibid., para. 71.
38 Ibid., para. 104.

Deccan very few peasant debtors actually ended up behind bars. Wedderburn noted of his experience in Ahmednagar District in the 1870s that 'on no occasion have there been more than seven civil prisoners out of a population of over 700,000 souls'.[39] Similarly, decrees granted against debtors' land may not have been automatically executed.

At this point, the question of the mutual interest of creditor and debtor again arises. For the Vania and Marwari moneylenders, whose major concern was business and trade, the acquisition of agricultural land, even where its value was rising, was not necessarily an attractive investment. Hostility from the village community might accompany any move into landownership by a newcomer, but such a man would especially need to draw on the labour reserves of the village to cultivate the land, for he was unlikely to take to the plough himself. Thus, when Shivram Marwari acquired some land at Kheirgaon in Bhimthari taluk (not, considering the size of his lending business, a very large amount), 'the land is waste, as no one will cultivate for him except the original owners, and they only on condition of getting their bonds etc. given up'.[40] Many moneylenders must have preferred to avoid Shivram's predicament. It was tightening control over a debtor's produce which formed the most desirable goal. The way to secure this was to threaten everything against the peasant debtor, but to take the final step of dispossession would rob the creditor of any future major sanction. Indeed one Ahmednagar official argued that Brahmin moneylenders proved less unpopular in 1875 precisely because 'as they sometimes aim more at the land than at money, there is an end of the business when they have got that': the former 'proprietor' would now, of course, be a 'tenant', but often 'a good deal better off than the bond slave of the village usurers'.[41] Auckland Colvin, then, was probably summing up normal Deccan practice, when he commented that 'the Marwaris do not, as a rule, desire to possess themselves of the land . . . They usually prefer to keep the nominal occupier on, and sweat him.'[42]

This again suggests that the credit system possessed structural limitations on its capacity as an agent for extensive social change. To destroy the debtor socially and personally – as formal land loss would often entail – made little sense, for, as the old Maratha saying put it, 'the Kunbi is the

---

39 MARD, Vol. 11 of 1879, No. 422, W. Wedderburn to the Chief Secretary to Bombay Government, Revenue Department, No. 1713, 20 August 1879, para. 12.
40 Deccan Riots Report, Appendix C, p. 37.
41 Notes by Mr Sinclair, Assistant Collector of Ahmednagar, quoted in ibid., para. 39.
42 Ibid., Memorandum by Auckland Colvin, para. 22. 'It is only, as I understand, when a Kunbi has to resort to a second money-lender, or is from any cause unable to cultivate, or has in his khata exceptionally good lands, that the Marwari steps in and causes a transfer of the proprietary title to his own name.'

Marwari's cow... and is too valuable an animal to be allowed to perish'.[43] This is not to deny, however, that the use of court decrees to threaten and cajole debtors was often widely resented. It is striking that everywhere the 1875 riots had one common aim: 'the object of the rioters was in every case to obtain and destroy the bonds, decrees etc. in the possession of their creditors'.[44] In this sense, the legal documents formed a new formalisation of each peasant debtor's economic dependence, which might previously have been unwritten and undefined. There were now, then, actual physical objects to serve as the focus for tensions within the credit system. In this way, the legal system introduced by British rule went to make the Deccan Riots a more coherent and effective movement, but it had not created an agrarian revolution to cause the disturbances.

These suppositions could, of course, be best underpinned by some survey of the actual amounts of land acquired and held by moneylenders in the years before 1875. The ubiquity of the credit system, however, as well as limitations of the statistics, complicates this. Since every caste and group lent and borrowed, it is impossible to determine definitively movements of land caused by the effects of debt. Even so, holdings by the 'professional' moneylenders, whose origins lay outside the Deccan (and the case for social revolution is based on their alleged acquisitions), can be estimated. To generalise widely, it seems likely that Marwari moneylenders owned around 5 per cent of the cultivated land in most central and east Deccan localities in 1875, possibly rather less in the west Deccan. A survey in that year of three Ahmednagar taluks revealed that Marwaris owned 660 holdings comprising 17,885 acres in Parner taluk, 13,004 acres made up of 702 fields in Sangamner and 976 holdings of 7,771 acres in Akola:[45] since the cultivated acreage of the taluks varied between a quarter and a third of a million acres (the latter in Parner), this in no case represented more than 6 per cent of land in use and in the western taluk of Akola very much less. Independent investigations in Poona District by the Deccan Riots Commission, some of which are presented in Table 1, portray a very similar picture. In six villages of Indapur, Purandhar and Haveli taluks, total overall ownership by Marwaris in 1874–5 stood at around 6 or 7 per cent of the cultivated acreage.

What does this suggest about social change in the riots region? All

---

43 *Selections from the Records of the Government of India, Home Department*, No. 342, Papers Relating to the Working of the Deccan Agriculturists' Relief Act, 1875–94 (Calcutta, 1897), Vol. 2, No. 13, A. F. Woodburn to the Secretary to the Bombay Government, Judicial Department, No. 189, 27 April 1889, para. 32.
44 Deccan Riots Report, para. 12.
45 MARD, Vol. 3 of 1875, No. 971, A. F. Woodburn, Assistant Collector, Ahmednagar, to H. B. Boswell, Collector of Ahmednagar, No. 271, 5 July 1875, para. 11.

Table 1 *Ownership of land by Marwaris in certain villages of Poona District, 1874–5*

	Village	Total no. of holdings	Total area (acres)	Holdings owned by Marwaris		
				No.	Area (acres)	Percentage of total area
Indapur taluk	Indapur	386	19,443	36	2,189	11.3
	Bhowri	336	18,503	18	1,438	7.8
	Nimgaon Khedki	186	8,252	1	17	0.2
	Kullus	177	12,396	3	294	2.4
	Lasume	127	8,643	Not known	584	6.8
	Palasdeo	178	7,417	11	467	6.3
	Total, six villages of Indapur taluk	1,390	74,654	69+	4,989	6.7
Purandhar taluk	Kasba Walhe	285	11,558	31	1,143	9.9
	Khalod	183	3,114	6	82	2.6
	Deweh	306	8,474	19	214	2.5
	Veer	162	7,184	10	240	3.3
	Kasba Sasur	384	5,692	44	555	9.8
	Belsar	201	4,170	19	236	5.7
	Total, six villages of Purandhar taluk	1,521	40,192	129	2,470	6.1
Haveli taluk	Wagholi	147	8,319	4	87	1.0
	Lonikund	81	4,211	9	311	7.4
	Lohogaon	132	9,106	5	704	7.7
	Loni Kalbhar	209	10,359	10	490	4.7
	Nandosi	61	2,736	17	573	20.9
	Hadapsar	204	5,066	23	467	9.2
	Total, six villages of Haveli taluk	834	39,797	68	2,632	6.6

Source: Deccan Riots Report, Appendix C, pp. 200–1.

ownership by Marwaris, it must be remembered, was not necessarily new acquisition. Some Marwari families, like that of Shivram Marwari of Kheirgaon, had been established in the Deccan for generations and had possibly acquired some land much earlier. Bearing this in mind, the levels of ownership established by the Marwaris, ubiquitous as they were throughout the central Deccan, do not seem high. Nor does ownership seem conspicuously greater in the riot taluks. In Ahmednagar District the proportion in Parner, where several riots occurred, was certainly higher than in quiescent Akola, but in Poona there is no meaningful contrast between Marwari landownership in the disturbed taluk of Indapur and in Haveli, unaffected by the 1875 troubles. One further feature is strikingly evident from Table 1 and the other evidence. Marwari ownership was not evenly distributed between villages. Instead, it was often relatively extensive in a few large market villages and extremely limited elsewhere. Thus over 40 per cent of Marwari ownership in the six villages of Indapur taluk quoted in Table 1 was in the taluk centre of Indapur village. In this way, there were, perhaps, a few villages where dramatic social change had apparently occurred, the classic example being Parner village, the administrative and market centre of the Ahmednagar taluk. Here one outsider moneylender, Chandrabhan Bapuji, owned land assessed at Rs. 923 revenue and another, Tularam Karamchand, had obtained holdings assessed at Rs. 604 in the thirty or forty years before 1875.[46] At the same time, the Kowray family, which had once held the Patelship of Parner, owned not a single acre of land in the village by the 1880s.[47] Yet Parner seems to have been a highly exceptional case. Whilst the Deccan Riots were often triggered by disturbances in market villages, they involved many smaller, more isolated villages and tension in 1875 has to be seen as general throughout the eastern areas of Poona and Ahmednagar Districts. To explain this phenomenon, acquisition of land by alien moneylenders seems inadequate.

The Deccan Riots, in contrast, were at once a more precise and more wide-ranging movement. The object was, typically, to obtain bonds concerning day-to-day debt. At the same time, the bulk of the village community was normally involved, not simply a coterie of dispossessed peasants. The final outbreak at Kukrur, Satara District, in September 1875 involved an agitation by 41 leading figures: 24 Kunbis, mainly peasant agriculturists, nine artisans including four potters and the village barber and carpenter and eight low castes, mainly labourers.[48] The latter, and some of the artisans, could hardly have been motivated by land loss or fears about it, since they would never have owned any land. Most of the Kunbis, however, were independent proprietors cultivating their own land. If we could construct a picture of the typical rioter of 1875, he would own some land, it would not be threatened in any legal sense, but he would be heavily in debt, sometimes to several moneylenders.[49] This is balanced by the specific character of the victims. 'The Marwari and

---

46 Deccan Riots Report, para. 76.
47 Ibid., Appendix C, p. 66.
48 Ibid., Appendix C, pp. 10–11.

Gujur sowkars were almost exclusively the victims of the riots.'[50] Not only did peasant creditors remain unmolested, but businessmen and trader moneylenders of other castes were also, normally, left alone. This, of course, gives firmer relevance to our remarks about Marwari landownership and the whole thesis about land loss, since it was the Brahmin lenders, apparently, who 'have a much greater taste for getting land into their own hands and names than the immigrant traders'.[51] Nevertheless, this also presents a picture, familiar to many students of peasant societies, of a tight-knit village community reacting against outsiders who are, simply, disliked personally. At once, this makes explaining the Deccan Riots a much more manageable problem, for, as Holderness notes of Europe, 'the "foreignness" of the local financier is as much a cause of social unrest as his alleged rapacity'.[52]

Even so, Marwari and Gujarati immigrants settled and ran rural businesses throughout western India. It is still necessary to explain why their presence and behaviour apparently came to be resented more fully in the Deccan of the 1870s than elsewhere. One important possible influence here is the varying evolution of land tenure and social structure, discussed in Chapter 2. Poona and Ahmednagar Districts – the area affected in 1875 – formed the very region of Bombay Presidency where traditional village elites had fared worst under the impact of late Mahratta and early British administration. Deshmukhs and Patels, as we saw, lost privilege and power and one striking consequence was a perceptible social vacuum by the middle years of the nineteenth century. In turn, credit provision by established, recognised social elites – often a solvent of tensions in the debt situation – was, necessarily, much more limited. The role played by the khot in the Konkan, the talukdar in Gujarat was absent and in the poorer east Deccan, capital for lending by successful and wealthy agriculturists was also less readily available. As a result, immigrant 'professional' moneylenders were simply more important in Poona and Ahmednagar rural society, the leading suppliers of credit by 1875 rather than, as elsewhere, merely one group within a diverse system. There is some evidence, too, that, as the decline of the office holding class intensified in the years after the Wingate settlements, professional moneylenders were becoming progressively more numerous and started dealing with an increasing range of small day-to-day debts. The Riots Commission believed that 'the smaller class of sowkar... have increased

---

49 See the depositions of convicted rioters collected in ibid., Appendix B.
50 Ibid., para. 12.
51 Notes by Mr Sinclair, Assistant Collector of Ahmednagar, quoted in ibid., para. 39.
52 Holderness, 'Credit in English Rural Society before the Nineteenth Century', p. 109.

very considerably during the last ten years'[53] and Temple, the Bombay Governor, told the 1880 Famine Commission that 'the petty moneylenders... have increased, almost multiplied, under British rule'.[54]

This, it can be argued, was inherently an unstable situation. The degree of tension in the credit system was likely to depend most on the personal relations between creditor and debtor, the extent of common acceptance of values and attitudes. One official contrasted the situation in the pre-British Deccan, when the moneylender 'belonged to the village and to its quasi corporation... speaking the same language, worshipping at the same temple' with the picture revealed by the 1875 riots: 'now into almost every Maratha village four or five moneylenders from Marwar have penetrated, having no sympathy and no connection whatever with the villagers in language, race or religion'.[55] Like so many such official pronouncements, this greatly exaggerated the innovatory role of British rule, but the basic analysis of change is accurate. In turn, however, we need to consider what this explanation says about the Deccan Riots as a phenomenon. At one level, the riots should, perhaps, best be seen as a safety valve for local tensions. Like previous occasional violence against moneylenders, they enabled peasants to work off their resentments and, in addition, they achieved successfully the genuine aim of limiting temporarily the moneylender's power and potential. Many moneylenders, faced with an angry mob of their debtors, did surrender their court decrees or at least negotiate with the leaders of the riot for a general reduction of claims.[56] Others, frightened but unmolested, like those of Kardeh village in Sirur taluk of Poona who fled to the taluk headquarters for police protection,[57] would at least tread more warily in the future. In a situation where alien moneylenders were major providers of credit, there were precise advantages to be gained from a periodic show of village force and unity.

All this relates to my earlier argument that the social significance of the Deccan Riots has been greatly exaggerated.[58] Actual rioting, it needs to be remembered, was confined to just 33 villages[59] and, even in these cases, the violence was limited and controlled. Overall, 'the most remarkable

---

53 Deccan Riots Report, para. 63.
54 Temple Papers, Vol. 208A, Evidence before the Indian Famine Commission, Answer to Question 40.
55 MARD, Vol. 3 of 1878, No. 140, Note on the Deccan Riots Commission Report by Sir Erskine Perry, 1 December 1877.
56 Deccan Riots Report, para. 12.
57 Ibid., paras. 1–2.
58 See Charlesworth, 'The Myth of the Deccan Riots of 1875', pp. 401–21.
59 Deccan Riots Report, paras. 8–9.

feature presented is the small amount of serious crime'[60] and in one village, Damareh, a Marwari moneylender who had broken his leg in the uproar was saved from death when some of the rioters dragged him from his burning house.[61] The riots, too, died out quickly, for only 'a show of force sufficed to restore the ... country to its normal state of peace and quiet in a short space of time'.[62] These features, again, put the arguments about social breakdown in their proper perspective. Nevertheless, more can now be said about the character of the riots and what they, and the immense amount of official comment they generated, reveal about this region and its problems in 1875.

Perhaps the most striking general feature of the Deccan Riots is that they stand as action by, at root, a poor peasant economy. This is still rural politics in a relatively uncommercialised world and, as we shall see, the contrast is potentially immense with the Gandhian agrarian campaigns of the twentieth century. The unity of the village in 1875 provides evidence for this, for in a more developed peasant economy the interests of richer, middle and poorer peasants may diverge, diversifying their political reaction and performance. Again, the slow spread of the riots, often requiring the contact of village market-days – a point well made by Catanach[63] – illustrates the nature of the local economy and the character of the action. So, too, does the very specificness of the rioters' aims, the limited objects and the ease with which the authorities were able to suppress the disturbances. All these aspects link the Deccan Riots with their equivalents in other less developed peasant societies, the 'grain riot' in medieval Europe or traditional China.

Once more, this stresses the unremarkable nature of the riots of 1875, but the question remains why they erupted at the particular time and place they did. Official action and attitudes may have played a role here. One suggestion, made originally by Carpenter of the Riots Commission,[64] was that the revision of land revenue assessments, which had begun in Indapur taluk in 1867, greatly exacerbated the situation. Certainly the enhancements, in quantitative terms, seemed high, but, as most of Carpenter's colleagues agreed, there was, in fact, very little actual correlation between the riots area and the progress of the revision. By the

---

60 Ibid., para. 16.
61 Ibid., para. 6.
62 *Bombay Presidency Annual Police Report, 1875*, Southern Division, p. 25.
63 As Catanach well points out, the riots 'took a fortnight to spread over an area of only about forty miles from north to south and sixty from east to west'. Catanach, 'Agrarian Disturbances in Nineteenth Century India', p. 72.
64 See Deccan Riots Report, Minute on the Connection of the Recent Enhancement of Land Revenue with the Riots, by Mr Carpenter.

summer of 1875, land revenue rates had been reassessed in four taluks of Sholapur District, totally undisturbed in 1875, and four in Poona District: but two of the latter were in the quiescent west. Only a minority of the reassessed areas, then, were involved in any riot outbreaks and Ahmednagar District, where most of the Deccan Riots occurred, had experienced no revenue revision. Only 5 of the 33 villages with recorded rioting had, in sum, seen any change in their land revenue rates.[65]

The real influence of officialdom on the timing and scope of the riots may be more subtle than the traditional scapegoats of land revenue and legal systems. The riots area was the subject of considerable official enquiry during the 1870s. Besides any preparations for revenue revision, research and investigation were in full swing for the compilation of the *Gazetteers* of Ahmednagar and Poona Districts, the latter with its three volumes much the fullest of the *Bombay Gazetteers*. Official formalisations of the local credit system, with frequent talk of 'fraud' and 'exploitation', may have, in some cases, fuelled tensions and given new weight to peasant grievances. Catanach notes how many of the rioters appear to have thought that government, fundamentally, supported them, tacitly approving attacks on Marwaris.[66] This, of course, is, again, a feature of the traditional peasant demonstration, shared by the rebels of 1381 in England and many Chinese peasant insurgents. In the Deccan of the 1870s, however, peasants may have been given much firmer grounds than usual for such assumptions.

Nevertheless, this could only have been of minor significance, though it is ironic that, amidst so much contemporary heart-searching over the alleged innovations of British rule, the impact of official investigations and formalisations themselves was rarely considered. The problem of the specific geographical nature of the riots region remains: why did tension become so much more acute in the eastern areas of Poona and Ahmednagar Districts? The factor which most clearly united the riots region within the districts was, as one member of the Riots Commission noted, 'the characteristic of poor soil and precarious climate'.[67] All the villages where disturbances occurred in 1875 were firmly within the Deccan famine belt. Supa, the village where the riots began, was over 50 miles east of the Ghat spurs and 20 inches of rainfall in a year here would amount to better than average climatic conditions[68] but, within the neighbouring area, freaks of climate and soil created even more disadvan-

---

65 Ibid.
66 Catanach, 'Agrarian Disturbances in Nineteenth Century India', pp. 70–1.
67 Deccan Riots Report, Minute on the Connection of the Revision of Assessment with the Riots, by Mr Richey.
68 *BG, Vol. 18, Poona,* Part 1, p. 16.

taged local conditions. One official wrote in 1843 of an area which would prove highly turbulent in 1875: 'there is a tradition amongst the people, that the country between Jejuri and Baramati was formerly cursed, the ill-effects of which remain to this day in the failure of the periodical rains'.[69] With such problems, agriculture throughout the riots region was marked by a high concentration on the basic millets, bajra and jowar and in Bhimthari taluk, apart from a little wheat, there was next to no cultivation of the more varied cereals grown further to the west.[70] Ownership of resources for effective tillage was undoubtedly limited. Throughout the Supa area, 'cultivation is very carelessly conducted', 'the crops overrun with weeds', and 'the use of manure on dry-crop lands ... is apparently unknown'.[71]

Nevertheless, the riots region had shared fully in the expansionist economic cycle of the mid-nineteenth century. Waddington's settlement report on Bhimthari taluk of 1871 was an almost lyrically optimistic account which catalogued substantial alleged increases in 'prosperity'.[72] Numbers of cattle, carts and ploughs had increased and there was evidence of extensive new well construction. At the same time, the most striking feature, as in many areas of the Presidency, was the rapid rise in population, for the numbers of inhabitants of Bhimthari taluk had grown from 20,401 to 28,467 over the thirty years of the settlement.[73] The Deccan Riots, then, broke out in a society where population was still steadily expanding. Even the year between July 1873 and July 1874 saw 2,000 added to Poona District's total population.[74] Despite our earlier scepticism about the applicability of Malthusian models, it may be that, in this very poor area, the population increase was generating, by the early 1870s, serious pressure on the means of subsistence. Land available for new cultivation had always been described as extensive in this region but the problems of soil and climate necessitated relatively large holdings – 10 acres and more – if a peasant family was to make a reliable living. Hence take-up of land had, necessarily, been substantial to meet population growth and by the 1870s, even the official statistics suggest that

---

69 *SBG*, NS, No. 151, Settlement Report on Supa Petha, Poona District, by E. Evans, 8 July 1843, para. 6.
70 For details of cropping patterns in the different areas of Poona District, see *BG, Vol, 18, Poona*, Part 2, pp. 35–46.
71 *SBG*, NS, No. 151, Settlement Report on Supa Petha, Poona District, by W. Waddington, 5 September 1873, para. 4.
72 See *SBG*, NS, No. 151, Settlement Report on Bhimthari Taluk, Poona District, by W. Waddington, 12 July 1871.
73 Ibid., para. 6.
74 MARD, Vol. 7 of 1875, No. 960, Administration Report on Poona District 1874–5, para. 5.

land shortages may have been threatening. In the Supa sub-division, the amount of 'available assessed waste' had fallen from 26,302 acres at the beginning of the 1840s settlement to only 1,843 acres by its third decade: in Bhimthari taluk, no land was recorded as 'available assessed waste' by the 1870s, compared with over 43,000 acres thirty years before.[75]

In part, then, the Deccan Riots may have been a response to population pressures. They can be seen as the political reaction to be followed immediately by the economic response of the great famine of 1876–8. Yet, once again, we cannot assume the action of some simple Malthusian cycle. Whilst population continued to rise in the riots region into the mid-1870s, the expansionary era had firmly ended and contraction begun in most other aspects of the local rural economy. Firstly, after 1870 some land even started to fall out of cultivation. Over Ahmednagar District, official statistics suggested, 92 acres of hitherto cultivated land were abandoned in 1871–2, as much as 12,002 acres in 1872–3, a further 6,793 acres in 1873–4 and another 16,266 acres in 1874–5.[76] More fundamental and widespread was a decline in general price levels, beginning in the late 1860s and stretching down to 1875. In Poona market, average prices of the basic foodgrain, jowar, slumped from 13 seers per rupee in 1870–1 to 21 seers per rupee by 1873–4.[77] The administration report on Poona District for the riots year of 1875 confirmed that prices of all commodities except wheat and rice had continued to drop.[78] With population still rising and demand for produce to that extent presumably still buoyant, these may seem, superficially, perplexing trends. They make the point, again, that population patterns are not indissolubly linked to and determinant of other basic trends. Indeed, different types of economic cycle may be in operation at the same time. Possibly, in the 1870s, with harsh effects for the peasantry, contraction in the price cycle happened to coincide with the culmination of the demographic cycle's expansionary thrust. Price falls were general to many areas of India in the 1870s, a response, many officials argued, to the rapid and largely unreal boom of the American Civil War era. In addition, Bombay officials pointed to the effects of temporarily good harvests in the early 1870s and to improved facilities for transporting grain in lowering the prices of produce.[79]

These are complex issues to which we must return but, whatever their

---

75 Deccan Riots Report, Memorandum by Mr Auckland Colvin, para. 31.
76 MARD, Vol. 3 of 1875, No. 971, Administration Report on Ahmednagar District 1874–5, para. 14.
77 Deccan Riots Report, para. 68.
78 MARD, Vol. 7 of 1875, No. 960, Administration Report on Poona District 1874–5, para. 10.
79 These were the reasons given in ibid.

explanation, the features of contraction, in a situation of continued population growth, must have created substantial difficulties for the bulk of the peasantry in the riots region. Even the poor peasant's familiar resort – widely available in the 1860s – of selling his labour to earn additional cash now offered limited returns. Not only was labour work more difficult to find, but wage levels in the riots region seem to have been declining steadily along with prices.[80] In sum, it is hard to escape the conclusion that the riots region faced, even by its standards of poverty, an economic crisis by 1875. Like most peasant protests in most historical situations, the Deccan Riots were probably, at root, a reaction to economic deprivation rather than any more complex social trend. Faced with severe economic problems, the Deccan peasantry reacted by lashing out at a group who were identifiably outsiders, an ideal target for dissension.

Yet there is another, overlooked dimension to this situation. Moneylenders themselves were far from unaffected by it. Those who were also agriculturists were hit by the low prices and the end of the easy expansion in land use, but in particular the 'professional' moneylenders were threatened by developments. For them, the advance of credit to peasants with very poor security (and as outsiders, they often lacked the effective local knowledge about peasants' true economic position which many agriculturist lenders would automatically possess) was always a risky business, but normally justified by the opportunities this offered to control the marketing and sale of produce. In the situation of the 1870s, however, agricultural produce was, temporarily, a deteriorating asset. Hence many Marwari and Gujarati moneylenders, so far from accomplishing an agrarian revolution in their favour, may have seen themselves as desperately struggling to maintain their share of the proceeds from agriculture. From their point of view, the situation was further exacerbated by the widening of the credit market which had accompanied the prior expansionary cycle in the rural economy: much greater competition among moneylenders had been created, evidenced by the fact that many of the 1875 rioters seemed to owe money to more than one creditor.[81]

How could the moneylender react to such pressures? Land acquisition, for the reasons outlined earlier, made little sense. Most, therefore, had to retrench by restricting credit supply in some way or other. Amidst the detailed official accounts of the specific background to the riots, some of the ways this was attempted begin to emerge. One obvious target was credit for land revenue payment. The sub-judge of Patus in Poona District 'heard complaints that the sowkars refused to pay the assessments

---

80 See ibid., para. 9.
81 See the depositions of convicted rioters in Deccan Riots Report, Appendix B.

after carrying away the whole of the produce of the last season with an undertaking to do so'.[82] Another Poona official thought he discerned a conscious strategy behind this: 'their aim this year was by refusing advances to the ryots to force Government to grant large remissions as they well knew that the whole Talooka could not be sold up'.[83] If the latter assumption was correct, this could have been a subtle and sophisticated attempt to ease the problem, for forcing land revenue remission would have been one clear way of loosening the squeeze on the region's agriculture. The bulk of the peasantry, however, could not see matters in this light. They were threatened enough by the economic depression and the social and cultural divide between themselves and the Marwaris prevented any mutual understanding of problems, much less the adoption of a common strategy. Throughout the depositions by convicted rioters, collected in the appendices to the Deccan Riots Commission Report, runs the strand of rioters' bewilderment at the behaviour of the moneylenders. At Kardeh village in Poona District, for example, the peasants believed that 'the Marwaris have commenced to ruin them'.[84] In particular, where moneylenders refused to pay land revenue, peasant debtors were frightened that the authorities would react by confiscating their holdings, as the Land Revenue Code laid down. 'Accordingly the ryots pressed the sowkars to pay the instalments and the refusal of the latter led to the disturbances which commenced in Bhimthadee and extended all around.'[85]

Like many contemporary officials and subsequent historians, the Deccan rioters probably assumed that the Marwari moneylenders in 1875 were seeking to increase their power within agrarian society. In reality, however, the economy of the riots region at that time hardly offered, to any group, novel opportunities for social or economic enrichment. In the end, many of the riots were precipitated by actions by moneylenders which were probably essentially defensive, designed to protect an established position rather than to enhance it. In one sense, the Deccan Riots remain evidence of social change. They do firmly reflect our earlier theme of the decline of the old office holding class in this region, for without a predominant moneylender group lacking close contact and community of interest with the bulk of the peasantry, the outbreaks probably would not have occurred. Yet, on examination, they

---

82 Ibid., Assorted Papers on the Enhanced Assessments of Land Revenue, Report by the Sub-Judge of Patus, Poona District, 28 September 1875.
83 MARD, Vol. 7 of 1875, No. 960, C. G. W. Macpherson, Assistant Collector, to G. Norman, Collector of Poona, No. 72, 17 July 1875, para. 11.
84 Deccan Riots Report, para. 1.
85 MARD, Vol. 7 of 1875, No. 960, C. G. W. Macpherson to G. Norman, No. 72, 17 July 1875, para. 11.

reveal nothing in the way of any revolution in landownership in the Deccan.

Indeed, we can go further. Some might now respond to the argument about ownership by suggesting that the formalities of proprietorship, as recorded in the village documents, matter little compared with the realities of economic power: that, even if the peasant agriculturist remained the owner of his traditional holding, traders and moneylenders, through the operation of the credit and marketing systems, were, perhaps, able to secure growing proportions of the output. This is a complex issue and discussion of it, over the second half of the nineteenth century, will form some of the dominant themes of the following chapters. We have already seen, though, that some surrender of produce was an established price to pay for credit for many peasants, particularly in a poor area like the riots region. This, again, makes degrees of possible change very difficult to estimate. What we might suggest, however, is that the Deccan Riots do not reveal, in the still relatively uncommercialised economy and area in which they occurred, any firm evidence of a shifting balance of economic power. Rather, the key to the situation was that both moneylender and debtor seem to have been caught in the squeeze of depression. The creditor's response, bringing down the wrath of the village on his head, suggests that he had no more secure means of maintaining his economic position than the peasant debtor. Once again, the Deccan Riots show an economy whose fundamental problem was shortages of capital and credit.

The official response

What, however, would be the official reaction to the outbreak of the riots? The initial response within the Bombay government was sanguine and balanced. Wodehouse, the Governor, informed the Viceroy, Northbrook, of the affair in a strikingly low-key report: 'no lives have been lost and they are of no political importance, but directed entirely against the Sowcars or moneylenders whose property has been plundered'.[86] Yet maintaining so balanced an assessment thereafter proved extremely difficult, for the outbreak of the riots had exactly coincided with the culmination of a major debate in Bombay over the problems of agrarian indebtedness. In November 1873, Wodehouse had written to the Secretary of State with some tentative proposals for legislation. The main suggestions were to exempt rural land from sale for debt and to provide

---

86 IOL, Papers of Sir Philip Wodehouse, Vol. 8, Wodehouse to Lord Northbrook, 18 May 1875.

for registration, by a European officer, of any debt bond containing an encumbrance on land.[87] The Secretary of State, the Duke of Argyll, firmly rejected the notion on grounds of laissez-faire principle,[88] but a somewhat more sympathetic audience was available when Argyll was succeeded by Salisbury in 1874. Wodehouse himself, however, was, as yet, a lukewarm reformer: it was pressure for legislation from grass-roots officialdom which seems to have been forcing the pace. He told Salisbury in the winter before the Deccan Riots: 'this Government is in the constant receipt of representations of increasing poverty (which I do not believe to be quite correct) and is urged to go to the rescue', but he still recommended adoption of his earlier proposals.[89] The discussion, then, was at a crucial stage when the Deccan Riots broke out. In this atmosphere, their impact was bound to be substantial, whatever the realities which lay behind them.

But what had created the pressure for legislative reform? Throughout India, of course, the question of agrarian indebtedness proved an issue supremely attuned to arousing the indignation and reforming zeal of the European rulers[90] and Bombay, like most provinces, boasted a long pedigree of official proposals for legislation.[91] At the same time, the development of the debate reflected attempts to resolve a major dilemma at the heart of British thinking about India's agrarian economy, a dilemma which had been particularly evident within the evolution of the Bombay revenue system. The Joint Report of 1847 had, in one breath, enshrined as a basic principle of the Bombay settlements 'the progressive development of the agricultural resources of the country, and the preservation of all proprietary and other rights connected with the soil'.[92] The possible contradiction here was evident even in the 1840s. To promote agricultural development, most officials then assumed, some degree of land transfer was imperative and Wingate promised firmly that the land of revenue defaulters would be put up for auction, for 'by thus bringing land into the market, traders, pensioners, and other parties having capital, would probably be induced to lay it out in the purchase of land, to the great benefit of agriculture and the community at large'.[93] Here, though,

---

87 Ibid., Vol. 12, Wodehouse to the Duke of Argyll, Secretary of State for India, 27 November 1873.
88 Ibid., Vol. 11, Duke of Argyll to Wodehouse, 29 December 1873.
89 Ibid., Vol. 13, Wodehouse to the Marquis of Salisbury, 20 October 1874.
90 For a study of official attitudes to the problem throughout India, see Dewey, 'The Official Mind and the Problem of Agricultural Indebtedness in India, 1870–1914'.
91 For proposals in the early 1840s to ameliorate rural indebtedness, see Kumar, *Western India in the Nineteenth Century*, pp. 190–2.
92 Joint Report, 1847, para. 4.
93 *PP*, 1852–3, Vol. 75, Official Correspondence on the System of Revenue Survey and Assessment in the Bombay Presidency, G. Wingate to E. H. Townsend, Revenue Commissioner, Southern Division, No. 239, 23 December 1848, para. 16.

already was change in proprietary rights. The contradiction was papered over, in the 1840s, by the comfortable assumption that only the morally undeserving would suffer land loss in this way and that the pursuit of agricultural development could, therefore, co-exist with the defence of the rights of the vast majority of landowners. However, greater familiarity with rural society showed that those threatened (once the fatal assumption had been made that debtors were 'threatened') were far more than a feckless minority. To many officials, therefore, 'the preservation of all proprietary and other rights' seemed, eventually, to demand legislative action. After 1858, too, the blood shed at Cawnpore and Lucknow added a new force to the argument. Not only was the undermining of traditional landowners contrary to all the principles of British rule, but it inevitably threatened the security of the whole operation.

Proposals for reform in Bombay, therefore, developed quickly from the 1850s. Wingate, who in 1847 had been content to let rural society go where it may, believed in 1852 it was 'absolutely necessary to take immediate steps for the protection of the cultivators or they will all be eventually reduced to the condition of those of Rutnagirree who are under the Khotes'.[94] He submitted a detailed scheme for legislation, designed to deal with abuses of the legal system.[95] Land, Wingate proposed, should be exempted from attachment for debt unless specifically pledged, there should be 'a simple and equitable insolvent law' and courts should have powers to fix suitable rates of interest. Other suggestions included allowing small suits to be dealt with summarily by revenue officers or panchayats and 'improved regulations' for the delivery of court summonses and orders. Wingate's proposals were to find frequent echoes in subsequent years. In the late 1850s, for example, C. Fraser Tytler, the Collector of Ahmednagar, presented a full-scale review of possible legislative palliatives for indebtedness.[96] Some of his suggestions, that land should be exempted from civil process and that panchayats might deal with some civil cases, directly mirrored Wingate's. Tytler also advanced the radical idea of abolishing any legal redress for all debts under a certain amount: he originally proposed Rs. 300 as the lowest amount, which, at a stroke, would have ended most of the business of the Bombay courts.[97] In addition, Tytler argued for a registry of all peasant debts as a measure against fraud.

Wingate and Tytler's ideas, it will be seen, formed one central tradition

---

94 Wingate Papers, Box 293/4, Diary, 1 April 1852.
95 Ibid., Box 118, G. Wingate to the Registrar of Suddur Dewanee Adawlut, Bombay, No. 319, 24 September 1852.
96 MARD, Vol. 167 of 1859, No. 322, C. Fraser Tytler to J. D. Inverarity, Revenue Commissioner, Southern Division, No. 1397, 9 December 1858, para. 18.
97 For this idea, see MARD, Vol. 164 of 1858, No. 934.

among official attitudes to the indebtedness issue, a tradition which was to dominate deliberations in the aftermath of the Deccan Riots. Its approach was to attempt legal reforms, for the working of the courts was seen as an innovatory and disruptive force for change. An entirely contrary programme was developed from the functionalist view of the credit system held by such as Wedderburn. If, as we saw Wedderburn argued, credit and moneylenders played an essential role in agrarian society, then the way forward might be to create much more credit within the village, developing resources and, hopefully, lowering interest rates. Wedderburn's panacea was, therefore, a mass of local agricultural banks, providing loans and encouraging the wealthier cultivators to save.[98] This was, in the 1870s, not a new concept. It had been one of the remedies reviewed by Tytler in 1859,[99] drawn by him from a proposal submitted by one of his Assistant Collectors, H. E. Jacomb. Jacomb was originally attracted to the village bank idea, because of its potential as a moral force for encouraging saving and thrift: his model was the English 'penny savings bank'.[100] However, he soon realised that savings and loan banks might be usefully amalgamated so that 'the Savings Bank would... supply at least a portion of the funds necessary to enable us to advance small sums of money on equitable terms to the cultivators'.[101]

The dichotomy between these different philosophies is clear. Wingate, in recommending his legislative provisions, had argued that 'what is wanted is to restrict credit' and 'to withdraw many of the facilities now afforded to the creditor for the realisation of his debts'.[102] Wedderburn, in contrast, believed that 'the more capital that flows towards the land the better'.[103] As a result, he rejected all legislative regulations suggested except, possibly, an insolvency law for extremely indebted peasants and the establishment of conciliation machinery, perhaps panchayats, to maintain amicable relations within the credit system.[104] Difficulties created by the differing approaches were enough to prevent any serious

---

98 These ideas were eventually presented most clearly in William Wedderburn, *Agricultural Banks for India* (Bombay, 1882).
99 See MARD, Vol. 167 of 1859, No. 322.
100 Jacomb's ideas are most fully developed in MARD, Vol. 116 of 1860, No. 497, H. E. Jacomb to C. Fraser Tytler, No. 13, 6 December 1858, para. 13.
101 Ibid., H. E. Jacomb to C. Fraser Tytler, No. 2, 29 January 1859, para. 2. For further discussion of Jacomb's proposals, see also Kumar, *Western India in the Nineteenth Century*, pp. 193–4.
102 Wingate Papers, Box 118, G. Wingate to the Registrar of the Court of Suddur Dewanee Adawlut, Bombay, No. 319, 24 September 1852, para. 12.
103 Ibid., Box 119/1, William Wedderburn, Report on the Indebtedness of the Ryot, 7 December 1876, para. 5.
104 See ibid.

steps towards action before the 1870s. Legislation on the lines proposed by Wingate, according to Jacomb and Wedderburn's arguments, might make peasants poorer – and more discontented – by restricting sources of credit. At the same time, as Wingate himself conceded, unless such legislation was firm and definitive, the moneylenders 'would experience no difficulty in evading the law'.[105] Programmes, however, which placed provision of new loan facilities in the van faced accusations of severe impracticality. With the limitations of private capital in the Bombay countryside widely acknowledged, most bank schemes envisaged some degree of government aid and control over their operations and many were opposed to this on principle.[106] There was also the grave political objection which Wodehouse put to Salisbury in 1874, when the latter betrayed some support for the bank concept: government-run agricultural banks would be 'calculated to transfer at once to the shoulders of the Government all the odium which now falls on the sowkars'.[107]

For these reasons, as pressure for action began to grow after 1870, proposals based on remedial legislation came to seem, as Wodehouse put it, 'more prudent'.[108] Yet, what was the driving-force behind the 'constant receipt of representations' which Wodehouse now described his government as facing? Frequently, it is seen as ideological. Ravinder Kumar describes a strong conservative reaction, inspired by the ideas of Maine, fuelling, in the 1870s, the progress towards legislation 'to restore that fine balance . . . in the villages which had been undermined by the advocates of reform'.[109] This view is based, to a large extent, on the writings of theorists like Raymond West, whose *The Law and the Land in India*, published in 1872, provided a powerful denunciation of supposed agrarian change based on deeply held and thought out conservative convictions. For West, indebtedness and land transfer were repugnant, not just for practical reasons, but for their moral and social product, 'a widespread and corroding political demoralisation'.[110] Yet the actual impact on the formation of policy of such broader metaphysics is always a matter for debate. West was a judge not a revenue man and it was within the revenue department that pressure for action was growing. In 1874 W. G. Pedder,

---

105 Ibid., Box 293/8, G. Wingate to H. Green, Professor of Literature, Poona College, 5 January 1852.
106 The Bombay government, for example, met Jacomb's suggestions with a firm statement that 'such a Bank could not be established by the State'. MARD, Vol. 116 of 1860, No. 497, Bombay Government Revenue Department Memorandum, No. 932, 8 March 1860, para. 1.
107 Wodehouse Papers, Vol. 13, Wodehouse to Salisbury, 21 June 1875.
108 Ibid.
109 Kumar, *Western India in the Nineteenth Century*, p. 203.
110 West, *The Law and the Land in India*, para. 8.

soon to become Bombay's Revenue Secretary, submitted a full scheme of legislation. It included the main Wingate palliatives of a law of insolvency and the exemption of land from sale for debt as well as proposals for reform of court procedure, such as ensuring that courts and not creditors recovered debts.[111] It was demands such as this which led Wodehouse to open direct negotiations with the Secretary of State.

Of course, Pedder, like West, might be seen as a conservative ideologue. Certainly, if utilitarianism is regarded as the assumed philosophy of the Bombay revenue department, Pedder was, at best, a lukewarm advocate and as a young settlement officer in Gujarat, he had been keen, as we saw earlier, to make exemptions within the ryotwari system for the narwa and bhagdari systems. Yet the central problem in linking demands for reform with one consistent ideology is that, with so complex an issue, precise attitudes often cut across wider beliefs. In practice, there were as many particular views on the indebtedness problem within the revenue department as there were revenue officials. Already, in distinguishing the Wingate–Tytler tradition and that typified by Wedderburn as two distinct philosophies, we have been guilty of inevitable over-simplification. Theoretically, it is true, the approaches appeared to differ fundamentally over issues such as credit provision, but that did not stop some officials wedding different aspects from each to form their own proposed reform package. Thus Tytler, in his survey of possible measures, included agricultural banks along with his legal remedies.[112] Further, there was often a divide between general philosophical views on indebtedness and specific attitudes to reform. West's, for example, was a moral denunciation of the effects of British rule, not, in any sense, a specific call to action. Although *The Law and the Land* did contain some suggestions for legislation, West later became highly sceptical about the advantages of legal reform and he proved a fierce critic of the eventual measure which emerged, the Deccan Agriculturists' Relief Act. On the other hand, some firm opponents of the work of the conservative law codifiers strongly supported legislation for the Deccan.[113] In this way, any link between the Bombay reform movement and a conservative reaction in the 1870s can only be tenuous.

In practice, the Bombay government in the early 1870s was feeling its way forward, not advancing on any firm ideological foundations. It was, fundamentally, deteriorating economic conditions in the 1870s which

---

111 IOL, Papers of Sir William Lee-Warner, Deccan Ryot Series, No. 2, W. G. Pedder, Note on the Indebtedness of the Indian Agricultural Classes, 29 July 1874.
112 See MARD, Vol 167 of 1859, No. 322.
113 One example was A. K. Connell. See A. K. Connell, *Discontent and Danger in India* (London, 1880).

brought possible legislation on indebtedness back into the forefront of debate. During the 1860s with its buoyant prices and, in particular, the American Civil War cotton boom, discussion had been largely in abeyance and even Tytler wrote that the Ahmednagar peasantry 'have extricated themselves from the meshes of which formerly they used so greatly to complain'.[114] After 1870 so sanguine a view was no longer possible. This was why the impact of the Deccan Riots was so electric: they seemed to provide positive evidence not just of the extent of debt but that its operation was a source of serious discontent. In addition, they had the effect of unifying the hitherto highly disparate reform campaign. Earlier proponents of legislation, like Pedder, took the opportunity to press their case more vigorously[115] and the Deccan Riots, also, provided the focus to attract a much wider audience, in Britain as well as in India. Pedder popularised the problems of the Deccan economy in an article in *The Nineteenth Century* of September 1877[116] and, far more influential, Florence Nightingale was drawn into the debate. With her skills as a keeper of the imperial conscience, she drew a poignant contrast between the different treatments received in the court by the indebted undergraduate in England and the indebted peasant in Bombay.[117] Throughout the late 1870s, Florence Nightingale used her not inconsiderable influence at the India Office (it was, by now, as Lytton Strachey put it, 'de rigeur for the newly appointed Viceroy, before he left England, to pay a visit to Miss Nightingale'[118]) to press for legal reforms as far-reaching as possible for the Deccan.[119]

Meanwhile, the terms of reference of the Deccan Riots Commission – necessarily carefully chosen to avoid offending the Bombay government by seeming to call into question its basic principles and system – had aided the process. The Commission's brief was to investigate the grievances of the Deccan rioters and to examine the problem of indebtedness in Poona and Ahmednagar Districts. This was clearly likely to direct it towards the supposed legal abuses of indebtedness rather than any daring flights of fancy towards a Wedderburn type of analysis. The Riots Com-

---

114 MARD, Vol. 30 of 1862–4, No. 492 of 1862, C. Fraser Tytler to W. Hart, Revenue Commissioner, Southern Division, No. 242, 14 February 1862, para. 4.
115 This aspect of the riots, which greatly enhanced the publicity and attention they received, is stressed in my 'The Myth of the Deccan Riots of 1875', pp. 401–21.
116 W. G. Pedder, 'Famine and Debt in India', *The Nineteenth Century*, 2, August–December 1877, pp. 177–97.
117 See Florence Nightingale, 'The People of India', *The Nineteenth Century*, 4, July–December 1878, p. 220.
118 Lytton Strachey, *Eminent Victorians* (Folio Society ed., London, 1967), p. 164.
119 Her correspondence on this with the new Secretary of State, Lord Cranbrook, is mentioned in Alfred E. Gathorne-Hardy, ed., *Gathorne Hardy, First Earl of Cranbrook. A Memoir* (London, 1910), Vol. 2, pp. 29–80.

mission Report, presented in spring 1876, duly obliged. Its proposals proved strictly in the Wingate–Tytler tradition. The Commissioners advocated abolishing imprisonment for debt, exempting a debtor's necessaries from attachment and limiting the validity of court decrees to no more than six years.[120] In addition, they hoped to prevent fraud by providing for the registration of all debt transactions and by entitling any debtor to receipts for payments and a pass book for his account.[121] The Commission also recommended a range of administrative reforms, making the revenue system less rigid and ensuring the courts were more accessible to a defendant and his witnesses.

Despite this unanimous verdict it was more than another three years before legislation was finally enacted and the Deccan Agriculturists' Relief Act received the Viceroy's assent in October 1879. The tortuous wrangling over the specific evolution of the measure in 1878 and 1879 can be followed elsewhere[122] but the major ingredients which maintained the pressure for legislation can be easily outlined. One was undoubtedly Wodehouse's replacement in April 1877 by Richard Temple, unlike most Bombay governors, a professional India man of great experience.[123] Temple regarded the reform movement as sheer pragmatic sense and his first proposals for legislation, made in November 1877, incorporated a fair balance of Wodehouse and the Riots Commission's schemes – the exemption of land from sale for debt unless specifically pledged; the enactment of an insolvency law – together with his own pet proposal, the strict examination by the courts of all debt transactions for which they were asked to pass decrees.[124] At the same time, the intensification of the economic downswing in the Deccan with the onset of the famine of 1876–7 created a further force for action. Temple later commented that 'such misfortunes as these caused us to reconsider the question whether anything could be done by legislation to mitigate the evils which arose from ... indebtedness'.[125] Most significant, the build-up of influences like

---

120 Deccan Riots Report, paras. 62–5.
121 Ibid., paras. 60–2.
122 See Kumar, *Western India in the Nineteenth Century*, Ch. 6; Charlesworth, 'Agrarian Society and British Administration in Western India 1847–1920', Ch. 3. For a collection of the official papers, see *SBG*, NS, No. 157, Papers and Proceedings Connected with the Passing of the Deccan Agriculturists' Relief Act.
123 Temple had worked in the North-Western Provinces and the Punjab before becoming successively Chief Commissioner of the Central Provinces, Finance Minister to the Government of India, and Lieutenant-Governor of Bengal. He had earned a reputation as a famine troubleshooter, which largely influenced his appointment to Bombay.
124 *SBG*, NS, No. 157, Papers and Proceedings Connected with the Passing of the Deccan Agriculturists' Relief Act, Minute by Richard Temple, 12 November 1877.
125 Richard Temple, *Men and Events of My Time in India* (London, 1882), p. 466.

these created unanimity of support for some action within the Bombay revenue department. During 1879, the Revenue Commissioner and all the Collectors within the Deccan region minuted their firm adherence to the principle of legislation.[126]

The Deccan Agriculturists' Relief Act was apparently a conclusive victory for those who had been pressing for strong legislation. The courts were to investigate carefully the background to debt transactions, as Temple had urged. Almost all the recommendations of the Riots Commission were enacted. Imprisonment in execution of a decree for money was abolished and any agriculturist with debts of Rs. 50 and more could now apply to be declared insolvent. Village Registrars were to be appointed to write out all instruments of debt involving agriculturists, and debtors were to be entitled legally to statements of account and receipts from their moneylenders. Land, unless specifically pledged, was to be exempt from attachment or sale for debt. Village munsifs would be extended to make the courts more accessible to the debtor and his witnesses. In addition, the Act introduced some procedures intended to prevent disputes ever reaching court. The period of limitation for suits was extended and a system of 'conciliators' introduced to try to settle disputes. This legislation was to apply to the four Deccan districts, Poona, Ahmednagar, Satara and Sholapur, but, it was declared in 1879, might eventually be extended to other Bombay districts.

So far from being grounded in specific ideology, the Deccan Act seemed to most Indian officials in 1879 simple practical politics. Once an essential assumption had been made – and that assumption could be shared by men of all ideologies and none – support for legislation would follow inevitably. The assumption was that high quantitative levels of indebtedness led inexorably to land transfer and social revolution; and most Bombay officials came, by the late 1870s, to make it. Wingate, as early as 1852, believed that 'the courts are beginning to sell land in execution of decrees, and when this measure becomes general, the whole land of the country will rapidly pass into the hands of the Marwarees'.[127] By 1858 Tytler was alleging 'a new and marked tendency on the part of moneylenders to secure land from their debtors'.[128] Such statements were, of course, generalisations based on a few particular cases, but the whole operation of British administration worked against the maintenance of a sense of proportion on issues such as these. Some spectacular cases of land transfer and social change undoubtedly did occur, as in

---

126 See MARD, Vol. 11 of 1879, No. 422.
127 Wingate Papers, Box 293/4, Diary, 1 April 1852.
128 MARD, Vol. 167 of 1859, No. 322, C. Fraser Tytler to J. D. Inverarity, No. 1397, 9 December 1858, para. 4.

Parner village: the local official existed to define and report problems and it was difficult to avoid the temptation to point to these isolated examples as probable symbols of much wider change. The Deccan Riots seemed to provide precise confirmation, and, in addition, they brought the security factor firmly to the fore. Many took them as a harbinger of the Maratha uprising which would inevitably follow continued government inaction. Ranade summed up the consensus of opinion among Bombay officialdom when he commented: 'the heavy indebtedness of the agricultural classes being admitted to be a great political evil, I do not see that there was any other alternative open to Government'.[129]

Yet, if it is easy to understand how events shaped the Deccan Agriculturists' Relief Act, this does not obscure an essential fact: the Act was based on a fundamentally misconceived view of the operation of the credit system and social change. This was, clearly, likely to complicate its impact. However, there was a further problem. Even on its own analysis of events, the Deccan Act did not dare to approach the problem centrally. It was social change represented by land transfer which had been identified as the dangerous problem but, unlike twentieth-century legislation on land transfer, there was no attempt to interfere with the conditions of landownership and exchange. The peasant was given every opportunity to avoid having to put his land at stake, but his right to sell it and a creditor's right to obtain it were both unimpaired. In sum, the Deccan Act was, as Gadgil later put it, 'in the main a rural Moneylenders' Act',[130] a regulatory measure designed to correct what were seen as obvious abuses of the legal system. Yet such regulation was a highly subtle matter, much more so than specific enactments about land transfer. Whether the Act would restrict credit availability or leave it unaffected, diminish or increase creditors' interest in land, remained to be seen. After 1879, at any rate, the operation of the Act introduced new artificial complexities into the working of the Deccan credit system.

---

129 *Selections from the Records of the Government of India, Home Department*, No. 342, Papers Relating to the Working of the Deccan Agriculturists' Relief Act during the Years 1875–94, Vol. 1, No. 139, M. G. Ranade, quoted in Dr A. D. Pollen, Special Judge, to the Secretary to the Bombay Government, Judicial Department, No. 60, 4 February 1882, para. 52.
130 Foreword by D. R. Gadgil to K. G. Sivaswamy, *Legislative Protection and the Relief of Agriculturist Debtors in India* (Poona, 1939), pp. i–ii.

5
Continuity and change in the rural economy, 1850–1900

Implicit in our whole discussion of the causes of the Deccan Riots and, even, the evolution of official attitudes to the problem of indebtedness has been the central issue of fluctuations within the rural economy. In a society which lived so close to the struggle for subsistence, these short-term changes in economic climate were likely to have substantial, deep-felt impact. Yet, also, they were the very stuff out of which any broader developments in economic performance might be constructed. If, for example, the rural economy could enjoy a prolonged period of buoyant demand and favourable climatic conditions, then some important steps might be taken towards extending local supply of capital and credit. On the other hand, even one season of serious famine might create more than short-term crisis, by killing off agricultural stock and hence intensifying crucial limitations of resources. So we need, at this stage, to try to delineate the pattern of economic fluctuations over the whole second half of the nineteenth century.

At the broadest level, this is not difficult. From the 1840s, as we have seen, population, land use and price levels seem to have been rising consistently in western India, although often from fairly low bases. The process then accelerated markedly during the first half of the 1860s. Tytler's comments, for that period, from Ahmednagar District might be applied to most regions of the Presidency: 'the seasons have been excellent, the market prices very high and the consumption of produce and consequent demand great and increasing'.[1] After about 1870, however, prices and, in some cases, the extent of cultivated acreage began to decline. This trend was most perceptible in the central Deccan but it also occurred elsewhere: average prices of raw cotton in Broach District of Gujarat, for example, were substantially lower over the period 1871–6 than the levels of 1869 and 1870.[2] The economic difficulties of the 1870s reached a

1 MARD, Vol. 30 of 1862–4, No. 492 of 1862, C. Fraser Tytler, Collector of Ahmednagar, to W. Hart, Revenue Commissioner, Southern Division, No. 242, 14 February 1862, para. 2.
2 *Index Numbers of Indian Prices, 1861–1926* (Department of Commercial Intelligence and Statistics, New Delhi, 1928), Table 5, p. 7.

culmination with the onset of major famine in 1876. The famine of 1876–8 mainly afflicted the traditionally vulnerable districts of the Deccan but, throughout Bombay Presidency, the growth in population experienced an abrupt halt and the 1881 Census revealed absolute decline in population levels since 1872 in Dharwar, Belgaum, Broach and Ratnagiri Districts as well as in the Deccan.[3] Yet recovery often seems to have been swift. In Indapur taluk of Poona, at the very heart of the famine destruction, numbers of people and cattle, despite calamitous falls in 1876–8, had recovered to well above the 1875 peak by the mid-1890s.[4] This illustrates the buoyancy of the new era of expansion which set in after 1880, marked as it was by consistently good climatic conditions and favourable opportunities for extending agricultural production for the market. The cycle, however, was, this time, more compressed. It ended in 1896–7 with the outbreak of a still more devastating period of famine, which, by 1900, had again substantially reduced levels of population and resources throughout western India.

This pattern, whereby periods of expansion appear to lead inexorably to crisis, might obviously be explained in Malthusian terms. Despite our reservations about the model for the mid-nineteenth century, it might be argued that after 1870 the Bombay agricultural economy faced some fundamental impasse. The economic system continued to operate within constant technical parameters and land availability was presumably not limitless: expansion could, therefore, occur to the limits of population pressure, before creating a crisis of subsistence which would itself restore the capacity for renewed growth on the same terms. In this classic form, the economic pattern would be cyclical and operating at relatively even levels. Yet it is at least possible that any cyclical pattern obscures overall movement in a positive direction, producing significant change in the levels and character of rural economic activity between 1850 and 1900. Was there, perhaps, some deterioration in the rural economy's capacity to support its population? The apparent shortening of the cycles – the gap of under twenty years between the occurrence of two great famines – might suggest this. It may be, in explanation, that the problem of land availability had now become a decisive obstacle. Possibly, the expansionary cycle of 1880–96 was always constrained and finally broken by lack of good, new cultivable land, for some official evidence suggests an impasse had been reached as early as the 1870s. Officially estimated

---

[3] *Imperial Census of 1881, Operations in the Bombay Presidency* (Bombay, 1882), Vol. 1, p. 28.

[4] The recorded population of Indapur taluk was 62,722 in 1875–6, 39,720 in 1877–8, 67,684 by 1895–6. SBG, NS, No. 379, Settlement Report on Indapur Taluk, Poona District, by F. B. Young, 13 October 1897, para. 22.

cultivable waste in the Deccan, as we saw, was now very limited; nonexistent in some taluks like Bhimthari. James Caird, the English agricultural expert, visiting India under the auspices of the 1880 Famine Commission, had concluded from this type of evidence that 'a great part of Madras and some part of Bombay seemed to me to have reached the margin of production under the present agricultural system'.[5]

Official estimates of overall land availability remained gloomy in the years after 1880. A major enquiry in 1888 concluded that waste land capable of cultivation came to only 9 per cent of the total area of Gujarat, 14 per cent of the Konkan, 7 per cent in the Deccan and 6 per cent of the Southern Maratha Country.[6] By 1892 another review of land availability reported that 'the margin is only sufficient to provide for a five to ten per cent increase of the agricultural population'.[7] And yet, it seems that actual cultivated acreage continued to extend in many localities in the boom of 1880–96 as in that of 1840–70, if not at so rapid a rate. Statistics on land in use remain unsorted before the twentieth century, but many are available for individual localities in the settlement papers. These often present a more sanguine impression than the general reports. To take, for example, the statistics for a central Deccan taluk traditionally heavily populated and intensively cultivated, Parner taluk of Ahmednagar: cultivated area seems to have risen here from 255,404 acres to 283,324 acres between the settlements of 1884–5 and 1913–14.[8] In practice, it seems that most localities were not reaching any absolute impediments to land expansion.

The question of the quality of the new land remains, of course, at issue and an impossible issue to judge definitively for this period. Nevertheless, even where land availability was strictly limited, there were other routes to the raising of overall output. We should not underestimate the capacity of Indian agriculture to wring yet more out of the land. Techniques of production, as we argued earlier, were by no means backward and, if the expansionary periods created sufficient capital for some limited extension of resources, then output per acre may have risen perceptibly. Again, gains could have occurred through the constant refining of techniques. These remarks are especially prompted by the case of one

---

5 IOL, Home Miscellaneous Series, No. 796, Indian Papers of Sir James Caird, Caird to Lord Northbrook, 19 September 1879.
6 IOL, Government of India Revenue Progs., December 1888, No. 18, Report on the Economic Condition of the Masses of Bombay Presidency, 27 August 1888, p. 2.
7 MARD, Vol. 258 of 1892, No. 749, Brief Memorandum on the Material Condition of the People of Bombay Presidency, 1881–91, para. 3.
8 SBG, NS, No. 567, Settlement Report on Parner Taluk, Ahmednagar District, by J. H. Garrett, 22 May 1916, para. 14.

Bombay region, the south Konkan. Here, in Ratnagiri District, there had been continued talk of intense pressure of population on land resources. Wingate described the district as over-populated as early as 1851[9] and the revenue surveys of the 1860s, before they were ended by khoti obstructionism, suggested to the settlement officers that the area might have reached the limits of numbers it could support.[10] Yet population continued to rise and at a substantial rate. Ratnagiri taluk's population, for example, increased by 46 per cent between 1866 and 1897[11] and by a further 19 per cent down to 1911.[12] This strikingly upward demographic pattern in the Bombay district traditionally highlighted for its overpopulation and limited agricultural capacity emphasises again that there was no simple link between population and resources. Of course, the process may have involved decline in per capita living standards or augmentation from non-agricultural sources, for emigrants from many Ratnagiri families found employment in the Bombay mills or even abroad. Nevertheless, the trends surely indicate some capability in the agricultural system to support increased numbers. One of the techniques whereby this might have been achieved was outlined in a report on Ratnagiri in the 1920s: 'the improvement of land whereby a level plot of rice is won from a sloping hill by stone levelling the land and erecting stone dams which collect silt deposit over a run of scores of years must be seen to be understood'.[13]

Despite, then, the absence of any decisive technological breakthrough, the possibility of significant rise in agricultural output over the second half of the nineteenth century cannot be discounted. However, the official statistics on agricultural output remain, in this period, much too limited and patchy to offer serious assistance in making judgement. Government estimates do exist for the expansionary era of the 1880s and 1890s. These broad figures declare that the total value of agricultural produce in the Presidency proper and Sind rose from Rs. 36,93 lakhs in 1881 to Rs. 48,69 lakhs in 1898;[14] but the statistical basis for this assessment was, to say the

---

9 *SBG*, No. 2, Report by Capt. Wingate on Introducing a Survey in Ratnagiri District, No. 44, 30 January 1851, para. 43.
10 See, e.g., *SBG*, NS, No. 377, Settlement Report on Saitavda Petha, Ratnagiri District, by W. Waddington, 27 April 1867, para. 5.
11 *SBG*, NS, No. 377, Settlement Report on Ratnagiri Taluk, Ratnagiri District, by J. L. Lushington, 30 March 1897, para. 12.
12 *SBG*, NS, No. 574, Settlement Report on Ratnagiri Taluk, Ratnagiri District, by J. A. Madan, 11 September 1914, para. 36.
13 IOL, Bombay Land Revenue Progs., Vol. 11540, July 1926, p. 405, Report on Ratnagiri District by L. J. Mountford, 26 January 1926, para. 8.
14 IOL, Bombay Land Revenue Progs., Vol. 10556, January 1919, Memorandum on the Value of Gross Agricultural Produce, 5 February 1917, para. 22.

least, thin and, even as the figures stand, their meaning in real terms is difficult to evaluate. There is, though, a consistent attitude evident in the major official statements on the Bombay rural economy over the last two decades of the nineteenth century. In contrast to the gloomy view taken of the agricultural economy and the effects of indebtedness in the 1870s, the tone now was of some measured optimism. In 1888 Bombay replied to the Dufferin Enquiry on the standards of living of the Indian masses that 'there is no large proportion of the population of this Presidency suffering from insufficiency of food'.[15] Two years later, the Bombay government responded to yet another Government of India initiative – this time, an enquiry on extent of population pressure – by insisting that, despite the statistics, 'the question is not a pressing one so far as this Presidency is concerned'.[16] Even the problem of indebtedness now received a relatively sanguine response. In 1903, despite the recent impact of famine, officialdom declared that 'there is not now in the Deccan a larger proportion of the agricultural population in debt than there was on the accession of British rule'.[17]

Does this change in official attitudes, evident at many levels of the Bombay administration, genuinely reflect a more substantially based economic expansion in the era 1880–96? The problem with this type of evidence is that, as in the 1870s, official pronouncements on rural economic performance tend primarily to serve the wider needs of policy evolution. In the 1870s, emphasis on the evils of debt promoted the cause advocated by the activists within the revenue department. This very process, however, released a Pandora's Box of criticisms of Bombay and, especially, its revenue system at all-India level. In particular, the 1901 Famine Commission singled out Bombay's as 'an agrarian system under which the cultivators fail to reap the full fruits of their industry and are kept in a state of indebtedness'.[18] Against such attacks, the Bombay administration needed to close ranks and rebuild the image of the province's stability. The 1903 response on indebtedness was one deliberate attempt in this vein.

Advancing the argument on detailed economic trends, therefore, throws us back to the specific data on population, cultivated acreage and

15 IOL, Government of India Revenue Progs., December 1888, No. 17, Minute by J. B. Richey, 27 August 1888.
16 MARD, Vol. 263 of 1890, No. 706, J. Nugent, Chief Secretary to Bombay Government to the Secretary to the Government of India, Revenue Department, No. 1655, 28 February 1890.
17 IOL, Government of India Revenue Progs., October 1905, No. 37, W. T. Morison, Secretary to Bombay Government to the Secretary to the Government of India, Revenue Department, No. 719, 2 February 1903, para. 4.
18 *PP*, 1902, Vol. 70, Report of the Indian Famine Commission, 1901, para. 325.

prices. The first two of these indicators reveal, within the cycles of expansion and contraction, a perceptibly upward movement. Population, and in most areas land in use, was rising discernibly in overall terms. Thus, the total population of the Bombay Presidency increased from 15.7 million in 1881 to 19.3 million by 1921.[19] This, however, was not especially impressive in comparative terms and it is clear that Bombay's population growth was less rapid than the all-India norm during the late nineteenth and early twentieth centuries.[20] Does this, then, indicate that the province's agriculture was more stagnant than elsewhere in the subcontinent? The Ratnagiri case – where population increase *was* at least as great as the all-India norm – counsels against that simple assumption and so, too, does a closer examination of the demographic material. Bombay's population growth was cut back by her relative susceptibility to mortality crises. The two great famines were the most important of these but disease and some major epidemics, notably the influenza pandemic of 1918–19, had deep-seated impact.[21]

The prevalence of these problems, of course, might indicate low levels of resistance among the peasant population and hence prior deterioration in standards of nutrition. Other explanations are, however, equally possible. Diseases may have their own life-cycles independent of human activity and vulnerability certainly rests on more than just low living standards. One might expect, for example, the world-wide influenza outburst to bite deep in the relatively cool plateau of the Deccan where the complications of pulmonary congestion could more easily develop.[22] Again, climatic factors, in the shape of some abrupt disruption of rainfall patterns, might always, in India, create sudden famine in a region where per capita output trends had previously been favourable. Alternatively, famine could reflect primarily distributional changes and difficulties. More extensive commercialisation of peasant agriculture, as is often argued,[23] might be the basis of more frequent famine. A fuller response to the market may, when shortfalls in production occur, swiftly drive up prices, placing food beyond the pockets of those in the village – the poorer peasants and the labourers – who need to buy it to supplement family consumption. Reports in 1897 suggested that 'the character of the season

---

19 *Census of India, 1931* (New Delhi, 1937), Vol. 8, Part 2, Bombay Presidency, Imperial Table 2.
20 L. and P. Visaria, 'Population', in Kumar, ed., *The Cambridge Economic History of India, Vol. 2*, p. 490.
21 Ibid.
22 I owe this suggestion to Mr Donald J. Morse of Edinburgh University, who has been working on the influenza epidemic.
23 B. M. Bhatia, *Famines in India. A Study in Some Aspects of the Economic History of India, 1860–1965* (London, 1967).

does not of itself account adequately for the large demand for relief': 'the distress has mainly arisen from the pressure on the poorer classes of the abnormally high prices'.[24] Occurrence of more regular and serious famine, then, is not automatic evidence of long-term decline in aggregate per capita foodgrain production. In general, too, the demographic evidence answers only quite limited questions about the overall economic trends.

What, though, can we learn from the price data? There is a mass of material available at local level here and, from this, Table 2 presents the long-term price trends in the leading foodgrain, jowar, in three widely separated taluks of the Bombay Presidency. It should be emphasised that the statistics are not inter-comparable, because of the existence of subtle variations in local weight measurements. Nevertheless, the material reveals more than just confirmation of the basic trends which we have already outlined. The most striking features are the extreme short-term fluctuations. The years of the great famines – 1877, 1878, 1897 and 1900 – produced sharp leaps in foodgrain prices, even in Ankleshwar taluk which was far from the famine heartland. Perhaps this is to be expected, but there were also in each locality continuous marked annual fluctuations in prices, presumably consequent on variations of production. At this level, the data re-emphasises the dependence of the agricultural economy on the quality of the season and its quick vulnerability to shortfalls in annual output. Clearly, this was an economy without substantial reserves to counterbalance short-term declines in production. The statistics also indicate strong local variations, but there is some tendency after 1890 for closer correlation between annual trends across the three taluks, suggesting a growing market integration. Furthermore, it is at least arguable that some overall upward momentum is evident. Certainly during the early years of the twentieth century, amidst the aftermath of the last great famine, prices seemed to be stabilising somewhat at levels perceptibly higher than those of thirty years previously.

This raises some interesting questions. Firstly, it is contrary to international price movements. Some would argue that the secular trend of prices in the industrialising West was downward over much of the nineteenth century[25] and undoubtedly this was the case during its last quarter. The 'Great Depression' of 1873–96 witnessed sharp declines of agricultural prices on world markets, which especially afflicted grains. The Bombay Presidency's avoidance of this phenomenon might perhaps

---

24 IOL, Bombay Revenue (Famine) Progs., Vol. 5326, January 1897, No. 122, Report on the Scarcity in the Bombay Presidency, 12 January 1897, para. 5.
25 See, e.g., S. B. Saul, *The Myth of the Great Depression, 1873–1896* (London, 1969), p. 13.

Table 2 *Prices of jowar in three taluks, 1866–1910*

Figures in numbers of seers per rupee

	Haveli taluk, Poona District	Bagalkot taluk, Bijapur District	Ankleshwar taluk, Broach District
1866/7	16.0	17.0	26.0
1867/8	19.0	26.0	24.9
1868/9	23.0	36.0	24.1
1869/70	18.0	28.0	26.0
1870/1	18.0	24.0	N.A.
1871/2	14.5	32.0	22.3
1872/3	18.5	18.0	21.4
1873/4	26.0	29.5	20.7
1874/5	30.0	25.0	19.5
1875/6	26.0	26.75	19.5
1876/7	12.0	10.0	14.8
1877/8	10.5	12.0	10.4
1878/9	10.5	11.5	10.4
1879/80	12.5	13.25	16.3
1880/1	20.5	28.5	28.6
1881/2	27.25	32.0	22.3
1882/3	23.5	N.A.	17.3
1883/4	17.5	N.A.	18.3
1884/5	16.5	N.A.	19.5
1885/6	17.75	N.A.	19.5
1886/7	22.75	N.A.	20.7
1887/8	22.5	N.A.	16.8
1888/9	14.75	22.5	15.6
1889/90	16.3	20.3	18.3
1890/1	17.5	23.5	21.5
1891/2	17.6	19.4	15.1
1892/3	15.1	19.8	17.6
1893/4	17.4	26.8	18.3
1894/5	18.3	22.9	18.3
1895*	20.2	27.2	N.A.
1896	16.3	20.0	18.3
1897	9.8	10.8	11.1
1898	15.9	20.75	15.3
1899	14.75	17.4	14.4
1900	8.8	10.4	9.1
1901	14.1	11.8	15.9
1902	14.4	15.6	14.0
1903	21.1	24.75	22.7
1904	20.4	26.6	18.9
1905	15.8	22.4	14.4
1906	11.8	13.0	12.8
1907	13.6	17.2	12.6

	Haveli taluk, Poona District	Bagalkot taluk, Bijapur District	Ankleshwar taluk, Broach District
1908	11.3	13.7	10.5
1909	14.1	13.6	12.0
1910	13.8	15.1	14.0

*At this point classification changed to calendar years. Figures are therefore only for the latter part of 1895.
Sources: *SBG*, NS, No. 577, Settlement Report on Haveli Taluk, Poona District, by R. D. Bell, 7 March 1916, Appendix N; *SBG*, NS, No. 176, Settlement Report on Bagalkot Taluk, Bijapur District, by W. M. Fletcher, 30 June 1883, Appendix N; *SBG*, NS, No. 564, Settlement Report on Bagalkot Taluk, Bijapur District, by G. C. Shannon, 3 November 1914, Appendix N; *SBG*, NS, No. 529, Settlement Report on Ankleshwar Taluk, Broach District, by H. Denning, 15 August 1912, Appendix N.

be regarded as an index of its limited incorporation within the international economy. Internal factors, notably harvest performance, at this stage exercised more influence. The force of the international price trend, however, would grow stronger, for, as we shall see, price fluctuations in the 1920s and 1930s would correlate much more closely with developments in the world economy. Yet influences external to agriculture still had some weight during the late nineteenth century. The steady devaluation of the rupee over the period acted as a force for inflation within the Indian economy and probably had some influence even on the prices of non-export crops.[26]

The price statistics, then, like the other broad indicators, are difficult to interpret. Nevertheless, we might attempt some preliminary hypotheses from this overall review of trends. Our long-run quantitative data, however complex and occasionally contradictory, does tentatively suggest some expansionary tendency within the agricultural economy of the Bombay Presidency over the second half of the nineteenth century. For all the fluctuations, the agrarian system was typically supporting larger numbers in the early years of the twentieth century than in the middle years of the nineteenth century. Prices even of foodgrains, too, within their sharp variations, had typically some upward momentum. This should have provided some stimulus to market production, especially if railway construction was easing access to wider markets and cutting costs

26 Charlesworth, *British Rule and the Indian Economy, 1800–1914*, p. 49.

of transportation. Testing these notions, however, requires much more specific investigation and, in particular, grappling with the issue of qualitative change. Assumptions of totally static technique, we have suggested, cannot be made. Shifts, too, may have been occurring in the organisation of agriculture, in cropping patterns, in the whole extent and nature of commercialisation. To advance the argument here, we need to look more closely at the short-term trends.

Expansion and contraction, 1850–1880

The expansionary era which ended with the onset of lower prices in the early 1870s was notable for its great vigour. For twenty years, population, cultivated acreage and prices had all risen relatively rapidly in western India. This, some would argue, had already introduced substantial structural change within the agrarian economy and society. Ravinder Kumar's rich peasants have 'emerged' from the village mass by the 1870s, thrown up by both the opportunities for acquisitive individualism presented by British rule and favourable economic conditions throughout the middle years of the nineteenth century.[27] The new revenue settlements from the late 1860s certainly hinted at the existence of a pioneering element within the village, developing new resources and techniques. Indapur taluk, at the heart of the famine belt, had apparently witnessed the construction of 625 new wells and the repair of 184 old ones in the thirty years before 1865–6[28] and numbers of carts, over the same period, had risen more than three times.[29] Similar increases were evident in neighbouring areas. Carts had more than doubled in number and wells in working order risen by over a third in Haveli taluk during the thirty years of the settlement revised in 1872.[30] Bhimthari taluk boasted, over a similar period, 141 new wells and a more than doubling of per capita cart availability.[31] At the same time, such developments seem likely to have been the work of a minority of peasant entrepreneurs with access to capital resources. A new well in the Deccan of the 1860s, it was estimated, cost around Rs. 400,[32]

---

27 See Ravinder Kumar, 'The Rise of the Rich Peasants in Western India', in D. A. Low, ed., *Soundings in Modern South Asian History* (London, 1968), pp. 25–58.
28 *SBG*, NS, No. 107, Settlement Report on Indapur Taluk, Poona District, by J. Francis, 12 February 1867, para. 95.
29 Ibid., para. 101.
30 *SBG*, NS, No. 151, Settlement Report on Haveli Taluk, Poona District, by W. Waddington, 30 November 1872, para. 17.
31 *SBG*, NS, No. 151, Settlement Report on Bhimthari Taluk, Poona District, by W. Waddington, 12 July 1871, paras. 6–7.
32 This was Francis' estimate. See *SBG*, NS, No. 107, Settlement Report on Indapur Taluk, Poona District, by J. Francis, 12 February 1867, para. 95.

but the Deccan Riots Commission was to assess the value of the typical Kunbi's total possessions at little more than Rs. 200.[33]

This is striking evidence for Kumar's thesis, drawn as it is from these relatively poor areas of the central Deccan. Yet it is hard to square with the situation revealed by the Deccan Riots, where the prevalence of the alien moneylender seems to rest on the relative absence of an indigenous elite with surplus capital to lend. In general, the evidence of these new settlement reports of the late 1860s and early 1870s may be limited by their timing: they came at the very peak of the recent boom and many of the increased resources enumerated may have been novel, and possibly even temporary, acquisitions. Nevertheless, the boom of the 1860s – a marked quickening of the pace within a period of expansion – may have had widespread effects, for it rested on the coincidence of a number of favourable factors. Climatic conditions were good, public works construction was extensive, creating new irrigation and communications facilities, and the steady buoyancy of demand was dramatically boosted, in the first half of the decade, by the effects of the American Civil War.

The outbreak of the American War at once greatly ameliorated the problems of demand which we earlier outlined as a major restriction on Bombay cash-crop production. However reluctantly, British manufacturers were forced to turn to India for much more of their raw cotton supplies and the massively enhanced demand had its immediate impact on price levels. Prices of western Indian cotton shot up from an average of Rs. 128 per candy in 1860 to Rs. 314 in 1863 and a peak of Rs. 627 in 1865.[34] The response to this incentive was considerable. The amount of Indian cotton shipped to Britain in 1861 exceeded the previous year's level by over 80 per cent[35] and the extent of land under cotton cultivation increased substantially throughout western India.[36] Prices of the foodgrains followed in the trail and they roughly doubled in most Deccan regions between 1860 and 1865.[37] Coincidentally, the public works construction programme in western India increased momentum substantially. This was possibly important not just in constructing new resources like irrigation canals but also in providing, through wages paid for labour work, cash and employment for the poorer elements within

---

33 Deccan Riots Report, para. 53.
34 MARD, Vol. 25 of 1875, No. 1855, Table of Cotton Prices.
35 *PP*, 1862, Vol. 55, Statistics on Indian Cotton, p. 627.
36 See Peter Harnetty, 'Cotton Exports and Indian Agriculture, 1861–1870', *Economic History Review*, 24, 3, 1971, pp. 414–29.
37 See, e.g., *SBG*, NS, No. 151, Settlement Report on Bhimthari Taluk, Poona District, by W. Waddington, 12 July 1871, Appendix C.

village society. In Poona District in the year 1868–9 alone, about Rs. 31 lakhs were spent on public works[38] and at one stage, the great Sahyadri irrigation works were employing 40,000 labourers on a 14-mile stretch of continuous construction.[39]

What were the effects of these developments? Many contemporaries argued that sources and supply of credit extended rapidly with the result that 'the ryot's powers as a borrower were those of a capitalist rather than a labourer'.[40] Later, in the official atmosphere of the late 1870s, this came to be seen as a snare which had ultimately increased indebtedness and tightened peasant economic subservience to moneylenders: from Gujarat, one settlement officer blamed 'extravagance engendered by gains made in the season of abnormally high-priced cotton'[41] for new levels of indebtedness and one official commented of the Deccan Riots situation that 'most of the ryots appear to have got in debt during or immediately after the cotton mania'.[42] Whatever the long-term consequences, however, these developments provided in the short run the exact prescription advocated by Wedderburn and others by extending credit availability within the village and this may well have offered new opportunities to increase ownership of resources like wells and carts. Harnetty also argues that the expansions in cotton cultivation of the American Civil War era generated both increased wealth and the qualitative change of extending commercial initiative within the village economy.[43] In these ways, the boom may have had more than temporary impact on levels and types of economic activity.

The price explosion in cotton presumably produced some substantial returns at some level, for the rise was swift enough to represent a marked increase in real terms. The cotton boom, too, affected many areas of the Presidency, not just the traditional cotton-growing centres of Gujarat, Belgaum and Dharwar. Even in Indapur taluk, just over 30,000 acres out of a cultivated area of around 200,000 acres were growing cotton in 1862.[44] This certainly demonstrates, as Harnetty stresses, the ability of Indian peasant agriculture, even in a period where the extent of prior commercialisation might be challenged, to respond rapidly to changes in world demand. Yet the distribution of the returns from cotton cultivation

---

38 Deccan Riots Report, para. 66.
39 *BG, Vol. 18, Poona,* Part 2, p. 109.
40 Deccan Riots Report, para. 66.
41 *SBG,* NS, No. 227, Settlement Report on Dhandhuka Taluk, Ahmedabad District, by T. R. Fernandez, 8 December 1888, para. 17.
42 MARD, Vol. 7 of 1875, No. 960, C. G. W. Macpherson, Assistant Collector, to G. Norman, Collector of Poona, No. 72, 17 July 1875, para. 11.
43 See Harnetty, 'Cotton Exports and Indian Agriculture'.
44 *BG, Vol. 18, Poona,* Part 2, p. 48.

within the village may have been uneven. Where peasant debtors handed the bulk of their produce over to their creditors – and in an inflationary climate it made sense for moneylenders to try to enforce and extend this situation – then most of the value from the price increase would go to the middleman. In addition, food where a peasant needed to buy it was expensive during the 1860s. This latter aspect of a generally rapid inflation puts into perspective the value of wages gained on the public works. Wages apparently rose swiftly from an average for unskilled labourers of Rs. 7¾ per month over 1860–2 to Rs. 13½ by 1863,[45] but foodgrain prices were increasing at about the same rate and those with little or no land may have been worse off, certainly in relative terms compared with cotton producers and middlemen.

This interpretation still accords with Kumar's thesis of a thrusting rich peasantry monopolising the benefits of agricultural expansion. However, the nature of the 1860s boom itself suggests reservations about the widespread and stable existence, before 1880, of any newly formed commercially orientated peasant elite. The cotton boom relied on special circumstances. Even at its height it was evident that 'the position of American cotton is not shaken in the English market by the present scarcity, and the Indian staple is not established on any more solid basis, because necessity has driven manufacturers to use it'.[46] As a result the boom was extremely short-lived. Prices fell away rapidly after 1865, from an average Rs. 627 per candy then to Rs. 343 in 1867 and Rs. 235 by 1869.[47] After 1870 cotton participated in the general price fall and by 1874 its typical price level was not greatly above those of the late 1850s.[48] In turn, just as the response to enhanced world demand had been rapid, so too was the reaction to contraction. Cotton cultivation declined spectacularly in the areas which had leapt on the bandwagon of the Civil War boom and in Indapur taluk, for example, there were a mere 100 acres under the crop by 1870–1. By the time of the Deccan Riots, cotton cultivation in Bombay Presidency was fundamentally confined to the regions which had grown the crop in the mid-nineteenth century: north Gujarat, particularly within Ahmedabad and Broach Districts, the frontier district of Khandesh and the Southern Maratha districts, primarily Belgaum and Dharwar. The four central Deccan districts now totalled under 100,000 acres growing cotton.[49]

---

45 Deccan Riots Report, para. 66.
46 Cassels, *Cotton: an Account of its Culture in the Bombay Presidency*, p. 346.
47 MARD, Vol. 25 of 1875, No. 1855, Table of Cotton Prices.
48 The average price for 1874 was Rs. 184 per candy: the average level for 1859 had been Rs. 148. Ibid.
49 MARD, Vol. 25 of 1875, No. 15, Administration Report of the Cotton Department for the Year 1873–4, Appendix A.

This not only serves to challenge Harnetty's emphasis on the importance of the cotton boom. It also makes one wonder about the nature of any Deccan commercial rich peasantry in this period. In parts of Gujarat and the Southern Maratha Country, such a group with a clearly differentiated, permanent and widespread range of market operations may indeed have been significant, for in Broach District, even by 1874, 44.7 per cent of the cultivated acreage was growing cotton.[50] This, however, was untypical of Bombay Presidency in this period. Indapur taluk, with its highly limited cash-crop cultivation by the early 1870s, was a much more usual case, certainly for the Deccan. Here, a small group of pacemakers possibly did draw ahead in the 1860s, building the wells and acquiring the carts which were so lauded in Francis' 1867 settlement report, but, if so, it was a brief and feeble flourish confined to the few years of the cotton boom. Even the character of these temporary kings – were they primarily agriculturists or alien moneylenders? – remains obscure, since few signs of their existence can be found in the welter of official investigations arising from the Deccan Riots. The Riots Commission painted a picture of a relatively uncommercialised economy with little evidence of social differentiation: 'the railway is not used for the export of produce ... A little produce is sent to Poona in carts for local consumption, but the foodgrain of the region is consumed by the inhabitants.'[51]

This impression is supported by the extent of use, by the 1870s, of the new facilities developed during the cotton and public works boom. The potential released by the new government irrigation works was not realised at this stage. The Collector of Ahmednagar wrote in 1875 that, although three major works could now supply water to a total of 41,510 acres in his district, 'such is the condition of the people that they cannot be brought to avail themselves of the water thus provided ready for them, except to the trifling extent of 457 acres'.[52] Throughout the Bombay Deccan in the famine year of 1876–7 'the area irrigated, in a season almost unprecedented in its severity, was less than one-twelfth of that actually under command at the time and only one-twentieth of the extent the works were designed to command'.[53] The difficulty was that use of these government facilities required paying a water rate and preparing land for irrigation also involved considerable work and prior capital expenditure.[54] For any securely based and entrepreneurial rich peasantry,

50 *BG, Vol. 2, Broach and Surat* (Bombay, 1877), p. 392.
51 Deccan Riots Report, para. 57.
52 MARD, Vol. 3 of 1875, No. 971, H. B. Boswell, Collector of Ahmednagar, to the Commissioner, Southern Division, No. 2132, 20 July 1875, para. 14.
53 Caird Papers, Minute by H. E. Sullivan, Madras Civil Service, 11 April 1879.
54 Sullivan estimated that preparing land for irrigation might cost as much as Rs. 20 per acre. Ibid.

however, this might have seemed a reasonable investment, as would the new advances government offered for agricultural improvements. During the 1870s legislation provided for much fuller use of 'takavi' loans, in the shape of official grants for irrigation, drainage and reclamation projects. Yet William Wedderburn could comment by 1881 that 'the system ... has proved a failure ... during the famine year of 1877–78, the total advances hardly exceeded £1,000 for the whole Presidency of Bombay'.[55] In the year before the outbreak of the Deccan Riots, no takavi advances whatsoever had been made in Poona District[56] and in Ahmednagar just Rs. 1,550, 'very little for a district of this size and so urgently in want of wells and of general agricultural improvements'.[57]

The extension of facilities like these may have been handicapped by institutional weaknesses and government shortcomings in popularising them, but the fundamental problem of the 1870s – as we saw in an extreme case in the Deccan Riots region in 1875 – was the familiar story of essential lack of capital and credit. This seems the reason why the government schemes were not taken up, for even takavi loans were intended for improvement projects and a peasant would require wider sources of credit to support them. In one sense, the official view with hindsight of the 1860s expansion – 'the fatal gift of almost unlimited credit destined soon again to collapse'[58] – was correct, for the credit contractions of the 1870s must have seemed extremely sharp and burdensome after the extensions of the previous decade. The retrenchment process calls into question some of the other qualitative innovations claimed for the 1860s boom. One was the creation of a much greater market value in land. Tytler quoted one case in Ahmednagar District in 1862 whereby 'fifty two times the year's assessment was bid for the occupancy of a field'[59] and from Surat District in the late 1860s came reports of prices 'for one acre of land ... varying from £40 to £50 (Rs. 400 to Rs. 500)'.[60] By the 1870s, however, as we saw in the Deccan Riot villages, the local land market seems to have been much more limited and less buoyant.

Yet the 1870s saw not just the ending of an era of expansion but the eventual onset of a precise crisis, particularly in the areas affected by the famine of 1876–8. How should we interpret this? If we assume that many of the new private wells described in the optimistic settlement reports of

55 Wedderburn, *Agricultural Banks for India*, p. 27.
56 MARD, Vol. 7 of 1875, No. 960, G. Norman, Collector of Poona, to J. E. Oliphant, Commissioner, Southern Division, No. 1625, 9 August 1875, para. 11.
57 MARD, Vol. 3 of 1875, No. 971, H .B. Boswell, Collector of Ahmednagar, to the Commissioner, Southern Division, No. 2132, 20 July 1875, para. 11.
58 Deccan Riots Report, para. 66.
59 MARD, Vol. 30 of 1862–4, No. 492 of 1862, C. Fraser Tytler, Collector of Ahmednagar, to W. Hart, Commissioner, Southern Division, No. 242, 14 February 1862, para. 2.
60 *BG, Vol. 2, Broach and Surat*, p. 183.

1867–72 were, like the irrigation facilities provided by government, not in widespread use by the mid-1870s, then a picture does begin to emerge, in the Deccan at least, of an economy under some strain. For examining other resources apart from wells and carts, the evidence, even of the settlement reports, is of a struggle to maintain levels of per capita ownership. Auckland Colvin noted that 'while with increasing population there has necessarily been an increase in plough-cattle, and the total number has grown, it seems that individual property in stock... has decreased'.[61] This seemed to be true of Bhimthari taluk where the increase in number of working bullocks over the thirty years of the settlement from 11,568 to 13,792 in no way matched the nearly 40 per cent growth of population.[62] Again, in Haveli taluk a 12 per cent rise in bullock numbers compared with a leap of 42.8 per cent in total population.[63] The real comparison, however, is not with population but with the extent of land required to be cultivated and this had certainly grown more rapidly than numbers of working cattle. As Pedder noted of such statistics, 'the smallness of the increase of plough cattle in proportion to the increase of cultivation is an argument' that cultivation is not as good as it was'.[64] In the central Deccan, at any rate, the fundamental problem of shortage of working cattle and the limitations this placed on effective tillage may well have been worsening in the 1870s.

This suggests that, despite our reservations about simple Malthusian models, the onset of famine in the region in 1876 might be the consequence of pressure on resources. It is noticeable that the outbreak afflicted the traditional famine belt, the eastern areas of the Deccan and the Southern Maratha Country. Its major victims were those such as labourers and some artisans, who enjoyed no access to land but could not afford to pay the higher prices demanded for food. Temple commented – and the statistics appeared to bear him out – that 'the mass of the ryots or peasant proprietors, who constitute the real agricultural community, never came on relief at all'.[65] The severity of the climatic failure of 1876, however, was such that shortage and price dislocation would have arisen temporarily whatever the economic trends, for only 5 inches of rain fell at Indapur during the year compared with 14 inches in the next-worst year of the 1870s.[66] This was enough, in an economy under considerable popu-

61 Deccan Riots Report, Memorandum by Mr Auckland Colvin, para. 29.
62 *SBG*, NS, No. 151, Settlement Report on Bhimthari Taluk, Poona District, by W. Waddington, 12 July 1871, para. 6.
63 Ibid., Settlement Report on Haveli Taluk, Poona District, by W. Waddington, 30 November 1872, para. 17.
64 Caird Papers, W. G. Pedder to Sir James Caird, 2 August 1879.
65 Temple Papers, Vol. 193, Minute by Richard Temple, 24 December 1877, para. 17.
66 *BG, Vol. 18, Poona*, Part 1, p. 16.

lation pressure, to create immediate high mortality and emigration. In turn, this limited opportunities for renewed cultivation and extended the crisis into the much better climatic seasons of 1877 and 1878. Despite the scale of the outbreak, there is no evidence here of any especial new type of famine problem.

Yet it might be argued that our interpretation, stressing the long-term economic inconsequentiality of the 1860–80 cycle, works best for the Deccan where both the riots of 1875 and the famine provide confirming evidence of a society facing basic difficulties of subsistence. Were the effects of the 1860s boom more durable elsewhere in western India? One might expect that parts of Gujarat, in particular, had advanced more fully over the period on the road towards commercialisation, enjoying, for example, a more consistent village land market. Nevertheless, the downswing of the 1870s was evident here too. Throughout western India, the expansionary cycle had been essentially of a once-for-all nature, fuelled by special, strictly temporary factors: rapid population growth and expansion in land use, linked in the 1860s with the particular, unique effects of the American Civil War. The special vitality of the process was something of an illusion, owing more to the low levels of economic activity under early British rule than to any marked momentum behind most of the cycle itself. Genuine improvement in per capita availability of resources occurred only patchily. Extensions in cart ownership perhaps represented the most durable gains, but carts remained the prerogative of a minority of cultivators and in Bhimthari taluk, for example, even the steady growth in cart numbers of the thirty years before 1870 left cart ownership at around 1 per 28 of population.[67]

Particularly important, however, is the absence of any significant qualitative changes upon which a more secure and permanent commercialisation than that of the 1860s might have been based. The 1880 Famine Commission estimated that the total irrigated area in the Bombay Presidency then amounted to barely a half million acres, compared with over 11 million acres in the United Provinces and over 5 million acres each in the Punjab and Madras.[68] This reflected not merely the limited use of facilities on the canal schemes, but also the comparative level of private well construction and utilisation. It was true that, in predominantly dry-crop Bombay, sophisticated methods of irrigation were not required for the traditional foodgrains and would therefore probably demand and indicate switches in cropping patterns. Yet to this extent the process of

---

67 *SBG*, NS, No. 151, Settlement Report on Bhimthari Taluk, Poona District, by W. Waddington, 12 July 1871, para. 6.
68 *PP*, 1880, Vol. 52, Report of the Indian Famine Commission, 1880, Part 2, Ch. 1, Section 5, para. 14.

the spread of irrigation is one useful index of the scale of commercialisation.

A more commercialised agricultural economy also required greater ease of communication between the village and the wider market. Special conditions of intense new demand, as in the early 1860s, would meet a quick response but regular channels of contact between the village market and the world beyond the locality were, before 1880, difficult. Railways in western India were still predominantly trunk routes with little local, commercial significance. In the late 1860s two major lines existed, the Bombay Baroda and Central India, stretching north for 305 miles through Baroda to Ahmedabad and the Great India Peninsula Railway, its two branches totalling 875 miles and running north-eastwards towards Nagpur and south-east to Sholapur.[69] Desire to break through to the cotton fields was an early influence on the construction of both lines, but after the Mutiny the main priorities of Indian railway building became political and administrative: improving channels of communication between officialdom, increasing internal security and guarding against the Russian frontier threat.[70] Under these diktats, new railway construction in Bombay Presidency lagged. By 1880 the Famine Commission could report that, although 'few villages in Gujarat are 30 miles from the railway', throughout the south – both the south Konkan and the Southern Maratha Country – there were no lines at all. Belgaum, despite its significance as a cotton centre, was 212 miles from the nearest station at Poona and, even in the north Deccan, parts of Nasik and Khandesh were 90 miles from a railway.[71] Limitations of communications, too, limited the durability of the 1860s boom.

A renewed expansionary cycle, 1880–1896

If the cycle ending in 1880 represented no significant breakthrough in either the scale or scope of the agricultural economy's activities, then any claims for the renewed expansion which set in after 1880 might seem necessarily limited. Superficially, the expansion was grounded on the same short-term trends as before: population growth, extensions in land area, price buoyancy and favourable climatic conditions. Yet their operation, after 1880, was apparently much less vital than in the earlier period. There was no special spur to cash-crop cultivation like that provided by the cotton price explosion of the early 1860s and the expansion

---

69 For the pattern of early railway construction in western India, see Nalinaksha Sanyal, *The Development of Indian Railways* (Calcutta, 1930), p. 35.
70 W. J. Macpherson, 'Investment in Indian Railways, 1845–1875', *Economic History Review*, 8, 2, 1955–6, pp. 177–86.
71 *PP*, 1881, Vol. 71, Part 2, Report of the Indian Famine Commission, 1880, Appendix 3, Ch. 1, Question 19, Statement by J. Peile.

of the cultivated acreage, although an evident phenomenon, was a much less spectacular process, little commented upon in the settlement papers. This might suggest that the whole dynamic of the new cycle was weaker. On the other hand, the demographic trend was sharply upwards: the total recorded population of the Bombay Presidency increased from 15,697,911 in 1881 to 18,035,280 by 1891.[72] This could indicate increasing capacity within the agricultural system to support growing numbers or the possibility of accumulating pressure on land.

Rapid population growth, however, must have been setting certain parameters for the agricultural economy. Granted that quantitative expansion was going, in part at least, to feed more mouths, the possibility of significant 'improvement' would depend, largely, upon qualitative innovations; in particular, some amelioration of that fundamental problem of basic shortages of resources. In the poorer areas of the Deccan, there is little evidence that per capita availability of most resources had changed much by the mid-1890s. Ownership, for example, of plough cattle apparently remained fairly constant. In Indapur taluk numbers of agricultural cattle rose from 21,972 to 23,231 over the twenty years after 1875-6, whilst population increased from 62,722 to 67,684.[73] In another area of the east Deccan famine belt, Malsiras taluk of Sholapur District, numbers of plough cattle actually diminished marginally between 1875 and 1890 despite a 6 per cent rise in population.[74] This seemed to mirror the familiar story of earlier periods. In some areas, however, there are indications that availability of cattle relative to land may have been improving perceptibly over the later nineteenth century. Revision settlements of Satara District during the early 1890s provide a valuable insight into the workings of the renewed expansion in a Deccan region of more climatically assured conditions. Here, in many cases, plough cattle numbers do seem to have been extending steadily over the thirty-year period since the previous revenue settlement: by 22.4 per cent in Khanapur taluk,[75] 18.3 per cent in Satara taluk,[76] 32.4 per cent in Patan taluk.[77] Such rates of growth at least matched and possibly exceeded the pace of expansion in the cultivated acreage. However, per capita availability of

---

72 MARD, Vol. 258 of 1892, No. 749, Brief Memorandum on the Material Condition of the People of Bombay Presidency 1881-91, para. 1.
73 *SBG*, NS, No. 379, Settlement Report on Indapur Taluk, Poona District, by F. B. Young, 13 October 1897, para. 22.
74 *SBG*, NS, No. 247, Settlement Report on Malsiras Taluk, Sholapur District, by W. M. Fletcher, 27 December 1890, para. 17.
75 *SBG*, NS, No. 259, Settlement Report on Khanapur Taluk, Satara District, by W. M. Fletcher, 15 March 1891, para. 14.
76 *SBG*, NS, No. 280, Settlement Report on Satara Taluk, Satara District, by R. B. Pitt, 13 May 1893, para. 17.
77 *SBG*, NS, No. 293, Settlement Report on Patan Taluk, Satara District, by R. B. Pitt, 24 April 1894, para. 21.

the specialist agricultural resources – wells and carts in particular – seems to have been extending firmly in both the Satara and the famine belt taluks. In Parner taluk of Ahmednagar, for example, numbers of irrigation wells apparently rose from 1,626 to 2,980 over the thirty years after 1883.[78]

This already suggests spatial differentiation as a major theme of the expansion of 1880–96. The expansionary trend may have varied considerably in force from region to region, from locality to locality and, within every geographical unit, from group to group of agriculturists. In many areas, particularly those where the downswing of the 1870s had been most intense, the cycle may have been operating at similar, or even lower, levels of economic performance than that before 1880. Elsewhere, the process of greater resources leading to higher levels of output and economic activity and additional capital and credit for investment may have been firmly under way. Yet resources like wells and carts could still have become far from general property even in the most relatively prosperous area. If Parner taluk boasted nearly 3,000 irrigation wells by the outbreak of the First World War, these were distributed between over 20,000 holdings.[79] Spatial distinctions in economic potential and performance, it is likely, were occurring at every level.

Why, though, should such spatial differentiation occur? We will argue that, during the last two decades of the nineteenth century, the process of commercialisation in Bombay agriculture extended steadily, but it was a commercialisation which, by its very nature, varied intensely in its power and extent. With no special leaps in international prices and demand during the 'Great Depression', the extension of export crops like cotton depended on growing competitiveness by Indian produce. In the absence, however, of any great technological breakthrough in methods of agricultural production, the sources for this were limited: the most striking and that with the largest cost-cutting potential would be communications improvement. As a result, the rate of expansion of the railway proved, perhaps, the fundamental determinant of the whole process of commercialisation in late nineteenth-century India. In western India, this made the 1880s a particularly consequential era, for between 1881 and 1891 railway mileage over the Bombay Presidency increased from 1,562 miles to 2,661 miles, a rise of 70 per cent.[80] In addition, the character of the new lines made the local impact often even greater than this might imply. The early railways in western India had been trunk routes, linking major administrative centres and the placement of the main junctions and stations

[78] SBG, NS, No. 567, Settlement Report on Parner Taluk, Ahmednagar District, by J. H. Garrett, 22 May 1916, para. 28.
[79] Ibid.
[80] MARD, Vol. 258 of 1892, No. 749, Brief Memorandum on the Material Condition of the People of Bombay Presidency, 1881–91, para. 26.

reflected this priority. The new lines of the 1880s, however – the Southern Maratha Railway and the West, South and East Deccan Railways – linked the district centres within the interior, creating more immediate effect on local commerce and agriculture.

The advent of these railways in a locality provided the opportunity not only to reach wider markets more cheaply and efficiently, but also to cut the extent of participation by the range of middlemen, hence potentially increasing the proportion of the returns secured by the agriculturist. Conveying produce to the railhead might offer significant gains over selling within the village, but this required access to some means of conveyance, usually a bullock cart. The significance of increases in numbers of carts has been pointed out by Amalendu Guha for earlier periods of the nineteenth century.[81] The rate of increase, however, seems in general to have been far faster during the expansion of 1880–96 and, strikingly, seems often to be linked to patterns of railway construction, suggesting that agriculturists were attempting to connect up with the new facilities. In Satara District, through which the Southern Maratha Railway was opened during the 1880s, there was nothing short of a cartage revolution. As Table 3 shows, the settlement reports of the 1890s recorded increases in cart numbers over the thirty years of the settlement of at least 85 per cent in every taluk, reaching up to 500 per cent in some.

The thesis can be demonstrated, too, for other very different areas of the Bombay Presidency. Viramgam taluk of Ahmedabad District was one of the more outlying areas of northern Gujarat but during the 1870s and 1880s it acquired good new railway communications with Ahmedabad, Kathiawar and the north. Here the settlement report of 1890 estimated that cart numbers had risen from 1,424 to 2,522 over the previous thirty years.[82] Wider ownership of carts enabled more peasants to reach the larger local markets as well as the railheads. One Satara settlement officer noted what he saw as subtle changes in patterns of marketing: 'many of the rayats prefer to sell in their villages to itinerant Vanis and agents of merchants residing in the trade centres, but a good few utilize their carts and bullocks in the slack season for carrying produce to the larger markets'.[83] In turn, this trade offered opportunities to sell a more extensive range of goods within the home village.[84]

81 See Guha, 'Raw Cotton of Western India: 1750–1850', pp. 1–42.
82 *SBG*, NS, No. 242, Settlement Report on Viramgam Taluk, Ahmedabad District, by T. R. Fernandez, 13 February 1890, para. 26.
83 *SBG*, NS, No. 345, Settlement Report on Valva Taluk, Satara District, by W. S. Turnbull, 24 June 1895, para. 15.
84 In Valva taluk in 1895 'I met a string of four or five carts on the Uran–Ashta road, the owners of which were rayats of Saspada, a village in the south of Satara Taluka, noted for its cultivation of ginger, who had taken ginger for sale to Sangli and after selling it at a profit of 8 or 9 rupees a cart-load had gone on to Athni and bought oilcake there for home consumption.' Ibid.

Table 3 *Increases in numbers of carts in Satara District, 1860s–1890s*

Taluk	Numbers of carts at the first settlement	Numbers of carts at the revision settlement	Percentage increase
Man	174	1,070	514.9
Khatav	377	1,250	231.6
Khanapur	430	1,560	262.8
Koregaon	1,025	1,907	86.0
Wai	309	1,257	306.8
Satara	938	1,735	85.0
Javli	171	409	139.2
Patan	169	1,055	524.3
Karad	941	3,267	247.1
Valva	1,018	3,930	286.0
Total, Satara District	5,552	17,440	214.1

Sources: SBG, NS, No. 240, Settlement Report on Man Taluk, Satara District, by T. M. Ward, 28 February 1890, para. 18; No. 241, on Khatav Taluk, by T. M. Ward, 29 April 1890, para. 14; No. 259, on Khanapur Taluk by W.M. Fletcher, 15 March 1891, para. 14; No. 267, on Koregaon Taluk by W. M. Fletcher, 12 August 1891, para. 12; No. 270, on Wai Taluk by W. M. Fletcher, 7 November 1892, para. 14; No. 280, on Satara Taluk by R. B. Pitt, 13 May 1893, para. 17; No. 285, on Javli Taluk by R. B. Pitt, 6 November 1893, para. 18; No. 293, on Patan Taluk by R. B. Pitt, 24 April 1894, para. 21; No. 299, on Karad Taluk by W. S. Turnbull, 29 November 1894, para. 18; No. 345, on Valva Taluk by W. S. Turnbull, 24 June 1895, para. 18.

Yet if these features reveal some of the local effects of railway construction, the railways also exerted a more fundamental influence on the overall shape of western India's commerce. The traditional export route for the produce of the interior was invariably the most direct way to the coastal port. Thus Dharwar cotton, exported during the 1860s boom, came westwards, down over the Ghats, to Coompta or some other south Konkan port. The railways, however, developed new routes, involving, in particular, the by-passing of local ports and much greater centralisation on Bombay City. The process had most dramatic impact on the south Konkan, for the region, with the technical problems presented by its heavily indented coastline, acquired no railway. Indeed, communications here remained strongly dependent on coastal shipping and in Ratnagiri taluk as late as 1897 'there is only one really good continuous cart-road'.[85]

[85] *SBG*, NS, No. 377, Settlement Report on Ratnagiri Taluk, Ratnagiri District, by J. L. Lushington, 30 March 1897, para. 7.

The result was that the new railways which from the 1880s fed the south Deccan and Southern Maratha Country eventually began to deprive the region of the traditionally valuable above-Ghats trade. In the expansionary climate ruling down to the mid-1890s, the south Konkan commercial centres experienced no immediate crisis: in Ratnagiri taluk up to 1895 'trade has not only not declined but shows some rise in spite of the diversion caused by the Southern Maratha Railway'.[86] Thereafter, however, the decline was sharp. The ports of Ratnagiri taluk exported produce worth over Rs. 11 lakhs in 1894–5, but under Rs. 5 lakhs' worth by 1906–7.[87] Dharwar cotton and Satara groundnuts now went direct to Bombay City via the rail link to Poona.

The establishment of rail-borne patterns of commerce tended to reduce the importance of all the great ports which had long been the major entrepots of western India. Even in Gujarat, where ports and producers had been more efficiently linked than the Southern Maratha Country and the south Konkan had ever been, the process was evident. In Broach District 'since the opening of the B.B. and C.I. Railway in 1861, the export and import trade of the ports of Broach and Jambusar has greatly fallen off'.[88] Further north in Ahmedabad District the ports of the Gulf of Cambay suffered a similar fate. Gogha was traditionally a major centre here, doing nearly Rs. 16 lakhs of trade per year in the late 1860s,[89] but the construction of a railway only 10 miles to the north-west through Bhavnagar considerably reduced the port's role. By 1889, Gogha's total trade amounted to under Rs. 8 lakhs a year.[90] Even commercial centres which remained on the rail-borne trade routes would be seriously affected, if the railway's organisation did not require them as major junctions. This was true of several of the markets of Thana District, notably Kalyan and Bhiwndi, through which much of the Ghats trade had traditionally filtered to nearby Bombay City. After 1880 the railway thundered across the plains of Thana, taking Deccan produce direct to the metropolis and its advent 'has deprived Bhiwndi of much of its old trade importance'.[91]

These events were of wider significance than as items of commercial

---

86 Ibid., Papers on the Settlement of Ratnagiri Taluk, Ratnagiri District, E. J. Ebden, Commissioner, Southern Division, to Government, No. 360, 2 February 1898, para. 4.
87 *SBG*, NS, No. 574, Settlement Report on Ratnagiri Taluk, Ratnagiri District, by J. A. Madan, 11 September 1914, paras. 31–2.
88 *SBG*, NS, No. 407, Settlement Report on Broach Taluk, Broach District, by P. R. Mehta, 11 August 1899, para. 13.
89 *SBG*, NS, No. 255, Settlement Report on Gogha Taluk, Ahmedabad District, by T. R. Fernandez, 23 March 1890, para. 15.
90 Ibid.
91 *SBG*, NS, No. 343, Settlement Report on Bhiwndi Taluk, Thana District, by C. W. Godfrey, 30 January 1895, para. 18.

history. Remembering our argument that communications improvement represented the major source of cost-cutting and hence a predominant influence on the process of commercialisation, the chance of where the railway ran had considerable influence on degree of production for the market and the whole pace of economic expansion in the 1880s and early 1890s. The impact on merchant groups of the shifts in trade routes was often by no means as important as the differential effect on agriculture of new access to wider markets or the continuance of traditional restrictions on demand. Agriculture in the south Konkan, as we noted, was able to support growing numbers, but it experienced little diversification over the late nineteenth century and it seems likely that living standards increasingly depended on money made elsewhere and remitted home.[92] Ratnagiri District remained overwhelmingly the main single supplier of labour to the Bombay cotton mills[93] and by the winter before the First World War nearly 10,000 people a month were embarking from the Ratnagiri ports for Bombay City.[94] In this situation, not even the most optimistic revenue officer could find any evidence of agricultural development. Despite population growth, 'I am personally not of the opinion', wrote the settlement officer of Ratnagiri taluk in 1897, 'that the prosperity of the cultivators has advanced to that stage it should have, after such a long period as thirty years.'[95]

In many localities, commercial decline induced by new rail-borne patterns of trade was accompanied by much more precise evidence of agricultural stagnation. In Ahmedabad District, for example, despite overall steady population growth, the population of Gogha taluk, with the decline of its port, actually fell between 1872 and 1890.[96] In this way, the impact of modern communications could produce marked spatial differentiation even within a district. Thus in Satara District, although the advent of the Southern Maratha Railway had a broadly stimulating effect, its vitality varied widely in direct proportion to proximity to the track. The central taluk of Koregaon, where every village was within 7 miles of one of the five stations, was hymned as enjoying 'a well-

92 In Ratnagiri taluk in the year 1913 alone, Rs. 12,693 worth of foreign money orders and Rs. 3,93,015 worth of inland money orders were paid out. *SBG*, NS, No. 574, Settlement Report on Ratnagiri Taluk, Ratnagiri District, by J. A. Madan, 11 September 1914, para. 40.
93 Morris D. Morris, *The Emergence of an Industrial Labour Force in India* (Berkeley and Los Angeles, 1965), p. 63.
94 *SBG*, NS, No. 574, Settlement Report on Ratnagiri Taluk, Ratnagiri District, by J. A. Madan, 11 September 1914, para. 39.
95 *SBG*, NS, No. 377, Settlement Report on Ratnagiri Taluk, Ratnagiri District, by J. L. Lushington, 30 March 1897, para. 29.
96 *SBG*, NS, No. 255, Settlement Report on Gogha Taluk, Ahmedabad District, by T. R. Fernandez, 23 March 1890, para. 15.

established continuous prosperity' when its settlement was revised in 1891.[97] In the outlying taluk of Man, however, its headquarters 28 miles from the nearest stretch of railway line, the official belief in 1890 was that 'the productive power of the soil has not increased'.[98] Whilst settlement officers frequently resorted to this type of explanation to account for growing local disparity, the pattern of access to modern communications was, by 1890, the most striking difference between the central and outlying areas of Satara District.

So far, our account of the nature of the late nineteenth-century expansionary cycle has depicted a commercialisation process of greatly varying local impact, primarily shaped by communications improvement. But what of the ecological dimension, evident in the comment about Man taluk? It is clearly possible that the differing extension of commercialisation and of expansionary forces in general depended, also, on the production of particular crops enjoying an especially favourable output and demand situation and requiring precise soil and climatic conditions. Against this, no one crop seemed to be the pioneer as cotton had been in the 1860s. Cotton remained the most extensively grown export crop in western India but its price history in the 1880s was the slow, steady upward movement which the foodgrains also enjoyed. Export prices per hundredweight rose from around Rs. 25 in the early 1880s to an average of just over Rs. 28 a decade later, not sufficient, probably, to represent any substantial improvement in real terms.[99] Movement in the total area of land under cotton cultivation mirrored this trend, for there was a slow increase reaching just over 3 million acres in the Bombay Presidency by 1890–1.[100]

Even so, cotton cultivation was clearly extending in overall terms, for the latter figure represented a substantial rise on the levels of the American Civil War cotton boom.[101] More significant still, the evidence suggests that expansion in cotton cultivation was now concentrated in the areas, especially within Gujarat and the Southern Maratha Country, particularly suited for the crop. The 1860s cotton boom was intensely unstable partly because it involved localities, like Indapur taluk, where soil and climatic conditions were too unreliable for continuous high out-

---

97 *SBG*, NS, No. 267, Settlement Report on Koregaon Taluk, Satara District, by W. M. Fletcher, 12 August 1891, para. 29.
98 *SBG*, NS, No. 240, Papers on the Settlement of Man Taluk, Satara District, H. E. Winter, Collector of Satara, to the Commissioner, Central Division, No. 1460, 15 March 1890, para. 2.
99 MARD, Vol. 75 of 1892, No. 733, Comparative Statement on the Progress of Cotton Cultivation.
100 Ibid.
101 See Harnetty, 'Cotton Exports and Indian Agriculture', p. 417.

put production. In this sense, the sudden, rapid explosion of international demand for one crop had produced serious distortion within the agricultural economy. During the 1870s, however, production in areas like the Deccan famine belt, as we saw, fell away and was not resumed on any scale after 1880. The steady overall expansion of the cotton acreage, therefore, was now based on rapid growth in some localities within the specialist regions. In Gujarat, for example, the western taluks of Ahmedabad District, now well served by rail links, substantially increased cotton cultivation. In Viramgam taluk, by 1890 cotton covered 45.4 per cent of the cultivated acreage (which had more than doubled since the late 1850s) compared with 32.6 per cent in 1877–8[102] and throughout north-western Gujarat by the end of the nineteenth century 'cotton has now become the great alternate crop to all the cereals in dry-crop land'.[103] The other area of Gujarat to experience rapid expansion in cotton cultivation was in the south, in the Tapti plain region of eastern Surat District. Here in Bardoli taluk the proportion of, again, a quickly extending cultivated acreage producing cotton increased from 17.5 per cent at the time of the *Gazetteer* enquiries of the mid-1870s[104] to 24.8 per cent by 1895.[105]

Favourable natural conditions may have been the main dynamic behind this type of expansion. The Tapti plain, in particular, was an ideal site for cotton cultivation, for here 'the finest, richest and most perfect black soil is met with, probably equal in quality to any in Gujarat, if not superior'.[106] In this way, unusually good output performance may now have been the spur to expansion which previously in the 1860s was provided by rapid price increase. The spectacular examples, however, of ecological conditions encouraging the growth of specialist cash-cropping came in the canal irrigated areas. In a sense the great Bombay canal schemes – as one official, H. E. Sullivan, argued during the late 1870s[107] – were, in terms of cost effectiveness, a highly inefficient means of stimulating agriculture in a dry-crop region. The problem of limited 'take up' remained. What the Deccan canals did achieve, however, were pockets of highly specialised production within areas of largely subsistence agriculture. Haveli taluk, for example, the taluk surrounding Poona City,

102 *SBG*, NS, No. 242, Settlement Report on Viramgam Taluk, Ahmedabad District, by T. R. Fernandez, 13 February 1890, para. 19.
103 Ibid., para. 20.
104 *BG*, Vol. 2, *Surat and Broach*, p. 280.
105 *SBG*, NS, No. 359, Settlement Report on Bardoli Taluk, Surat District, by T. R. Fernandez, 2 December 1895, para. 16.
106 Ibid., Settlement Report on Bardoli Taluk, Surat District, by C. J. Prescott, 11 November 1865, para. 15.
107 See Caird Papers, Minute by H.E. Sullivan, 11 April 1879.

remained little influenced by new patterns of cultivation during the nineteenth century. As late as 1916 cereals still covered around 81 per cent of the gross cropped area, 'oil seeds are little grown' and 'cotton is practically unknown'.[108] Around 10,000 acres within the taluk, however, were irrigated, mainly by canal works, and of these favoured lands nearly 4,000 acres were producing sugarcane.[109] In nearby Purandhar taluk, the works of the valley of the Karha river created another major centre of sugar cultivation.[110]

The thesis of specialist cash-cropping extending in particularly favourable localities in the late nineteenth century can be supported by other examples. In central Gujarat, tobacco cultivation expanded considerably in the eastern parts of Kaira District. Its area in Borsad taluk, for instance, increased from 4,259 acres in 1876–7 to 7,404 acres in 1886 and by 1895 it was 'the most important crop of the taluk'.[111] In Satara District, the communications expansion produced a substantial growth in cultivation of groundnuts which now could be exported swiftly to Bombay City. In Patan taluk, for example, the area under groundnuts rose from a minimal 570 acres to 8,811 acres in the thirty years before 1894.[112] The settlement report on Karad taluk in 1894 described groundnuts as the area's main export 'in which transactions took place last year (1893) to the value of Rs. 4,76,570'.[113]

This differential expansion of commercial agriculture in the localities of western India in the late nineteenth century needs, however, to be viewed sceptically in terms of its wider economic impact. Cash-crop cultivation would not, of itself, automatically create more wealth within the village unless output and price conditions were favourable. In addition, land, as we have seen, was often now described as relatively scarce. It is therefore possible that commercial crops were competing heavily with foodgrains in some areas, their expansion consequently threatening food supply to the poorest peasants and labourers. On the first issue – that of the rates of return to the village on commercial agriculture – it is clearly impossible to make any firm assertions. Estimating 'profit' on the

---

108 *SBG*, NS, No. 577, Settlement Report on Haveli Taluk, Poona District, by R. D. Bell, 7 March 1916, para. 10.
109 Ibid., para. 11.
110 See *SBG*, NS, No. 571, Settlement Report on Purandhar Taluk, Poona District, by R. D. Bell, 24 January 1916, para. 10.
111 *SBG*, NS, No. 337, Settlement Report on Borsad Taluk, Kaira District, by E. Maconochie, 19 January 1895, para. 14.
112 *SBG*, NS, No. 293, Settlement Report on Patan Taluk, Satara District, by R. B. Pitt, 24 April 1894, para. 18.
113 *SBG*, NS, No. 299, Settlement Report on Karad Taluk, Satara District, by W. S. Turnbull, 29 November 1894, para. 11.

cultivation of any particular crop depends so much on the variables, particularly the price of inputs. What we can do, however, is to record impressions of the value of some crops and compare this with the worth of many resources, particularly the Rs. 200 which the Deccan Riots Commission estimated as the value of a typical Deccan peasant's possessions.[114] Official estimates suggested that an acre of some crops might produce nearly as much as this every year. In 1880 the average yield of sugarcane per acre in the Bombay Presidency was estimated at 18 hundredweight at a time when the price of sugar averaged around Rs. 9 per hundredweight.[115] Kaira tobacco, equally, seems to have enjoyed a high price. Examples were quoted in the Nadiad taluk settlement report of 1894 of smallholdings producing crops worth Rs. 170 and Rs. 205.[116]

At the same time the cultivation of crops like these involved heavy costs, such as the employment of additional labour. Many cash crops may have required the gamble of initial large outlay for possibly high later returns. One settlement officer in Thana District, immediately to the north of Bombay City, tried to estimate rates of return, in the 1890s, on the production of hay, a major export of the area. He traced 27,000 bundles of hay, the harvest of about 15 acres of land, exported from a station in Mahim taluk to Bombay City.[117] The hay sold in Bombay for Rs. 378, clearly nothing like so lucrative a price as for the more specialist crops like sugarcane, but still representing well over Rs. 20 per acre. However, costs on freight, cartage and wages paid to harvesters totalled Rs. 168 and the annual land revenue on the 15 acres amounted to a further Rs. 4. Nevertheless, this still left a good profit of nearly Rs. 14 per acre. Cases like this suggest that much depended on keeping costs, so far as possible, under control. For the owner of a very large holding, considerable economies of scale may have been possible but, on the other hand, the small peasant family farm may have been able to economise substantially on additional labour costs. At any rate, between 1880 and 1896 prices and markets were stable enough to encourage prior investment in the hope of later profit.

What, though, of the issue of land availability and the possibility that expansions in commercial agriculture threatened foodgrain production?

---

114 Deccan Riots Report, para. 53.
115 MARD, Vol. 213 of 1882, No. 1574, Statement on Sugar Cultivation Prepared for the Government of India.
116 SBG, NS, No. 295, Settlement Report on Nadiad Taluk, Kaira District, by T. R. Fernandez, 17 April 1894, para. 17.
117 SBG, NS, No. 398, Settlement Report on Mahim Taluk, Thana District, by A. G. Hudson, 21 December 1896, para. 22.

Again, it is impossible to be dogmatic. In general, cultivated acreage was expanding but in some areas the rapid extension of cash crops must have involved some absolute diminution in extent of foodgrain cultivation. Yet the problem may have been mitigated by increasing opportunities for inter-regional trade. One striking feature of late nineteenth-century western India is growing flexibility in crop production. The most rapid extensions in cash-crop cultivation came, as we have seen, in specialist areas and, whilst foodgrain production here may have fallen, increasing regional specialisation could well have produced neighbouring grain surplus localities, able to export food to the deficient areas. The cotton-growing centre of Bardoli taluk, for example, was a significant food importer by the early twentieth century but its needs were supplemented by other localities within the south Gujarat region.[118] The most spectacular expansions of cash-cropping, however, do seem to have occurred in localities, like Bardoli and like Viramgam taluk of Ahmedabad, which were frontier areas, where cultivation as a whole was extending rapidly. This may have eased the problem of maintaining foodgrain cultivation. Thus in Viramgam taluk the proportion of the land growing cereals declined from 60.4 per cent to 38.7 per cent between 1877–8 and 1890,[119] but the cultivated acreage was doubling every thirty years.[120] Again, in Bardoli, the overall cultivated area was later estimated to have grown more than three times over the whole period between 1840 and the 1920s,[121] so that, although supplementary food imports were necessary, any absolute fall in extent of foodgrain cultivation was probably slow. This, on the other hand, may suggest that it was the cash crops which were pushed on to the peripheral land, grown mainly when foodgrain supply was first assured.

These reflections return us to the central issue of the dynamic behind the commercialisation process. Ecological factors may often have been vital. Speedy extension in cash-crop agriculture frequently depended on the availability of good waste land in frontier zones and sugarcane would not have been grown extensively in the Deccan without irrigation facilities. Yet one returns to communications improvement as the most important single determinant. More extensive production for the market in a peasant society depended on external forces and market organis-

118 *SBG*, NS, No. 648, Settlement Report on Bardoli Taluk, Surat District, by M. S. Jayakar, 30 June 1925, para. 11.
119 *SBG*, NS, No. 242, Settlement Report on Viramgam Taluk, Ahmedabad District, by T. R. Fernandez, 13 February 1890, para. 19.
120 Ibid., para. 27.
121 This was the Settlement Commissioner's estimate in 1928. IOL, Public and Judicial Department Papers, Vol. 1957 of 1928, No. 984, Report of Bombay Legislative Council Debate, 13 March 1928.

ations cutting their path to the village as well as the other way around. Railways not only enabled produce to be exported more cheaply and efficiently but also provided a new thrust and focus for merchants and trading houses to seek increased business in the mofussil. It was this latter process which frequently provided the spur to new production and trade in particular crops. In Satara District, for example, the extension of groundnut cultivation was stimulated by the opening, in the wake of the Southern Maratha Railway's arrival, of a Ralli Brothers depot in Karad town dealing with groundnut export.[122] The expansion of cotton cultivation in Viramgam taluk, according to its settlement report of 1890, was 'almost entirely due to the railway having made the town of Viramgam into a great cotton trade centre, at which at least one mercantile house of Bombay, dealing in cotton, has established a permanent European agency'.[123]

The central role played by the railway, however, underpinned the pattern of spatial differentiation in the commercialisation process. Uneven advantages conferred by communications improvement often enhanced old ecological divisions, for the railway could be constructed more easily across fertile plainlands, like the Tapti valley of Gujarat, than through broken or hilly country. Differentiation, created in this way, was particularly marked within Satara District, which serves as an apt microcosm of the character of commercialisation. Here the eastern villages of Valva taluk, in the south-west of the district, close to the Southern Maratha Railway, grew by the 1890s both groundnuts and tobacco and had enjoyed considerable expansion in local trade.[124] However, 17 of the 98 villages of the taluk, situated in the remote western hills, played no part whatsoever in commercial change. Isolated from the railway and protected from outside pressures by the absence of long-distance bullock carts, these villages remained entirely dependent on a primitive, hill-farming subsistence agriculture.[125] Again, to take the neighbouring taluk of Tasgaon: here, in the far south of the Deccan, cotton was the most important commercial crop, but most villages grew none. Eighty per cent of the crop – and a similar proportion of the tobacco grown in the taluk – came from just 14 villages in the west, where 'the rich alluvial lands of Samdoli and the Krishna villages' closely adjoined the railway.[126] Growing

122 *SBG*, NS, No. 299, Settlement Report on Karad Taluk, Satara District, by W. S. Turnbull, 29 November 1894, para. 11.
123 *SBG*, NS, No. 242, Settlement Report on Viramgam Taluk, Ahmedabad District, by T. R. Fernandez, 13 February 1890, para. 20.
124 *SBG*, NS, No. 345, Settlement Report on Valva Taluk, Satara District, by W. S. Turnbull, 24 June 1895, para. 18.
125 Ibid.
126 *SBG*, NS, No. 204, Settlement Report on Tasgaon Taluk, Satara District, by H. K. Disney, 26 March 1887, para. 11.

commercialisation, then, whatever its ultimate impact on overall economic performance, was typically pioneered, in the expansion of 1880–96, by precisely defined localities where communications facilities, underpinned by favourable ecological conditions, enabled cash crops to be produced competitively.

The famine problem, 1896–1900

We would argue that, for all the spatial differentiation, the process of agricultural commercialisation in western India nevertheless made more substantial strides during the expansion of 1880–96 than in any previous era. Even so, the period ended once more in contraction and devastation amidst the severe famines of 1896–7 and 1899–1900. Serious scarcity, in fact, continued in many areas until the next favourable season in 1902–3 and hence these five years represent a protracted and deep-seated agrarian crisis.

How can we explain the onset, at this stage, of the worst famine period in western India under British rule? Accumulated population pressure in the traditional famine belt regions would form an obvious explanation. We saw earlier that per capita availability of resources was probably changing little, possibly even declining in some east Deccan taluks. Renewed population growth here, therefore, may have quickly led to pressure on resources, particularly as land availability could hardly have been so extensive as during the middle years of the nineteenth century. The onset of famine in 1896–7 might certainly be interpreted in this way. The areas affected were all within the Deccan and eastern parts of the Southern Maratha Country. Yet the famine of 1899–1900 was a far wider disaster. It afflicted, not just the traditional famine belt, but also many areas of Gujarat, hitherto considered immune from serious famine, and parts of the Deccan and the Southern Maratha Country, which had avoided the worst calamities of the late 1870s. Indeed of the districts of the Bombay Presidency, only Dharwar and those of the south Konkan completely avoided famine in 1899–1900.[127] Recovery, too, was delayed in all affected districts except Thana, Nasik and Satara until 1903.

This makes the famine of 1899–1900 a much more difficult phenomenon to explain. Death rates were, in fact, at their highest in Gujarat, which had seemed to possess the most advanced and commercialised agricultural economy in western India.[128] Here areas seriously affected included parts of Kaira and Ahmedabad where 'there had then

---

127 *Report on the Famine in the Bombay Presidency, 1899–1902* (Bombay, 1903), Vol. 2, Appendix 1.
128 Death rates per 1,000 population in 1899–1900 stood at 127.8 in Gujarat, as against 64.4 in the Deccan. Ibid., Vol. 2, Appendix 61.

been no bad famine for more than 60 years'[129] and even twenty years later, one settlement officer would reflect of 1900 that 'this disastrous year is known throughout Northern Gujarat as Chappan (or fifty-six), and used as a land mark for reckoning dates and ages'.[130] The famine era also coincided with and stimulated a major bout of serious epidemic illnesses. An upturn in mortality from fevers and the virulence of the plague epidemic, which had erupted in the mid-1890s, contrived to reduce population levels even where the famine had been less severe. In Belgaum taluk, for example, which had a population of just under 120,000 in 1891, nearly 14,000 people were later estimated to have died from plague and fever between 1897 and 1901.[131] As in the late 1870s, then, the famine period administered a widespread check to population growth.

Recovery from the famine of the late 1870s, however, had been universally quick in western India and population and livestock numbers had everywhere quickly reasserted themselves. Yet with the epidemic mortality and the continuance of poor seasons in the first years of the new century, the famine of 1899–1900 seems to have had a more deep-seated and permanent impact in some localities. In Dholka taluk of Ahmedabad, for example, the population in 1911 was lower than at any previous census since the middle of the nineteenth century,[132] whilst, at the other end of the Presidency, in Belgaum taluk the population of 1911 was lower by over 8,000 than that of 1891.[133] These falls, however, were typically matched, and even exceeded, in the worst hit areas of the Deccan, by heavy decline in numbers of agricultural stock. In Sirur taluk of Poona, for instance, the proportionate loss in numbers of plough cattle – from 24,790 in 1895–6 to 19,828 in 1913–14 – was greater than the approximately 10 per cent decline in population between 1891 and 1911.[134] The consequences of this for cultivation may have been grave. One Poona official, writing in 1909 of Indapur taluk, spoke of 'a very serious decrease in cultivating capacity and consequent decrease in agricultural income . . . As eight to ten bullocks are needed to drag the heavy wooden plough, . . .

---

129 *SBG*, NS, No. 607, Settlement Report on Dholka Taluk, Ahmedabad District, by J. F. B. Hartshorne, November 1920, para. 32.
130 Ibid.
131 *SBG*, NS, No. 584, Settlement Report on Belgaum Taluk, Belgaum District, by M. Webb, 7 January 1918, para. 10.
132 *SBG*, NS, No. 607, Settlement Report on Dholka Taluk, Ahmedabad District, by J. F. B. Hartshorne, November 1920, para. 87.
133 *SBG*, NS, No. 584, Settlement Report on Belgaum Taluk, Belgaum District, by M. Webb, 7 January 1918, para. 10.
134 *SBG*, NS, No. 558, Settlement Report on Sirur Taluk, Poona District, by R. D. Bell, 25 September 1915, para. 2, Appendix F.

very much less land is ploughed and harrowed than formerly.'[135]

This type of evidence might suggest serious reservations about the whole trend of our previous argument. The famine of 1899–1900 – apparently so widespread and consequential – might seem to indicate a society universally on the verge of subsistence crisis, incapable of extensive agricultural diversification. Yet, in a number of ways, this assumption can be challenged. Firstly, the onset of famine in 1896 was not preceded, as was its counterpart in 1876, by any evident contraction of agricultural activity or any sense of impending crisis. So far as can be judged, the poor season of 1896–7 burst on an economy where the trend in land use and prices had remained firm. Again, although population decline, as we have noted, occurred outside the regions of severe famine, the sense of general demographic crisis is not quite so evident as in the late 1870s and in Ratnagiri District, for example, unlike in the late 1870s, population growth continued undeterred.

Further, in the late 1890s famine and its effects clearly could and did co-exist with the continuance of extending commercialisation. In both Dholka and Belgaum taluks, despite population and stock losses, commercial agriculture experienced no fundamental check. The Dholka settlement report of 1920, which declared the effects of the famine of 1899–1900 as 'the great outstanding feature of the period of settlement',[136] went on to describe 'the remarkable advance of cotton', its acreage in the taluk increasing from 7,250 in 1896–7 to 28,217 by the period 1912–16.[137] In Belgaum taluk, too, despite a decline of nearly 20 per cent in plough cattle numbers to match the population decrease,[138] cotton cultivation increased nearly three times in the thirty years before 1918.[139] This might intimate, nevertheless, a direct causal connection between famine and the spread of commercialisation. Not only is there the possibility that expansion in cash-crop production may have robbed the foodgrains of scarce land in these localities: commercialisation may also have created more frequent 'famine of price' on the lines outlined earlier. The Bombay famines might epitomise Bhatia's assessment that by the late nineteenth century 'the development of means of communications and transport

---

135 *SBG*, NS, No. 516, Papers on the Modification of Rates of Assessment in Indapur Taluk, Poona District, 1904–10, J. P. Brander, Assistant Collector, to the Collector of Poona, No. 2070, 10 November 1909, para. 10.
136 *SBG*, NS, No. 607, Settlement Report on Dholka Taluk, Ahmedabad District, by J. F. B. Hartshorne, November 1920, para. 32.
137 Ibid.
138 *SBG*, NS, No. 584, Settlement Report on Belgaum Taluk, Belgaum District, by M. Webb, 7 January 1918, para. 13.
139 Ibid., Report by L. J. Mountford, Commissioner, Southern Division, No. 3137, 19 July 1918, para. 8.

helped to distribute a local scarcity evenly over the country as a whole resulting in a rise in prices of foodgrains which rendered food beyond the reach of poorer sections of the community'.[140]

There is some evidence of the operation of the famine of price syndrome in the late 1890s. Prices certainly leapt up immediately: jowar prices in Poona market nearly doubled in the month between mid-September and mid-October 1896,[141] and official reports commented that 'in places where crops are poor but have not entirely failed and where in ordinary years there would have been no need for relief, the abnormally high prices press severely on persons who have no savings'.[142] If, however, this were the entire explanation for the extent of the famine, we might expect its victims to form a very precise group socially. Ordinary cultivators or at any rate anyone with access to sufficient land to produce some food should have been capable of survival. Indeed, in the 'famine of price' any surplus would be highly valuable on the market. In contrast, most labourers and poor peasants with very small holdings would have been in desperate straits. Yet by 1899–1900 those seriously affected seemed relatively diverse socially, certainly more so than in the famine of 1876–8. As one might expect, the poorest in the village suffered worst, but by 1900 the presence of many landowners was noticed on the famine relief works. In Ahmednagar District 'high class Marathas with their families and Musalman pardanishin women were observed on the works'[143] and here, at the famine's height in February 1900, 46.7 per cent of the relief workers were classified as 'cultivators'.[144]

The famines of the late 1890s, then, undoubtedly present complex problems, but the search for elegant new interpretations by historians and economists alike may have made matters a great deal worse. The simple realities of famine need reassertion. The massive outbreaks during the nineteenth century did not occur in just the poorest areas or even those where population pressure had become most intense. If that were so, the south Konkan which avoided famine throughout the British period must, at some stage, have been affected. The absence of severe famine in many areas of the Madras Presidency, so seriously afflicted in 1876–8, is also a curious circumstance. Alternatively, neither was famine inexorably linked with the progress of commercialisation, for the east Deccan – the area outside the Konkan where commercial agriculture had

---

140 Bhatia, *Famines in India*, p. 9.
141 IOL, Bombay Revenue (Famine) Progs., Vol. 5326, January 1897, No. 122, Report on the Scarcity in the Bombay Presidency, 12 January 1897, para. 5.
142 Ibid., para. 8.
143 *Report on the Famine in the Bombay Presidency, 1899–1902*, Vol. 1, para. 8.
144 Ibid., Vol. 2, Appendix 13.

made fewest inroads – remained the region where famine's impact was deepest and most permanent. Famine, at root, in nineteenth-century western India was caused because the rains failed on an extensive scale. The year 1899 was one of unique climatic failure throughout most areas of the Bombay Presidency, a season in which, across the whole northwestern quarter of the sub-continent, rainfall was drastically diminished. Gujarat, where rainfall was normally regular and ample, and where methods of cultivation had long adapted to these reliable conditions, received levels of rainfall which would have threatened scarcity in the east Deccan. Thus in Dholka taluk just 4.9 inches of rain fell during 1899.[145] This inevitably produced a massive shortfall in agricultural production: one official estimate was that 'the deficiency in the outturn of food crops alone in the years 1899–1902 as compared with the outturn of the normal year 1898–99 amounted to some 8½ million tons'.[146] In this situation, a crisis would have occurred whatever previous economic trends or the effects of the operation of the market.

This is not to overlook the human difficulties and misery caused by famine and certainly not to attack attention to economic and social problems to attempt to ameliorate it. Yet, in examining at this stage long-term economic trends, this does put the significance of famine in the late 1890s into perspective. Its existence does not call into question our earlier account of the commercialisation process, nor even suggest that the latter inevitably created problems of foodgrain supply to the poorest, although the social effects remain to be examined. Nevertheless, this focuses our attention on the consequences of the famine, which may have been considerable. Official estimates were that around 2 million cattle died in 1899–1900[147] and this must have affected methods of cultivation. Even so, our preliminary evidence suggests the speedy resumption of cash-crop cultivation in the more commercialised localities. It may be – and we will return to this point in examining the period after 1900 – that levels of performance remained static or even declined in some areas while elsewhere per capita ownership of many resources continued to expand.

Continuity and change in the agrarian economy, 1850–1900

The complex developments we have outlined in this chapter will benefit from some summary. Over the whole second half of the nineteenth century, the likelihood is that total output grew significantly in Bombay

---

145 *SBG*, NS, No. 607, Settlement Report on Dholka Taluk, Ahmedabad District, by J. F.B. Hartshorne, November 1920, para. 32.
146 *Report on the Famine in the Bombay Presidency, 1899–1902*, Vol. 1, para. 16.
147 Ibid.

agriculture, mainly through the mechanism of steady expansion in cultivated acreage. Opportunity for the continuation of the latter process seemed to be becoming more limited by 1900, but, in practice, there was never any crisis of land availability in the nineteenth century. Much of the increase in output went to support rising levels of population and before 1880 developments within the rural economy were dominated by a strong cyclical trend which seemed to involve little or no long-term change of substance in the structure or organisation of agriculture. Between 1880 and 1900, the pattern was again cyclical, ending in a climatic disaster which may have had very harmful effects on levels of resources in some localities. However, some qualitative changes are evident in the strongly expansionist era of 1880–95. Outside the famine belt, per capita ownership of resources like plough cattle may have been increasing in many areas, creating opportunity for more efficient cultivation. In contrast, the trend in some parts of the east Deccan may even have been downwards. Almost everywhere, however, ownership of the specialist resources – wells and particularly bullock carts – was increasing markedly, even if absolute levels still suggested that only a minority of peasants were concerned. At the same time, production of crops for the market was extending, now on a more secure ecological foundation than during the transient boom of the 1860s. The major impetus was the cost-cutting effect of communications improvement and the differential pattern of local railway construction created substantial disparities in the ease with which different regions and localities could react to the process.

The limitations of the commercialisation process, both in terms of extent and effect, need to be further stressed. It is by no means self-evident that cash-crop cultivation led to massive 'profits', nor, indeed, that returns were necessarily diverted back into agricultural investment on any scale. There is certainly some superficial correlation between areas of commercial agriculture and localities where ownership of resources was extending most, but determining cause and effect is a difficult matter. Returns from cash-crop agriculture may have gone to support social positions, been simply hoarded as insurance, or used in the service of that prime obligation of any man of capital, the lending of money. Yet in any case the agricultural economy remained, in many areas, highly localised. The sharp contrast between prices in the western Indian village and those in the outside world during the 'Great Depression' era provides itself evidence that the rural economy was still unintegrated within the wider sway of the international economy: the devaluation of the rupee can provide only part of the answer to the Indian price trend, for other currencies – the Russian rouble, for example – depreciated during this period without fully insulating internal agriculture from international price falls. In the twentieth century, however, such disparity would steadily disappear, and the next onset of depression in agricultural prices in the 1920s would affect the rural economy of western India as fiercely as any in the world. This, perhaps, provides the least flawed general evidence on the real timing and scope of commercialisation in the countryside.

Nevertheless, whilst elements of continuity remained strong, change had occurred within the agricultural economy. At Presidency level, the balance of economic power between different regions had started to shift. In 1879 Temple, giving evidence before the Indian Famine Commission, had remarked that Gujarat was probably the wealthiest area of the Bombay Presidency, followed closely by the Konkan.[148] Twenty years later, the south Konkan was becoming something of a backwater, with a high, possibly growing degree of concentration on subsistence agriculture. At the same time, pockets of rapid development, particularly within Gujarat and the Southern Maratha Country, had sprung up. By 1900, levels of economic performance and, certainly, degree of involvement in wider markets had probably become deeply differentiated from locality to locality as well as from region to region. This pattern was likely to have complex social effects and it is to the structure of agrarian society which we will turn in the next chapter.

---

148 Temple Papers, Vol. 208A, Evidence of Sir Richard Temple before the Indian Famine Commission, Question 6.

6
The Bombay peasantry, 1850–1900: social stability or social stratification?

The pattern of economic change we have described – marked, during the later nineteenth century, by a highly divergent process of commercialisation – permits, in turning to its social impact, the presentation of a familiar theme. During the 1970s it became almost established orthodoxy, so far as that ever exists in Indian history, to see the onset of a more extensive commercialisation within the peasant economy as creating significant social stratification. Access to the opportunities presented by commercialisation, it is often argued, depended on control of resources, such as bullock carts, irrigation facilities and, particularly, capital and credit, which were not widely held within village society. In turn, the cultivation of cash crops and their sale to wider markets allegedly gave the commercial producer enhanced wealth and power compared with the bulk of the villagers who remained, primarily, under the tutelage of subsistence concerns. In sum, the 'rich peasant' is frequently seen as the beneficiary of rural change in the late nineteenth century. Social stratification, of course, has long been a traditional theme of Indian agrarian history, antedating detailed economic analyses of rural markets and commerce, but its causes, in the historiography of the 1960s, were typically ascribed to government policy or socio-political processes.[1] This emphasis has steadily disappeared in the more recent literature, leaving differential patterns of commercialisation strengthening stratification as the interpretation of both economic and social historian. This connection has been stressed not only within our interpretation of western India but also, prominently, by Washbrook and the Rays from parallel developments in Madras and Bengal.[2]

---

1 Thus Ravinder Kumar's 'nouveau-riche class of kunbis' 'owed their prosperity to the policies pursued by the Government of Bombay'. Kumar, *Western India in the Nineteenth Century*, p. 330.
2 I have argued for economically based stratification in 'Rich Peasants and Poor Peasants in Late Nineteenth Century Maharashtra', in Dewey and Hopkins, eds., *The Imperial Impact*, pp. 97–113. Similar types of interpretation are developed in David Washbrook, 'Economic Development and Social Stratification in Rural Madras: the "Dry Region", 1878–1929', ibid., pp. 68–82, and in Rajat and Ratna Ray, 'The Dynamics of Continuity in Rural Bengal under the British Imperium', *The Indian Economic and Social History Review*, 10, 2, May 1973.

The Bombay peasantry, 1850–1900

Stratification, as an interpretation of agrarian change, has the advantage of being a broad church. Rich peasants may have drawn firmly ahead in a society where overall levels of wealth were moving upwards or they may have been rare escapees from agricultural stagnation. Again, the rich peasant may be an emergent nouveau-riche or a member of an established elite, merely using new devices to enhance his traditional supremacy. In these ways, stratification has, arguably, proved a flexible model for examining agrarian society.[3] Yet there is one major problem with the whole interpretation: as Stokes remarks, 'the case for the rich peasant ... rests on inference'.[4] Historians have typically assumed that certain patterns of economic change, like those depicted in the previous chapter, must have created social stratification, rather than demonstrated that it was actually extending on the ground. Positive proof, though, given the shortcomings of our data, is extremely difficult to secure. Demonstrating differentiation in, for example, ownership of land will not do, for it is clear that factors like this could have co-existed with overall social stability. Thus Chayanov's model, which highlights needs created by the life-cycle of the family as the dominant force behind peasant actions, explains substantial differences in size of holdings in terms of merely temporary divisions between individuals.[5]

This situation has prompted, perhaps inevitably, a recent reaction against the stratification thesis. Stokes on northern India, Mishra on Bombay and Robert for one district of Madras have challenged the notion of social change and reasserted the stability of the peasant mass.[6] At root, their argument is the old Russian Populist charge that agrarian society was too steeped in poverty, too oppressed and deprived of dynamism for any distinguishable elite to emerge. Thus Stokes asserts that, if social divisions did widen, it was 'more because of the slow impoverishment of the mass than the enrichment of the few'.[7]

---

3 For a fuller argument on this, see the conclusion of my 'The Russian Stratification Debate and India', *Modern Asian Studies*, 13, 1, February 1979.
4 Eric Stokes, 'The Emergence of the Rich Peasant under Colonial Rule: Some Northern Lights', paper presented to SSRC Seminar on Stratification and Social Mobility amongst the Indian Peasantry, Leicester, April 1977, p. 12.
5 Chayanov's ideas, too, have recently been directly applied to India. For a discussion of the literature, see my 'The Russian Stratification Debate and India'.
6 Eric Stokes, 'Agrarian Relations. 1. Northern and Central India', in Kumar, ed., *The Cambridge Economic History of India, Vol. 2*, pp. 36–86; Satish Chandra Mishra, 'Commercialisation, Peasant Differentiation and Merchant Capital in Late Nineteenth Century Bombay and Punjab', *Journal of Peasant Studies*, 10, 1, October 1982, pp. 3–51; Bruce Robert, 'Economic Change and Agricultural Organization in "Dry" South India 1890–1940: a Reinterpretation', *Modern Asian Studies*, 17, 1, February 1983, pp. 59–78.
7 Stokes, 'Agrarian Relations. 1. Northern and Central India', p. 65.

How can we advance the argument? What seems to be self-evident is that by 1900 very different types of economic activity and, probably, highly diverse levels of economic performance existed within agriculture in the Bombay Presidency and its regions. However, on this analysis, stratification may have been merely the spatial differentiation we have already postulated. Different localities could have come to perform at different levels without any permanent polarisation developing within their own individual societies. In the more commercialised areas, for example, the benefits from greater production for the market may have been spread more broadly than at first appears likely, for several of the leading Bombay cash crops – cotton and sugarcane – were relatively labour-intensive and hence probably provided additional sources of employment and wages for the poor peasants in villages where they were grown.[8] In similar ways, many of the forces traditionally described as productive of stratification undoubtedly contained implications with contrary effect. Famine in the late nineteenth century might seem, for example, an obvious source of stratification. Prices of all crops rose rapidly, to the apparent benefit of those with a surplus to sell and the detriment of those who had to buy food on the market. Yet famine's eventual legacy may have been the exact opposite of greater social polarisation, since if the poorest were killed off, the survivors might represent a less differentiated village.

Historians and economists, like the sources they use, are, by the nature of their trade, attracted by conspicuous change. A rich peasantry riding to wealth and power or the pauperisation of the weak is, where it occurs, a much more striking phenomenon than the maintenance of an approximate stability within a middle peasant mass. However, this problem is now much more widely recognised and a number of recent studies have highlighted the forces acting for social cohesion.[9] In this chapter, we will first examine the evidence for social developments as suggested by patterns of economic change, because the trends may still be decisive enough to make logical inference from them a valuable methodology. Then, in turn, the evidence of land tenure and social structure itself can be investigated.

---

8 See Michelle B. McAlpin, 'The Effects of Expansion of Markets on Rural Income Distribution in Nineteenth Century India', *Explorations in Economic History*, 12, 1975, pp. 289–302.
9 See Tom G. Kessinger, 'The Peasant Farm in North India, 1848–1968', *Explorations in Economic History*, 12, 1975, pp. 303–23; Clive Dewey, 'Social Mobility and Social Stratification amongst the Punjab Peasantry: Some Hypotheses', seminar paper, Institute of Commonwealth Studies, London, 2 March 1976; Stokes, 'The Structure of Landholding in Uttar Pradesh, 1860–1948', pp. 113–32; Kumar, 'Landownership and Inequality in Madras Presidency: 1853–54 to 1946–47', pp. 229–61.

The evidence: deductions from patterns of economic change

From where does the notion of a 'rich peasantry' arise? Stokes, in questioning the evidence for stratification, points to the relative absence of contemporary comment in north India on the phenomenon[10] and within the Bombay official literature, too, stratification was certainly never a developed concept. On the other hand, what are frequently described are distinctions, presented in social or even moral terms, between 'good' and 'slovenly' cultivators, each reaping their due rewards in different agricultural performance. Thus, on the canal lands of the Deccan, 'there is a small number of really skilled cultivators who grow good crops ... a fairly large class of ordinary good cultivators ... and a large class of cultivators who through lack of skill, care or capital grow considerably less'.[11] These divisions were often presented on a caste basis. Over Ahmedabad District, for example, 'where the Kanbi is found, then careful, and often high culture, large-sized and well-fed bullocks, and villages with substantial houses wearing a thriving appearance are visible ... On the contrary, where Kolis and Musalmans abound, the villages afford a picture the reverse of that.'[12]

Such comments, it might be argued, represent mere abstractions based on Victorian assumptions about the hierarchical nature of a caste society. Officials may have naturally believed that high castes automatically provided 'good cultivators' and low castes poor ones. Yet not all descriptions of the successful are of the groups one might expect. On the canal lands of Purandhar taluk, Poona District, it was a low-caste group, the Malis of the market town of Saswad, who 'have made fortunes in sugarcane growing' and received the accolade of official approval as being 'widely known for their skill and enterprise'.[13] Again, in the hill areas of western Poona, the pioneers in the cultivation of myrabolams, a fruit used for dyeing and tanning, were the tribal Kolis.[14] In descriptions of such less likely standard-bearers of commercialisation, an identifiable economically rather than socially based rich peasantry does fleetingly emerge in the official literature. Thus, among a small Christian sect in Thana District, noted for their cultivation of garden lands, we find an individual example

---

10 Stokes, 'The Emergence of the Rich Peasant under Colonial Rule: Some Northern Lights', p. 1.
11 G. Keatinge, *Agricultural Progress in Western India* (London, 1921), pp. 81–2.
12 SBG, NS, No. 214, Settlement Report on Daskroi Taluk, Ahmedabad District, by T. R. Fernandez, 20 February 1889, para. 19.
13 SBG, NS, No. 571, Settlement Report on Purandhar Taluk, Poona District, by R. D. Bell, 24 January 1916, para. 10.
14 MARD, Vol. 345 of 1888, No. 1813, H. F. Silcock, Assistant Collector, to A. Keyser, Collector of Poona, No. 593, 11 May 1888, para. 16.

like 'Nicholas Karwal . . . a fine substantial Khatedar, with a bluff manner, not unlike an English farmer'.[15]

Such remarks, however, advance the argument little. Stripped of their social prejudices, all comments about varying performance by different caste groups really only support spatial differentiation, for in any one locality one group tended to hold most of the land. This is still stratification of a qualitatively different type than is usually meant by the term. Nevertheless, at some point, geographical distinctions, repeated down to the very lowest level, might well involve polarisation of greater significance. Peasant horizons were not so limited: by 1900 many Bombay peasants must have known neighbouring villagers whose patterns of economic activity differed greatly from their own. In the canal zones, in particular, spatial differentiation at a highly local level was marked. In Haveli taluk 'the canals benefit the villages which they command to very different degrees, in accordance not only with the area irrigated but also with the kind of irrigation practicable; there is a vast difference between a village where sugarcane and high class garden cultivation is practised and one in which water is available only for a late crop': adjacent villages in the taluk enjoyed irrigated land comprising nearly half of their total cultivated area and only 1 per cent of it.[16]

However, economic stratification between neighbouring villages could be mitigated by migration. In the extreme case of the canal zone, labour and capital seeking investment in land moved swiftly into the more favoured villages and away from the relatively disadvantaged. Thus over Haveli taluk as a whole, population increased by 45 per cent between 1871 and 1911, but the rise was as great as 129 per cent in 11 villages commanded by the Mutha Canal whilst in 13 western villages, with few irrigation facilities, population actually fell by 7 per cent.[17] Such trends moderated stratification, for by 1911 over a quarter of Haveli's agricultural population lived in the 11 pacemaking villages.[18] Throughout the Presidency, whilst the more commercialised areas were able, anyway, to support more rapidly growing population, there exist marked differences in demographic performance between localities which can only be explained in terms of this type of local migration.

In these ways, only particular forms of stratification can be safely inferred from the evidence so far. Can, then, social stratification within the village in any way be logically deduced from patterns of economic

15 *SBG*, NS, No. 372, Papers on the Settlement of Bassein Taluk, Thana District, Diary by A. C. Logan, Collector of Thana, accompanying No. 4226, 15 June 1897.
16 *SBG*, NS, No. 577, Settlement Report on Haveli Taluk, Poona District, by R. D. Bell, 7 March 1916, para. 11.
17 Ibid., para. 21.
18 Ibid.

The evidence: deductions from patterns of economic change   167

change? One argument, which can be applied to Bombay, depends on the central issue of possession of resources, for even critics like Mishra seem to accept that skewed ownership will indicate stratification. In almost every Bombay locality in the 1880s and 1890s there are signs of growing expenditure on land, houses and agricultural facilities. The evidence of settlement reports, for example, indicates that numbers of privately constructed wells throughout western India probably nearly doubled between around 1870 and the mid-1890s. At the same time, simple division of these assets by local population levels gives us a rough guide to the extent of ownership. Thus, to take a relatively commercialised area like Karad taluk of Satara District: numbers of irrigation wells stood at just over 2,000 by 1892–3, nearly double the levels of the early 1860s, but these were distributed among a population of over 126,000.[19] Again, the dramatic rise in cart numbers in the taluk evident in Table 3 still meant levels of ownership of only about 1 per 40 people, and the documentary evidence suggests that those who carried to the railheads and wider markets often owned several carts. Settlement reports of the late nineteenth century also speak of growing expenditure in many villages on specialist items of consumption. In particular, there seems to have been a shift, evident in many areas of western India, from thatched houses to superior tiled or flat-roofed varieties. In the more commercialised Satara taluks, for example, increases in the better type of houses during the currency of the settlement ending in the 1890s amounted, to take two typical cases, to 32.9 per cent in Koregaon taluk,[20] and 49.7 per cent in Wai taluk.[21] This might well indicate expenditure by a rich peasantry.

More fundamental, however, is the behaviour of the land market. During the late nineteenth century, both the volume of land sales on the market and the prices paid seem to have grown rapidly in most areas.[22] Before 1880, as we saw in the Deccan Riots region, free exchange of land remained limited and the 1880 Famine Commission discovered that still only around 0.4 per cent of cultivable land was then being transferred each year in the Bombay Presidency.[23] Thereafter, land transactions

[19] *SBG*, NS, No. 299, Settlement Report on Karad Taluk, Satara District, by W. S. Turnbull, 29 November 1894, para. 18.
[20] *SBG*, NS, No. 267, Settlement Report on Koregaon Taluk, Satara District, by W. M. Fletcher, 12 August 1891, para. 12.
[21] *SBG*, NS, No. 270, Settlement Report on Wai Taluk, Satara District, by W. M. Fletcher, 7 November 1892, para. 14.
[22] Discussion in what follows is confined to free market sale and purchase of land. Land might also, clearly, have changed hands as the result of foreclosure of mortgages, court orders and other ramifications of the credit system. The significance of this will be considered in the later section on land tenure.
[23] See the figures presented in *PP*, 1880, Vol. 52, Report of the Indian Famine Commission 1880, Part 2, Ch. 3, Section 2, para. 7.

apparently became much more frequent. There were a total of 27,313 registered sales of land in the Presidency over the period 1878–83, but 36,329 by 1883–8 and 48,295 in 1888–93.[24] These figures, though, taken from the Registration Department statistics, are notoriously unreliable, for they are not comprehensive of all land transactions and, also, include sales of urban property.[25] The story can be followed more accurately in the four Deccan districts, where after 1879 the Agriculturists' Relief Act provided for the compulsory registration of all land transactions; although, even here, investigation is complicated by the categorisation of sales and mortgages together. Even so, the evidence for a growth in the land market seems clear. During the 1890s, for example, the number of land transactions in the four districts rose from 100,964 in 1892–3 to 121,816 by 1896–7.[26] The Bombay government reported, in reply to Elgin's all-India land transfer enquiry of 1895, 'a considerable and continuous increase in registered transfers'.[27]

Nevertheless, the extent of free exchange on the market still needs to be viewed sceptically. How much actual land was involved? Bombay's land transfer report spoke of 75,114 land sales involving 828,061 acres in the four Deccan districts between 1883 and 1891.[28] This would represent about 10 per cent of the total cultivated area changing hands in eight years, and it roughly accords with some occasional evidence provided by settlement officers. In Amalner taluk of Khandesh, for example, some 11,184 acres, around 4.5 per cent of the occupied land, were sold in three years during the late 1880s.[29] In all, a safe estimate would be that the amount of land sold each year increased, approximately, from the Famine Commission's 0.4 per cent of available land to around or just over 1 per cent during the expansionary era of 1880–96. This represents a steady expansion of the market rather than the commercial revolution sometimes implied by officialdom. Yet the buoyancy of the market clearly varied widely from region to region and from locality to locality. In the more commercialised areas of the Deccan, by the early twentieth century, 'everyone with savings seems to endeavour to convert them into

---

24 IOL, Government of India Revenue Progs., October 1895, No. 73, Bombay Presidency Land Transfer Note, p. 35.
25 Even the Bombay government conceded that the statistics were 'useful rather as a general guide as to the increase of transfers than as an indication of their actual extent'. Ibid.
26 *Bombay Presidency Annual Registration Department Reports, 1895–96*, para. 45; *1898–99*, para. 46.
27 IOL, Government of India Revenue Progs., October 1895, No. 73, Bombay Presidency Land Transfer Note, p. 62.
28 Ibid.
29 *SBG*, NS, No. 229, Settlement Report on Amalner Taluk, Khandesh District, by W. S. Turnbull, 13 February 1889, Appendix K.

The evidence: deductions from patterns of economic change 169

land'.[30] In parts of the isolated Panch Mahals District of north-eastern Gujarat, in contrast, 'sales of land are exceptional and due to peculiar circumstances' as late as 1909.[31]

This reiterates our constant theme of spatial differentiation and the highly uneven impact of the commercialisation process. A different type of evidence is provided by changes in land price levels within the areas where a broadly based market existed. After 1880, rises in price relative to levels of land revenue are widely reported in the revenue papers. In the early 1880s, to make a sweeping generalisation, average dry-crop land often cost around 20–25 times its revenue assessment: 21 times was the norm in Parner taluk of Ahmednagar,[32] and Lee-Warner in Satara District in 1882 spoke of 24 times the assessment as 'the average rate to which the vendors would agree'.[33] The settlement reports of the 1890s, however, typically present, in their statistical appendices, prices of around 40 or 50 times the revenue assessment as the new norm. Within the Satara taluks, for example, Karad's average land price of 51 times the assessment[34] was fairly typical and this was matched in the Konkan. Average sale prices were recorded at 43.6 times the assessment in Alibag taluk, Kolaba District, in 1893[35] and 45.6 times in Mahad taluk of the same district in 1898.[36] The statistics, of course, were presented in these terms to provide scientific justification for revenue enhancements and their impression would be qualified if, as is quite possible, the real burden of the land revenue was declining.[37] Even so, occasional references to actual money prices still suggest significant increase. The Parner price of the early 1880s, for example, represented little more than Rs. 10 per acre, whilst the Karad taluk average of the 1890s amounted to over Rs. 100 per acre. The direct comparison obviously is meaningless, for Karad taluk had been much more affected by commercial change. Nevertheless, some major rise in absolute price levels seems likely and even in the Panch Mahals the occasional land sales apparently involved prices of about Rs.

30 *SBG*, NS, No. 561, Settlement Report on Junnar Taluk, Poona District, by R. D. Bell, 12 April 1916, para. 22.
31 *SBG*, NS, No. 509, Settlement Report on Jhalod Taluk, Panch Mahals District, by R. B. Ewbank, 17 May 1909, p. 5.
32 *SBG*, NS, No. 177, Settlement Report on Parner Taluk, Ahmednagar District, by G. A. Laughton, 12 March 1884, Appendix C.
33 IOL, Lee-Warner Papers, Deccan Ryot Series, No. 1, W. Lee-Warner, Notes on the Bombay Land Revenue System, 16 December 1882, para. 1.
34 *SBG*, NS, No. 299, Settlement Report on Karad Taluk, Satara District, by W. S. Turnbull, 29 November 1894, Appendix K.
35 *SBG*, NS, No. 290, Settlement Report on Alibag Taluk, Kolaba District, by C. W. Godfrey, 31 October 1893, Appendix K.
36 *SBG*, NS, No. 430, Settlement Report on Mahad Taluk, Kolaba District, by A. G. Hudson, 7 September 1898, Appendix K.
37 This issue will be discussed directly in Chapter 8.

40 per acre by the late 1890s.³⁸ In the era between the two famines, then – marked by much lower rates of inflation than during the 1860s and after 1897 – land prices, almost certainly, were increasing substantially in real terms.

At the simplest level, this phenomenon might be attributed to mass peasant action under conditions of population growth and consequent pressure on land. However, if the bulk of the peasantry had been engaged in these land purchases, we would surely expect a much more widespread village market. In general, what we have described for the late nineteenth century is a still relatively modest land market, subject nevertheless to buoyant demand and increasing prices. This might suggest the activities of a minority of richer peasants, as indeed settlement officers did recognise in the most conspicuous examples. Thus in Junnar taluk of Poona District land prices became much higher than the average in just three villages, but 'it is not an accident that the most striking cases of inflated values come from these villages. Money is abundant in each of them; each is the home of prosperous fruit merchants in Bombay.'³⁹

Yet the most ubiquitous outlet for any spare cash remained lending operations. This guaranteed status and clients without the special problems of arranging cultivation which land purchase might often entail. Any rich peasantry, therefore, riding to greater economic power on the back of commercialisation should be increasingly conspicuous within the credit system. Judging the nature of credit supply amidst so vast a system is clearly difficult, but there are signs of widening activity by peasant cultivators during the late nineteenth century. This was particularly evident – and significant – in the central Deccan, where, as we saw, the relative absence of an indigenous elite, with spare capital to lend, was a major influence behind the crisis of 1875. By 1892, however, throughout the region 'dealings with fellow ryots have to a much greater extent than formerly taken the place of dealings with professional moneylenders'.⁴⁰ Amidst the areas most directly influenced by extending commercialisation, like Valva taluk of Satara District, it is no surprise to find that, by now, 'the moneylending was largely done by patels and rich Kunbis',⁴¹ but detailed enquiries into the working of the Agriculturists' Relief Act in

---

38 *SBG*, NS, No. 481, Settlement Report on Jhalod Taluk, Panch Mahals District, by C. V. Vernon, 3 March 1904, Appendix K.
39 *SBG*, NS, No. 561, Settlement Report on Junnar Taluk, Poona District, by R. D. Bell, 12 April 1916, para. 25.
40 MARD, Vol. 89 of 1893, No. 1038, Part 2, Report of the Deccan Agriculturists' Relief Act Commission, 11 June 1892, para. 11.
41 *SBG*, NS, No. 345, Papers on the Settlement of Valva Taluk, Satara District, J. Muir Mackenzie, Acting Survey Commissioner, to the Secretary to Bombay Government, Revenue Department, No. S2325, 27 September 1895, para. 6.

poor areas like Bhimthari taluk also revealed several villages where most credit operations were, by the 1890s, dominated by agriculturists.[42] The vacuum in the central Deccan's social structure, which had made the Marwari moneylender uncharacteristically important, seemed on the way to solution by 1900 and this was reflected by generally lower levels of tension within the credit relationship. In other regions, too, lending by peasants, although always relatively important, receives increased comment in the revenue papers of the late nineteenth century. In Kaira District, for example, by 1895 'a considerable business is carried on by well-to-do Patidars'.[43]

Nevertheless, much of this evidence, it might be objected, could indicate, not stratification, but a general rise in income levels throughout agrarian society in the 1880s and early 1890s. Officialdom typically explained increases in numbers of carts and the better houses as well as the expanding land market in these terms, and was often at pains to suggest that the poorer sections of the village community were also benefiting from the processes. Comparing the isolated, tribal areas of West Poona in 1888 to the 1878 situation, for example, one official remarked how 'then there were few or no cooking pots to be got – only earthen vessels; now it is quite natural to see brass and copper pots in the houses'.[44] Historians, presented with an agrarian economy which manifestly failed to achieve transformation, tend inevitably to baulk at assumptions of broadly based increases in 'prosperity' at any period and this is one major reason why social stratification has become so acceptable a thesis. Some developments and patterns of expenditure, however, do suggest a socially widespread involvement. The trend towards improved flat-roofed and tiled houses, for example, seems to have involved many peasants: in the Satara taluks, the inferior thatched varieties were very much in the minority by the 1890s[45] and the same feature is even evident in some of the famine belt areas of the Deccan.[46] Differentiation by house type in the 1890s, then, suggests a broad peasant mass, capable of improved living standards,

---

42 For example, in Bhandgaon in 1892 the peasants 'only borrow and lend among themselves'. In Deulgaon Gada 'all but one' of the village moneylenders were agriculturists and in Deulgaon Rasal the creditors were 'all Kunbis'. MARD, Vol. 89 of 1893, No. 1038, Part 2, Report of the Deccan Agriculturists' Relief Act Commission, 11 June 1892, Appendix 1.
43 *SBG*, NS, No. 337, Settlement Report on Borsad Taluk, Kaira District, by E. Maconochie, 19 January 1895, para. 18.
44 MARD, Vol. 345 of 1888, No. 1813, H. F. Silcock, Assistant Collector, Poona, to A. Keyser, Collector of Poona, No. 593, 11 May 1888, para. 20.
45 See, e.g., *SBG*, NS, No. 270, Settlement Report on Wai Taluk, Satara District, by W. M. Fletcher, 7 November 1892, para. 14.
46 See, e.g., *SBG*, NS, No. 177, Settlement Report on Parner Taluk, Ahmednagar District, by G. A. Laughton, 12 March 1884, para. 17.

with a notably poorer minority beneath them. This does not necessarily conflict with our description of the widening of credit supply, for there may have been a rapid increase in numbers of peasants lending very small amounts, rather than the formation of a distinct agriculturist moneylender elite with businesses to rival the Marwaris. Many official comments indicate that the 'moneylender' was becoming a much less socially precise entity and that 'agriculturists and the artisan classes borrow and lend amongst themselves'.[47]

Again, the operation of the land market could be explained in strictly Chayanovian terms. Even if only a minority of peasants were involved bidding high prices, the purchasers might still be, not ambitious entrepreneurs, but basically those on the upward section of Chayanov's cycle. Faced with the growing consumption needs of a large family (and amidst the buoyant demographic trends of the 1880s and 1890s this would have been a common situation), they must acquire more land. Yet in many localities by the late nineteenth century availability of good cultivable waste land was extremely limited so that any such peasants would necessarily have turned to the market, their competition inevitably driving prices up.

In these ways, then, social stratification within villages is not an unavoidable inference from the economic phenomena. This, though, does not mean that stratification did not occur. If some evidence, like that of house construction, suggests greater consumption within the mass of the peasantry, a richer peasant elite may still have drawn ahead at a faster rate. The most powerful case for stratification from the economic evidence depends, not on consumption patterns per se, but on access to resources which themselves, in turn, would exacerbate differences in economic performance. Acquisition of a better house or even more land, if family needs dictated it, might have few qualitative implications for a peasant's economic status. Yet, as most commentators in the debate concede, variations in ownership of carts and wells form a different case, for in the conditions of extending commercialisation they opened wider opportunities for the peasant producer.

At root, however, the most likely source of economically based differentiation was the question of the financing of agricultural operations, particularly those involving more extensive production for the market. The credit system remained the normal source of capital, for those with commercial ambitions as well as for poor peasants' subsistence needs, and during the late nineteenth century it seems likely that the western

---

47 *SBG*, NS, No. 561, Settlement Report on Junnar Taluk, Poona District, by R. D. Bell, 12 April 1916, para. 22.

Indian credit system widened considerably and quantitatively larger sums may have been involved. Yet was credit relatively freely available to all, on equal terms? Models of cyclical social mobility certainly seem to depend, to some degree, on this assumption.[48] As we saw, though, credit supply for some had traditionally involved surrender of control over the marketing of one's crops and this may have become a greater source of economic disadvantage in more commercialised conditions. Creditors, too, if their role and numbers did expand, may have acquired greater commercial ambitions and the attitude to the dispossession of the heavily indebted landowner may even have changed. This must remain speculation, until we can examine the tenurial situation in the next section. However, official comments seem to confirm that peasants, increasingly in the late nineteenth century, borrowed on greatly differing terms. Numerical interest rates mean little in an absolute sense, but where assessments of them reveal large relative variations they are, perhaps, suggestive. The Deccan Riots Commission had recorded up to 37.5 per cent as the normal rate for loans, to the typical Kunbi, based on personal security[49] but some other later assessments are of very much lower rates. In Ahmedabad District in 1904, for example, wealthy Patidars were borrowing from town moneylenders at estimated rates of between 4.5 and 9 per cent.[50] In particular, where peasants were able to borrow exclusively within their own small social group (and there is evidence that groups like Gujarati Patidars and the Saswad Malis consciously tried to develop this), much more favourable terms than the average may have been available.

Differential access to credit was most evident at times of economic crisis, particularly during the famine of 1899–1900. Many creditors were reluctant to lend money on personal security: 'people cannot raise money now', the Collector of Khandesh commented.[51] At the same time, the famines were, paradoxically, in some areas occasions of considerable capital outlay on agricultural improvement. One group of Poona landowners, for example, built three new irrigation tanks costing Rs. 22,000 during the famine year of 1905–6.[52] Clearly, for them the credit problem

48 Thus Chayanov seems to give little special attention to the financing of the upward section of his demographic cycle, apart from use of greater labour effort. For criticism of this, see my 'The Russian Stratification Debate and India .'
49 Deccan Riots Report, para. 83.
50 MARD, Vol. 74 of 1905, No. 428, Part 2, Lallubhai Samaldas to Sir James Monteath, 30 July 1904.
51 MARD, Vol. 14 of 1902, No. 1039, Collector of Khandesh to the Commissioner, Central Division, No. 2644, 20 March 1902.
52 *Report on the Famine in the Bombay Presidency, 1905–06* (Bombay, 1907), para. 14.

was surmounted and, if so, such large capital projects made economic sense, for labour was widely available and relatively cheap. Access to credit, however, was not just a sheer quantitative problem. Some peasants seemed able to borrow extensively with few restrictions on their freedom of activity whilst for others, as we shall see, land mortgage might be the only means of obtaining cash in a famine year.

The case for social stratification based on differential access to resources and credit still, of course, rests on logical inference. We can only assume that borrowing whilst retaining control over the marketing of one's harvest offered real advantages over the peasant who surrendered crops as part of his debt obligation. However, in the conditions of extending commercialisation which we have postulated, this might seem a not unreasonable deduction. To sum up our conclusions so far: it is the case for spatial differentiation which seems most secure. Often at a very localised level, marked differences in types of economic activity between villages had clearly developed by 1900, although some peasants were quick to recognise the condition and temper it by migration. In contrast, much of the economically based evidence for social stratification within the village seems inconclusive, for it might indicate increased wealth on the land not its automatic control by an elite. Differentiation is most evident – and the assumption that it would cause social stratification most reasonable – with regard to access to resources, notably credit. Yet the heart of Stokes' criticism, that social stratification cannot be proved indisputably on this type of evidence, remains intact.

The evidence: land tenure and social structure

After the necessarily inconclusive nature of so much of our economic evidence, even a superficial examination of the evolving tenurial situation in western India seems to indicate some unexpectedly dramatic developments. For there, recorded in the revenue papers' statistics, is the appearance of extensive tenurial change during the late nineteenth and early twentieth centuries: the extensive growth of tenancy throughout most parts of the Bombay Presidency.

Down to 1880, the ryotwari ideal had undoubtedly remained realistic and cultivating proprietors, outside the talukdari and khoti regions, still formed a significant majority of the agricultural population. The Census of 1891, however, suggested that some change was occurring. It recorded the existence of over 2½ million 'tenants and sharers' out of a total of around 9 million landholders[53] and, when landless agricultural labourers

---

53 *Census of India, 1891*, Vol. 7, Bombay and its Feudatories, Part 1, Report, Ch. 11, p. 180.

were added,[54] those who did not own land were now estimated at just over 40 per cent of the total agricultural population.[55] By the end of the century, they were probably the majority. In 1898 an experimental record of rights for 23 villages scattered throughout the Deccan and Southern Maratha Country revealed considerable change within the 13,722 separate interests existing. In 7,130 cases, the land was no longer owned by its original proprietors: on most of these lands, tenants now paid fixed rents to landlords.[56] The process is most strikingly presented in those areas where revision settlement reports of the 1910s can be contrasted with equivalents in the 1880s. In Junnar taluk of Poona District, for example, 27,963 of the total 41,435 survey units were recorded as being cultivated entirely by their proprietors in 1883.[57] Yet, at the settlement of 1916 only 1.8 per cent of the Junnar population were classified as 'land occupants' as against 68 per cent described as 'tenants and sharers'.[58] Again, in Parner taluk of Ahmednagar District, over two-thirds of the holdings were apparently cultivated entirely by registered owners in 1884,[59] but the 1916 settlement report enumerated almost all of the agricultural population as sharers, tenants or labourers.[60] Similar statistics are available for other regions of Bombay Presidency. In Belgaum taluk in the Southern Maratha Country, for example, in 1887 landowners cultivated 54.6 per cent of the holdings compared with 35.7 per cent which were sub-let.[61] By 1911 the statistics of the agricultural population recorded 77.7 per cent as tenants and sharers, 18.1 as labourers and only 4.2 per cent as cultivating landowners.[62]

How should we interpret this evidence? One obvious possibility, particularly in view of the Thorners' famous strictures about the reliability

---

54 The historical extent of landlessness will be considered later.
55 See MARD, Vol. 258 of 1892, No. 749, Brief Memorandum on the Material Condition of the People of the Bombay Presidency, 1881–91, para. 1.
56 IOL, Bombay Confidential Revenue Progs., Vol. 5777, October 1899, No. 15, J. W. P. Muir-Mackenzie, Secretary to Bombay Government, to the Secretary to the Government of India, Department of Revenue and Agriculture, No. 7100–168, 7 October 1899, para. 10.
57 SBG, NS, No. 205, Settlement Report on Junnar Taluk, Poona District, by G. A. Laughton, 7 May 1883, Statement A.
58 SBG, NS, No. 561, Settlement Report on Junnar Taluk, Poona District, by R. D. Bell, 12 April 1916, Appendix D.
59 SBG, NS, No. 177, Settlement Report on Parner Taluk, Ahmednagar District, by G. A. Laughton, 12 March 1884, Appendix A.
60 SBG, NS, No. 567, Settlement Report on Parner Taluk, Ahmednagar District, by J. H. Garrett, 22 May 1916, Appendix D.
61 SBG, NS, No. 215, Settlement Report on Belgaum Taluk, Belgaum District, by H. K. Disney, 14 February 1887, para. 26.
62 SBG, NS, No. 584, Settlement Report on Belgaum Taluk, Belgaum District, by M. Webb, 7 January 1918, Appendix C.

of later census data,[63] is to regard the phenomenon as illusion created by dubious statistical material. Changes in classification procedure are certainly evident. In the nineteenth century, as the statistics quoted show, the record of land tenure in Bombay was compiled by investigation of the actual survey units on the ground. Second revision settlement reports after 1900, however, relied on census data to classify the agricultural population by tenurial status. The latter practice may have involved occasional recording of the heavily indebted peasant as a 'tenant', whereas investigation of his land, assuming he was a hereditary cultivator, would have suggested tenurial consistency. In any case, there is the sheer random variation within these statistics: the difficulties of defining very complex proprietary arrangements were clearly considerable and the practice of each individual classifier must have varied.

Yet, although the statistics may well exaggerate the speed and extent of the process, it is hard to deny the breadth of support for the development of tenancy in Bombay. Documentary as well as statistical evidence exists within the settlement reports. In Junnar taluk, for example, the settlement officer of 1883 could describe 'the instances of sub-letting, which being very few, I have shown in detail'.[64] By 1916 in the same taluk 'sub-letting of lands is extensive, for holders classed as agriculturists frequently sub-let as well as non-agriculturists'.[65] The evidence, too, of western India's later history demands recognition of the development of a tenancy problem which would require major legislation in 1939 and 1948. Even after the introduction of the 1948 Bombay Tenancy Act – probably creating some initial diminution in the tenancy area – 'nearly 30 per cent of the cultivated area appears to be tenant cultivated'.[66] This provides a useful gauge of the situation, warning against both simple exaggerations that Bombay had become a region of total landlord domination[67] and equally simple assumptions of tenurial stability. The likelihood is that, starting during the late nineteenth century, the spread of tenancy was a steady development in most Bombay localities. In only two types of area was the process strongly delayed: firstly, in a few pacemaking highly com-

63 See Daniel and Alice Thorner, *Land and Labour in India* (London, 1962), Ch. 6.
64 *SBG*, NS, No. 205, Settlement Report on Junnar Taluk, Poona District, by G. A. Laughton, 7 May 1883, para. 31.
65 *SBG*, NS, No. 561, Settlement Report on Junnar Taluk, Poona District, by R. D. Bell, 12 April 1916, para. 22.
66 V. M. Dandekar and G. J. Khudanpur, *The Working of the Bombay Tenancy Act, 1948. A Report of Investigation* (Poona, 1957), p. 28.
67 E.g. the assessment of one Congress Committee in 1936 that in Maharashtra a 'top stratum of the community rents most of its land to the landless tenant'. *Report of the Peasant Enquiry Committee of the Maharashtra Provincial Congress Committee* (Poona, 1936), p. 48.

mercialised parts of Gujarat where land was still relatively extensive[68] and, more commonly, in backward localities where the extent of commercialisation remained extremely limited. Bijapur District in the southeast, a famine zone district poorly served by modern communications, formed the prime example of the latter category and most peasants here remained cultivating proprietors at the outbreak of the First World War.[69]

However, the relevance of tenurial change for the thesis of social stratification depends on the dynamic behind the expansion of tenancy. The traditional assumption has always been that extensive tenancy evidences the 'decline' of the old peasant proprietor. Thus, typically, in Sholapur taluk the explanation for tenurial change was that 'a large proportion of the hereditary cultivators have mortgaged or sold their rights in the land'.[70] As we saw, the free land market, although growing, seemed hardly intense enough to generate large overall shifts in landownership, but there is some evidence that the implications of debt for land tenure may have been more consequential after 1880. Mortgaging of land may then have become a more general condition for obtaining credit. Thus by 1915 one settlement officer in Poona District believed that 'nearly all the moneylending transactions are based on "poklist" sales, i.e. sales where the moneylender verbally agrees to resell the land on repayment of the consideration'.[71] Certainly numbers of mortgages seem to have been growing, as our earlier statistics of the growth of registered land transactions indicate. Mortgages of land in fact greatly outnumbered free sales on the market. Over the period 1883–91 in the four districts covered by the provisions of the Deccan Agriculturists' Relief Act, 75,114 land sales were registered but as many as 218,309 mortgages.[72] Most of these mortgages, in turn, seem to have given the creditor an extensive stake in the land, as A. F. Woodburn demonstrated in his report of 1889 on the

---

68 Thus in Olpad taluk of Surat District, in the south Gujarat black soil plain, even by 1896 'more than three-fourths of the occupants till their own fields'. *SBG*, NS, No. 361, Settlement Report on Olpad Taluk, Surat District, by T. R. Fernandez, 22 February 1896, para. 20.
69 E.g. in Badami taluk in 1912–13, over 68% of holdings were being cultivated by their registered owners. *SBG*, NS, No. 560, Settlement Report on Badami Taluk, Bijapur District, by G. C. Shannon, 3 November 1914, Appendix I.
70 *SBG*, NS, No. 443, Settlement Report on Sholapur Taluk, Sholapur District, by H. L. Painter, 27 July 1903, para. 36.
71 *SBG*, NS, No. 558, Settlement Report on Sirur Taluk, Poona District, by R. D. Bell, 25 September 1915, para. 21.
72 IOL, Government of India Revenue Progs., October 1895, No .73, Bombay Presidency Land Transfer Note, p. 62.

operation of the Deccan Agriculturists' Relief Act.[73] Of 4,528 mortgage transactions examined by him in both Deccan and Konkan districts, 57.2 per cent specified that if the debt was not paid within a fixed time-span the land automatically went to the mortgagee. A further 23.2 per cent of cases were 'poklist' sales: the mortgagee became the formal owner of the land until the payment of the debt. In all, 89 per cent of the mortgages investigated by Woodburn explicitly specified land, rather than its produce or further debt bonds as the sacrifice for a mortgagor's nonpayment.

In this way – the argument is familiar – encumbrances on land as a condition of credit could have been a major source of tenurial change. The peasant proprietor, unable to repay loans, would apparently become a tenant of his creditor. This, of course, conflicts directly with our explanation of the Deccan Riots situation where securing formal title to land seemed an ineffective strategy for most moneylenders. Yet a strong case can be made for different conditions after 1880. Peasants themselves may have been influenced by the atmosphere of extending commercialisation and the buoyant economic conditions of 1880–96 to offer land as security for more extensive loans, gambling that they would be able to prevent its permanent loss. Some officials, indeed, suggested that this greater flexibility to manoeuvre within the credit system was being achieved: in Ahmednagar, the Collector claimed, 'the same ryot whose land was apparently sold for ever in 1880 may have full possession of it in 1885 and again borrow for a marriage and go through a mock sale in 1890, and so on'.[74] Alternatively, the long famine years of 1896–1902 may have driven many peasants to offer land as their only security for desperately needed loans. Access to credit on personal security was, as we saw, severely cut back by the onset of famine, and the Collector of Ahmednagar predicted in December 1896 that this would inevitably mean that 'a large area of land will, for the time being at any rate, pass into the hands of sowkars'.[75] The statistical evidence appeared to support him. Registered land transactions, mostly mortgages, leapt up by over 28 per cent between 1895–6 and 1897–8.[76] In the latter year 'the value of the transactions aggregated about 1½ times the Land Revenue assessment, and... on an average one

---

[73] *Selections from the Records of the Government of India, Home Department*, No. 342, Papers Relating to the Working of the Deccan Agriculturists' Relief Act during the Years 1875–94, Vol. 2, No. 13, A. F. Woodburn to the Secretary to the Bombay Government, Judicial Department, No. 189, 27 April 1889.

[74] IOL, Bombay Confidential Revenue Progs., Vol. 5777, October 1899, No. 15, F. L. Charles, Collector of Ahmednagar, to J. K. Spence, Commissioner Central Division, No. 2027, 30 March 1896.

[75] Ibid., F. L. Charles to J. K. Spence, No. 8812, 26 December 1896, para. 12.

[76] *Bombay Presidency Annual Registration Department Report, 1898–99*, para. 46.

transaction occurred to every seven of the agricultural population'.[77] Prolonged famine could well have been a decisive instrument of tenurial change.

However, we still need to understand the motivations of creditors in these cases, for the impulse to make land common security must have come, at least partially, from them. Indeed, officials spoke, during the famine, of moneylenders 'pressing their clients' and 'doing their best to get possession of land'.[78] What, then, might have produced so fundamental a change since the 1870s? Land, especially in the more commercialised regions, had become a much more attractive investment by 1900, its value probably rising in real terms, but any growth of creditor interest in land may reflect, most of all, developments within the moneylending body. We argued earlier that the numbers and the extent of business of agriculturist moneylenders were both probably increasing during the late nineteenth century. Such creditors, in turn, may have had a much greater interest in land acquisition. They understood the business of agriculture more fully and typically would have enjoyed more intimate knowledge about the value and capability of different land. In Gujarat the connection between increased moneylending by the richer peasants and extensions in their landholdings was particularly noted. Thus in Broach District in 1896 'Patidars and Bohras are both much addicted to moneylending, and in Vagra villages (for instance) ... the Patidar inhabitants had got nearly all the lands of the Girassias into their possession.'[79] In parts of Kaira District, too, as a result of the Kanbi Patidars' control of the credit system, 'the tendency must be for the Kanbi to oust the Koli'.[80] In addition, the acquisition of a debtor's land by an agriculturist moneylender might have had further implications, for the agriculturist, unlike the Marwari, sometimes had his own firm plans for the organisation of cultivation. The Collector of Poona commented in 1896 that 'the professional moneylender prefers to keep the land in its owner's hands and get as much rent as he can, knowing he can always sell at the end, but the other [i.e. peasant lender] prefers the land itself as he can sub-let it at high rates'.[81] In the more commercialised areas of Satara District, it was

77 Ibid., Government Resolution No. 8942, 12 December 1899, para. 6.
78 MARD, Vol. 14 of 1902, No. 1039, Collector of Khandesh to the Commissioner, Central Division, No. 2644, 20 March 1902.
79 IOL, Bombay Confidential Revenue Progs., Vol. 5777, October 1899, No. 15, A. C. Logan, Collector of Broach, to the Commissioner, Northern Division, No. 3367/41, 26 November 1896, para. 6.
80 SBG, NS, No. 337, Settlement Report on Borsad Taluk, Kaira District, by E. Maconochie, 19 January 1895, para. 18.
81 IOL, Bombay Confidential Revenue Progs., Vol. 5777, October 1899, No. 15, H. F. Silcock, Collector of Poona, to J. K. Spence, Commissioner, Central Division, No. 2616/10, 27 March 1896, para. 12.

alleged, the 'patils and rich Kunbis' who lent money positively 'make it their business to add field to field in the hopes of increasing their garden cultivation'.[82]

In this way, a case can be constructed that the growth of tenancy in Bombay after 1880 reflected marked social and tenurial stratification: the formal dispossession of many peasant proprietors by a rich peasant creditor elite. Yet any thesis of abrupt 'peasant decline' would need to be kept within limits. There is no evidence in western India of any proletarianisation process, the creation of an extensive new problem of landlessness. In Bombay Presidency, as in Dharma Kumar's Madras,[83] a landless labouring class who worked for others had always existed. Sykes in the 1830s spoke of the 'husbandry or domestic servant' who received 'four rupees per month, with which he finds his own food and clothes'.[84] The ryotwari system, for all its ideals of making every cultivator a proprietor, could not cut through to such groups and in many areas formalised patterns of dependence and servitude by landless labourers were strengthened by the proprietary titles guaranteed to their masters by British rule. An excellent example was the 'hali system' of Broach and Surat Districts in south Gujarat.[85] The hali labourers worked as traditional servants of large farmers, 'usually an Anavla Brahmin or Kanbi':[86] possibly the system owed much to the early emergence of a relatively labour-intensive commercial agriculture in the region in which sugarcane was extensively grown.[87] At any rate, the *Gazetteer* of the 1870s estimated those 'who, for generations, have held the position of hereditary servants'[88] at 'about 27,500 souls, or one-sixth part of the entire strength of the aboriginal population'.[89] Many other tribal groups,

---

82 *SBG*, NS, No. 345, Papers on the Settlement of Valva Taluk, Satara District, J. Muir Mackenzie, Acting Survey Commissioner, to the Secretary to the Bombay Government, Revenue Department, No. S2325, 27 September 1895, para. 6.
83 See Dharma Kumar, *Land and Caste in South India. Agricultural Labour in the Madras Presidency during the Nineteenth Century* (Cambridge, 1965); an important study which demonstrated that, in south India at any rate, the landless labourer class was not a simple creation of agrarian change under British rule.
84 Sykes, 'Special Report on the Statistics of the Four Collectorates of Deccan, under the British Government', p. 323.
85 For a study of the hali system and its evolution in modern times, see Jan Breman, *Patronage and Exploitation. Changing Agrarian Relations in South Gujarat, India* (Berkeley, Los Angeles and London, 1974).
86 *SBG*, NS, No. 381, Settlement Report on Chikhli Taluk, Surat District, by E. Maconochie, 17 June 1897, para. 21.
87 *SBG*, NS, No. 2, Report on the Southern Districts of the Surat Collectorate by A. F. Bellasis, 15 October 1850, para. 42.
88 *BG, Vol. 2, Surat and Broach*, p. 198.
89 Ibid. Since in 1871 the 'aboriginal population' totalled almost exactly a third of Surat's population, the traditional hali classes probably represented, therefore, 5 to 6% of the district's population.

The evidence: land tenure and social structure

in south Gujarat as throughout the Presidency, were landless, traditionally hiring themselves out as labourers.

As a result, the assessment of 1891 – that 19.33 per cent of Bombay's agricultural population were then 'agricultural labourers'[90] – could hardly have represented a decisive extension of landlessness since the middle of the nineteenth century. And the evidence of settlement reports reveals little additional change thereafter, at any rate before the First World War. Landless labourers – to take some examples – were classified as 18.1 per cent of the agricultural population of Belgaum taluk in 1911,[91] 13.5 per cent in Parner taluk in 1916[92] and just 10.4 per cent in Maval taluk of Poona at the same date.[93] So much is mere confirmation for western India of a now well-established orthodoxy, but it suggests that any complete physical ousting of debtors from the land by ambitious agriculturist entrepreneurs must have been limited. In fact, official opinion in Bombay was that 'in most cases where a cultivating owner is expropriated he is not ejected'[94] and 'even when partition is resorted to . . ., the actual cultivation generally stays in the family'.[95] Tenurial change, therefore, on the analysis of 'peasant decline' through the operation of credit and debt, was clearly a subtle process. The former proprietor invariably remained the cultivator of the land and usually his previous debt obligations were merely translated into a more formalised payment of rent. For many, the decisive step in the conversion must have formed a relatively uneventful part of a whole series of loan transactions. Indeed, the typical 'victim' would probably not have seen himself as a 'tenant', viewing the alienation of his land, if he could place the precise moment of its occurrence, as merely a temporary measure. Nevertheless, on this explanation social stratification, in more than simple tenurial terms, may still have developed strongly, the regular rent demand offering greater opportunity to the creditor-landlord to enhance his return from the peasant's effort.

Yet an entirely contrary interpretation of the whole tenancy process is possible; one which would have radically different implications for the

---

90   MARD, Vol. 258 of 1892, No. 749, Brief Memorandum on the Material Condition of the People of Bombay Presidency, 1881–91, para. 1.
91   *SBG*, NS, No. 584, Settlement Report on Belgaum Taluk, Belgaum District, by M. Webb, 7 January 1918, Appendix C.
92   *SBG*, NS, No. 567, Settlement Report on Parner Taluk, Ahmednagar District, by J. H. Garrett, 22 May 1916, Appendix D.
93   *SBG*, NS, No. 565, Settlement Report on Maval Taluk, Poona District, by R. D. Bell, 26 June 1916, Appendix D.
94   IOL, Government of India Revenue Progs., October 1895, No. 73, Bombay Presidency Land Transfer Note, p. 83.
95   *SBG*, NS, No. 443, Settlement Report on Sholapur Taluk, Sholapur District, by H. L. Painter, 27 July 1903, para. 36.

social stratification thesis. The growth of tenancy on the Bombay land may be evidence, not of 'peasant decline', but of greater tenurial flexibility in a more commercialised economy faced, at the same time, with a serious problem of land availability: virgin land was strictly limited and the land sale market, although growing, was still relatively small. In this situation, peasants seeking to expand operations might often need to rent land and tenancy, then, would become a natural response to economic stimuli. On the Chayanov model, too, peasants needing to widen their activities because of increasing family consumption demands would readily rent land where opportunities for other types of land acquisition were few.

There is no doubt that expanding markets for all types of right and asset existed within Bombay rural society by the close of the nineteenth century. Inam land, for example, became subject to massive demand and transfer. The compromise evolved by official policy towards alienated lands in the mid-nineteenth century had left them as valuable assets and heavy buying in and consequent subdivision of the rights resulted. One official spoke, in 1906, of 'two main classes of Bombay inamdar'. One was 'the real squires of the country' – the old village officer class and the traditional inheritors of Mahratta military grants – who now, nonetheless, were 'very rare in the Deccan'. The second group, forming the vast majority of inamdars by the early twentieth century, were 'the absentee rent receivers who . . . look upon their estates as an investment pure and simple'.[96] From the official point of view, however, the real complication was the extensive, highly complex division of rights among both new rent receivers and payers on the inam lands. By 1910, the Collector of Satara remarked, 'the origin of sub-tenancies was often hopelessly obscured'.[97] In the same era, government set out to resume control over many service inam lands, on the grounds that the original holders had typically long since sold off their rights and the services, the original condition of the grant, had hence been discontinued.[98] Yet the process of trying to rationalise service inams remained slow and extremely limited, as well as subject to legal challenge by the new holders.[99]

The case of alienated land demonstrates, in an extreme form, the intense demand for rights and land, generated by extending commer-

[96] *Bombay Presidency Land Revenue Administration Report, 1905–06*, Part 2, p. 66, Report by M. C. Gibb, Commissioner, Central Division.
[97] IOL, Bombay Land Revenue Progs., Vol. 8851, January 1911, No. 4, Report by A. R. Bonus, Collector of Satara, No. 4082, 20 June 1910.
[98] Examples of cases can be read in IOL, Bombay Land Revenue Progs., Vol. 8851, January 1911, No. 7; February 1911, No. 10; May 1911, No. 22.
[99] For a typically involved case, see IOL, Bombay Land Revenue Progs., Vol. 11844, May 1930, p. 43.

cialisation. In such conditions poor peasants too could undoubtedly participate in the market and even experience upward social mobility. Attwood has demonstrated that a thriving land market in an extensively irrigated village in the west Deccan could permit smallholders to 'get richer' by making a large number of land purchases.[100] Tenancy might well have operated in the same way, allowing peasants more subtle satisfactions of their land needs. Even in the classic Leninist case of late Tsarist Russia it is now recognised, in recent work, that the growth of tenancy could help to support, not oppress, the middle and poor peasantry.[101]

In western India, we have already met classic examples of renting-in entrepreneurs in the shape of the Saswad Malis. Originally they owned little land and their success in sugarcane cultivation on canal irrigated land was based on renting from local owners.[102] By the First World War period, many Malis were set to repeat the performance further afield, 'adding to the wealth of their community by leasing and cultivating lands under the new Godavari Canal in the Ahmednagar District'.[103] In some cases, notably within the talukdari areas, the position of tenant became regarded as a highly valuable right because landlords were unable to enforce realistic rent levels and, where this occurred, a fiercely competitive market, like that for inam land, often developed. On the Mehlol estate in Godhra taluk of the Panch Mahals District, brought under government management in 1884, competition for tenancies – even though the status was only tenant-at-will – became intense during the early twentieth century.[104] Much of the land on the estate was waste with considerable potential for development and the authorities maintained low rents to encourage expansion. As a result, large prices were paid to buy the tenancies by such as Kanbi Patidars from 'distant places like Charottar and Kanam' in Kaira District. One such tenant-at-will, in the 1920s, would spend Rs. 5,000 in installing an irrigation plant. Clearly he at least was hardly a victim of social decline.

In these ways, a much more flexible tenurial structure may have come into existence by the early twentieth century, accompanied, too, by associated changes in social attitudes. One settlement officer in 1914

100 D. W. Attwood, 'Why Some of the Poor Get Richer: Economic Change and Mobility in Rural Western India', *Current Anthropology,* 20, 3, September 1979, pp. 495–508.
101 See, e.g., Elvira M. Wilbur, 'Was Russian Agriculture Really That Impoverished? New Evidence from a Case Study from the "Impoverished Centre" at the End of the Nineteenth Century', *Journal of Economic History*, 43, 1, March 1983, pp. 137–44.
102 *SBG*, NS, No. 571, Settlement Report on Purandhar Taluk, Poona District, by R. D. Bell, 24 January 1916, para. 10.
103 Ibid.
104 See IOL, Bombay Land Revenue Progs., Vol. 11844, April 1930, p. 37.

'heard a lament on the decay of morality and self-respect in regard to debt. The fathers of the present generation thought it terrible to have to contract a debt of Rs. 100. . . . Now rayats think nothing of debts running over several hundreds of rupees and land is lightly bought and sold as if such dealings are of no consequence.'[105] But, perhaps, in terms of the peasant's real economic status, they were not. Many Bombay peasants by 1900 must have been owners on some land and tenants on other. As the 1911 Census summed up the situation, the landlord class 'is not marked off by any clearly defined limits from the tenant class, and even from the coolie class; a man receives rent from one person and pays rent to another'.[106] This flexibility, too, may have extended to the use of part-time labour work, for many peasants, by the early twentieth century, migrated to the towns and cities to find wage employment during the slack agricultural season. Investigating the Mahar caste in Saswad in 1912, Harold Mann found 112 women members of the caste, 101 children but only 60 men: 37 male Saswad Mahars were away working, mostly in Bombay City.[107] If, as some officials alleged, real wages for all forms of labour were rising in the early twentieth century,[108] this may have represented a deliberate, profit-maximising allocation of effort. Some even argued that cash earned in labour work formed capital for new land acquisition or at any rate to redeem mortgaged holdings. The Commissioner of the Central Division noted in 1912 the paradox that 'after a year of acute famine the area held by the agriculturists of Malsiras [taluk, Sholapur District] increased by 7,174 acres'.[109] The explanation, he claimed, lay in the ability of peasants, forced to labour work by famine conditions but then earning high wages, to purchase land and the desire of non-agriculturists, after the difficulties of the famine year, to sell any recently acquired lands quickly.

The growth of tenancy – and the expansion in part-time labouring – can, then, be explained without any assumption of social stratification. The Bombay peasantry, by the early twentieth century, may have been allocating its work and status flexibly according to which seemed to bring the best return at any one time. Different peasants would still make different decisions as to how much land to rent or how much time to spend in wage employment according to differing perceptions of the local situ-

---

105 *SBG*, NS, No. 560, Settlement Report on Badami Taluk, Bijapur District, by G. C. Shannon, 3 November 1914, para. 18.
106 *Census of India, 1911*, Vol. 7, Bombay, Part 1, Report, Ch. 12, para. 240.
107 Harold H. Mann, 'The Mahars of a Deccan Village', in Daniel Thorner, ed., *The Social Framework of Agriculture* (London, 1968), pp. 75–6.
108 This important issue will be considered in the next chapter.
109 *Bombay Presidency Land Revenue Administration Report, 1911–12*, Part 2, p. 48.

ation and changing individual circumstances and needs, but permanent polarisation within the peasantry would not be occurring. Alternatively, the growth in part-time labouring can be used as reinforcement for the 'peasant decline' thesis. Peasants may have been forced to resort to wage employment as a desperate response to a fall from proprietor to tenant status and, even if real wages could be proved to be rising, the dynamic may still be need, not peasant perception of the situation. How can we make the vital adjudication between these two interpretations? Clearly, both were, to some extent, operating. The credit system, as the case of the Ahmedabad talukdars back in the mid-nineteenth century had revealed, always carried a potential as an instrument of downward social pressure. Some undoubtedly did lose rights and status as a result of debt, and in so far as commercialisation increased the value of those rights, one would expect the process to have extended. Against this, we can demonstrate that others, like the Saswad Malis, consciously rented land as an economic decision. What we are searching for, then, is the predominant overall influence behind developments and what this reveals for the evolution of agrarian social structure.

Superficially, tenancy in Bombay Presidency came in in the wake of agricultural commercialisation for it was the least affected areas, like Bijapur District, where the development of tenurial change was most conspicuously slow and limited. Yet, if this initially suggests tenancy fundamentally as a flexible response to economic processes, we soon encounter difficulties. The growth of a tenantry was a striking phenomenon in most parts of western India by 1914, but the progress of commercialisation, as we saw, was more uneven and localised. Areas like the Deccan Riots region experienced tenurial change whilst the nature and methods of its agriculture appeared to be altering only very slowly. In overall terms, the precise correlation between tenancy and the spread of commercialisation seems slight.

But what was the nature of the relationship between landlord and tenant? In northern India, the active 'rental market' which Neale describes was stimulated by legislative protection for many tenancies following the Bengal Rent Act of 1859.[110] In Bombay Presidency, in contrast, the bulk of tenants – there were exceptions in talukdari and khoti regions and on inam land – enjoyed, throughout our period, no legal status or special protection. Men like the Saswad Malis, then, were making a calculated gamble in committing so much effort and investment to land which, legally, they cultivated only while the owner willed. As a result, much of

---

110 See Walter C. Neale, *Economic Change in Rural India: Land Tenure and Reform in Uttar Pradesh, 1800–1955* (New Haven and London, 1962), pp. 64–72.

the renting in may have occurred only where particular conditions underpinned the tenant's security. On the canal lands, for example, output may have been so high that tenants could offer satisfactory, even improving, terms to the landlord whilst still securing to themselves the lion's share of returns: it is significant that Attwood's case of the poor 'getting richer' occurs in an extensively irrigated village. Again, the case of the Mehlol estate, quoted earlier, was untypical, for the fierce competition for tenancies developed only after the estate came under government management and the new tenants then had grounds for assuming that the authorities, anxious to promote the rehabilitation of the estate's finances, would prove benevolent and helpful landlords. Nevertheless, their security was continually threatened by some official attitudes, resentful of the dispossession of the traditional tenants: in 1926 a temporary change of policy produced refusal by the local authorities to recognise transfer of tenancies and some evictions of the newcomers.[111] Some of the behaviour of the Saswad Malis, too, suggests that they would not ideally have chosen their role as tenant entrepreneurs. In the early twentieth century they committed massive sums of cash to land purchase, on one occasion paying Rs. 12,000 for just 21½ acres at Khalad, a few miles east of Saswad.[112] In some such cases, inflation in land values may have been the product of a desperate quest for the much greater security guaranteed by landownership as against tenancy.

Even without legal protection, however, many tenants may still have been able to secure a relationship with their landlord favourable to themselves. Musgrave argues that in the United Provinces tenants were often, in practice, the effective 'lords of the land'[113] and the realities of local power which underlay this situation sometimes existed in western India too. Social constraints invariably inhibited the landlord's freedom of action: 'landlords are as much in need of tenants as the latter are for the former. In many cases the landlord dares not become too exacting for fear of having no one to cultivate his lands.'[114] In particular, as in Musgrave's United Provinces, any absentee landlord or one, for any reason, lacking effective local knowledge could be frustrated by tenant obstructionism. This often applied to incomer holders of inam rights. Rights and obligations on the alienated lands were a continual source of dispute and ten-

---

111 IOL, Bombay Land Revenue Progs., Vol. 11844, April 1930, p. 37.
112 *SBG*, NS, No. 571, Settlement Report on Purandhar Taluk, Poona District, by R. D. Bell, 24 January 1916, para. 21.
113 P. J. Musgrave, 'Landlords and Lords of the Land: Estate Management and Social Control in Uttar Pradesh, 1860–1920', *Modern Asian Studies*, 6, 3, July 1972, pp. 257–75.
114 *Bombay Presidency Land Revenue Administration Report, 1908–09*, Part 2, p. 68, Report by J. B. Vachha, Deputy Collector of Thana.

sion – as early as 1882 Lee-Warner spoke, in Satara District, of 'the chronic contest between landlord and tenant in inam villages'[115] – but by the First World War era official opinion was that 'as a general rule . . . inamdars and tenants are a fair match for one another'.[116] The position of many inamdars was, by now, fatally weakened by excessive divisions of rights between numerous sharers, which created internal dissension, ill-organisation and an inability to present a united front to the tenants. In addition, absentee right-holders relied entirely on the village officers to collect their dues and, clearly, this often led to arrangements with cultivators to reduce and withhold collections.[117] Constant resort by inamdars to legal action – in 1906–7 inamdars in Satara District alone had to file as many as 6,669 suits for official assistance to collect their dues[118] – suggested that tenants were capable of resisting demands and hence, presumably, securing an acceptable economic position for themselves.

Yet it may only have been some tenants on inam land who were capable of such action. Rack-renting was also a charge levelled at absentee inamdars and resistance may have been the prerogative of an elite of tenants whilst others faced severe oppression.[119] Tenants on inam lands faced some special difficulties, particularly in years of poor harvests. From 1896 proprietors on ryotwari holdings enjoyed suspensions and remissions of land revenue in bad seasons, but many inamdars were loath to sacrifice their dues by making equivalent grants to their tenants. It is significant that the years of intense legal action, like 1906–7 mentioned earlier, were often scarcity or famine years. Hence, by resisting payment in such conditions, tenants of inamdars may have been, not aggressively attempting to enhance their economic position, but desperately trying to mitigate serious hardship. The authorities sought to deal with the problems of the inam lands by attaining greater information. The 1879 Land Revenue Code introduced a new procedure, permitting, if the inamdars wished, full revenue survey, including measurement and classification, of inam lands.[120] Inamdars were offered the incentive, to allow the survey

115 IOL, Lee-Warner Papers, Deccan Ryot Series No. 1, W. Lee-Warner, Notes on the Bombay Land Revenue System, 16 December 1882, para. 5.
116 *Bombay Presidency Land Revenue Administration Report, 1914–15*, Part 2, p. 54, Report by the Commissioner, Central Division.
117 See *Bombay Presidency Land Revenue Administration Report, 1907–08*, Part 2, p. 59, Report by K. J. Agashe, Deputy Collector of Satara, where the allegation that village officers 'wilfully neglect to collect inamdars' dues' is discussed.
118 *Bombay Presidency Land Revenue Administration Report, 1906–07*, Part 2, p. 68, Report by S. R. Arthur, Collector of Satara.
119 For some cases which appear to indicate this, see IOL, Bombay Land Revenue Progs., Vol. 8851, July 1911, No. 30.
120 For details, see R. G. Gordon, *The Bombay Survey and Settlement Manual* (Bombay, 1917), Vol. 2, Part 3, Appendix 5, p. 538.

in, of more effective legal redress against recalcitrant tenants and, in return, tenants were given specific rights: in particular, from 1907 inamdars were obliged to allow remissions and suspensions of land revenue on the same scale as in ryotwari villages.[121] The new official involvement on the ground in inam villages may, in this way, have eased many tenants' positions but survey remained an optional measure for the inamdar, possibly resorted to in villages where, in any case, the balance favoured the tenants. Many inam villages were still entirely unsurveyed at the outbreak of the First World War.

The official aim which regarded the extension of the revenue survey to inam villages as 'an object to be attained wherever possible'[122] was part of a general policy, in the late nineteenth and early twentieth centuries, of mitigating the circumstances of tenants of the formal overlord tenures. The concessions of the mid-nineteenth century, if they could not be undone, were now at least widely regretted. In 1904 a new Khoti Settlement Act provided machinery for Ratnagiri khoti tenants to commute their traditional crop share to cash rents, at new fixed scales to be agreed under official supervision. Officials claimed that this had produced, by the 1910s, much more reasonable rent levels, though it is hard to adjudicate on the validity of this. Yet any such rights secured for khoti, talukdari and inam tenants contrasted with the lack of official recognition of tenancy on ordinary ryotwari lands. Even if the inamdar after 1907 granted automatic remission and suspension of rent to his tenants – and it must have been easy, in practice, to evade the regulation – the landlord on the formally ryotwari lands was under no such obligation. Whilst his own land revenue payment would be lowered in a season of famine or scarcity, the rent he received would be still subject to the normal practical laws of what he could obtain.

These, in turn, must have been considerably influenced by the economic conditions of production. If, for example, land had been universally extensive relative to population, landlords would presumably have been forced to moderate demands to keep their tenants. Throughout western India, however, at any rate at the population peak of 1896, per capita land availability seemed to be coming under threat. There is evidence that where land was scarce and population pressure particularly acute, as in parts of the Deccan and the Konkan, the tenant's position was especially weak. Thus, for all the attempts to moderate the effects of the khoti system, investigations by the 1920s suggested that the tenant's relative economic and social status had worsened over the previous half century and many tenants, given occupancy rights at the compromise of

---

121 Ibid., p. 539.   122 Ibid., p. 538.

1880, had apparently lost them.[123] Khoti tenants in Ratnagiri had 'to pay to the khot one-half of the gross produce and in some cases more' and 'have to perform all kind of labour that is demanded ... on a pittance of one anna and a meal a day'.[124] They were apparently forced to accept these conditions because 'they said there was no room for them anywhere, that all landlords had their full complement of tenants'.[125] Similar conditions affected tenants in the non-khoti areas of the Konkan. In parts of Kolaba District, typical rents by the 1920s amounted to a half or even up to three-quarters of the harvest and 'the practice of demanding straw as well as grain has extended, with the effect of raising rents in practice'.[126]

These remarks, however, bring us closer to the nub of the tenancy issue. We can most easily and effectively judge the balance of economic power between landlord and tenant by investigating the character and levels of rent payment. In what forms, then, was rent rendered? In a few areas, notably in the south Deccan and Southern Maratha Country, tenants paid cash rents. In Tasgaon taluk of Satara District in 1887, for example, over two-thirds of tenants were recorded as paying a money rent[127] and, again, in Belgaum taluk in 1918 'leases ... are usually for cash in the case of dry crop lands'.[128] Elsewhere, however, crop share rents were overwhelmingly the norm. In the Konkan, as we have seen, rents were typically shares of the harvest in the 1920s and this was a long-established situation. In Alibag taluk of Kolaba in 1893, for example, 1.4 per cent of holdings were being cultivated by tenants paying a money rent, but 52.3 per cent by those rendering grain.[129] Throughout Poona District 'tenants usually pay their rents as a share of the produce',[130] and the same practice seems to have predominated even in the more commercialised areas of the central and northern regions of Bombay Presidency.

---

123 'The statement of the Gazetteer that 90 per cent of the tenants have occupancy rights cannot be accepted today. In 45 villages of Dapoli, I find there are 3,770 permanent tenants and 1,174 upri tenants or 24 per cent.' IOL, Bombay Land Revenue Progs., Vol. 11540, July 1926, p. 405, para. 8.
124 Ibid., para. 4.
125 Ibid., para. 7.
126 *SBG*, NS, No. 624, Settlement Report on Pen Taluk, Kolaba District, by J. R. Hood, 31 October 1923, para. 21.
127 *SBG*, NS, No. 204, Settlement Report on Tasgaon Taluk, Satara District, by H. K. Disney, 26 March 1887, Appendix I.
128 *SBG*, NS, No. 584, Settlement Report on Belgaum Taluk, Belgaum District, by M. Webb, 7 January 1918, para. 16.
129 *SBG*, NS, No. 290, Settlement Report on Alibag Taluk, Kolaba District, by C. W. Godfrey, 31 October 1893, Appendix I.
130 *SBG*, NS, No. 571, Settlement Report on Purandhar Taluk, Poona District, by R. D. Bell, 24 January 1916, para. 20.

Thus in Dholka taluk of Ahmedabad, a thriving cotton-growing area of Gujarat, 'cash rents cannot be said to be common' as late as 1920.[131] Sixteen years later, a Congress committee made the identical remark for Maharashtra as a whole,[132] and this characteristic of western Indian rent apparently remained stable into modern times. Immediately after Independence, it was estimated, 63.5 per cent of rents across the Bombay Presidency were paid entirely as crop share and a further 7.2 per cent involved some grain component.[133]

This is a striking phenomenon, for across northern India money rents apparently developed as early as Akbar's reign[134] and by 1900 kind payment would have been unusual in most United Provinces districts. Crop rents, however, in terms of their widest significance, are no automatic evidence of backwardness or lack of commercialisation for they were the norm, for example, throughout Meiji Japan.[135] Their significance, arguably, is for what they indicate about the balance of rural power. In Japan, it is sometimes suggested, landlords secured a substantial transfer of income in their favour by selling crop rents on the market at appreciating prices whilst the burden of the land tax which they paid to the government was steadily declining in real terms. It would be over-simple to assume that this is what happened in western India. Not only have we yet to demonstrate any fall in the real burden of the land revenue, but prices were rising only steadily between 1880 and 1896. Nevertheless, even in an entirely stable overall price situation, there were advantages for the large receiver of grain who could afford to time his sales, for price fluctuations within any year were considerable. Any landlord who received, as some did, 'a definite amount of grain'[136] should have been obtaining an appreciating asset.

Most Bombay rents, though, were shares of the crop. Share rents had the advantage for the landlord that they secured him an automatic stake in increasing agricultural output as well as in rising prices. Only in famine years was there danger of diminished advantage from the system and even then the scale of price increase might compensate for absolute

---

131 *SBG*, NS, No. 607, Settlement Report on Dholka Taluk, Ahmedabad District, by J. F. B. Hartshorne, November 1920, para. 106.
132 *Report of the Peasant Enquiry Committee of the Maharashtra Provincial Congress Committee*, p. 48.
133 Dandekar and Khudanpur, *The Working of the Bombay Tenancy Act, 1948. A Report of Investigation*, p. 112.
134 Neale, *Economic Change in Rural India: Land Tenure and Reform in Uttar Pradesh, 1800–1955*, p. 65.
135 See R. P. Dore, *Land Reform in Japan* (London, 1959), pp. 42–3.
136 *SBG*, NS, No. 584, Settlement Report on Belgaum Taluk, Belgaum District, by M. Webb, 7 January 1918, para. 16.

shortfall in amounts of grain received. One half of the harvest is generally recorded as the norm throughout the Deccan[137] and 'the old recognised proportion' in the Konkan, too.[138] In practice, however, as we saw earlier, population pressure creating competition for tenancies enabled landlords to force up rent proportions in the south Konkan and even as early as 1899 the landlord's share was described as 'being as much as 60 to 70 per cent in many cases'.[139] Making any more precise statement about a landlord's 'profit' on leasing out land is, of course, a forlorn task in the absence of reliable output statistics. Some settlement officers, however, attempted rough assessments of rates of return by comparison of average land price and rent levels relative to the revenue assessment. R. D. Bell, working in Poona District in the 1910s, found that 'on average the landlord, after paying the Government assessment, had a return of 5.3 per cent on his invested capital' in Maval taluk[140] and that '5 per cent on the purchase price was generally considered a fair return' in Sirur taluk.[141] Typically, this meant land selling at around forty to fifty times its land revenue and rent charges averaging the equivalent of three to four times the assessment.[142] In other areas, notably, as we might expect, in the south Konkan, this type of calculation suggested approximate rates of return on leasing land of up towards 10 per cent per annum.[143]

These rates were considerably below most arithmetical assessments of interest rates on lending money but land boasted many other advantages as an investment, besides the social prestige which ownership conferred: 'the chief are', Bell commented, 'the absolute security of the investment and the tendency of land to appreciate in value'.[144] Yet, in any case, such calculations may be highly misleading, since most tenancies imposed

137 See, e.g., SBG, NS, No. 577, Settlement Report on Haveli Taluk, Poona District, by R.D. Bell, 7 March 1916, para. 23; SBG, NS, No. 571, Settlement Report on Purandhar Taluk, Poona District, by R. D. Bell, 24 January 1916, para. 20.
138 MARD, Vol. 10 of 1902, No. 331, Acting Superintendent, Land Records, Southern Division to J. Muir-Mackenzie, Acting Secretary to Bombay Government, Revenue Department, 19 February 1899, para. 2.
139 Ibid.
140 SBG, NS, No. 565, Settlement Report on Maval Taluk, Poona District, by R. D. Bell, 26 June 1916, para. 23.
141 SBG, NS, No. 558, Settlement Report on Sirur Taluk, Poona District, by R. D. Bell, 25 September 1915, para. 23.
142 This was exactly the situation in Maval taluk.
143 E.g., the situation in Alibag taluk, Kolaba District, in the 1890s, where land sales averaged 43.6 times the assessment and rents as much as 4.5 times the revenue charge, would indicate, on Bell's technique, a rate of return of just over 8%. SBG, NS, No. 290, Settlement Report on Alibag Taluk, Kolaba District, by C. W. Godfrey, 31 October 1893, Appendices K, L.
144 SBG, NS, No. 558, Settlement Report on Sirur Taluk, Poona District, by R. D. Bell, 25 September 1915, para. 23.

'from above' seem to have been the product of the operation of the credit system rather than land purchase on the market. What needs to be compared in these cases is the regular rent received (minus the land revenue charge) with probable income from the earlier supply of credit. In the change, there may have been firm advantages of formalisation and regularity for the new landlord.

The evidence of the growth of tenancy in western India can, then, superficially be taken in a number of ways, some indicative of social stability as much as stratification. However, the typical character of the landlord–tenant relationship, in particular the prevalence of kind rents, suggests that the landlord was probably the usual protagonist of the evolving pattern of change. Tenancy as an act of expansionary entre-preneurship undoubtedly did occur, but typically needed to rest on cash rents if the landlord was not to secure a major proportion of peasant initiative. Tenants were probably most assertive and successful on the canal irrigated lands and on holdings, usually talukdari or inam, subject to traditionally weak landlord authority. In general, though, the steady expansion of tenancy seems to underpin the case for social stratification occurring during the late nineteenth century. Nevertheless, this need not have been a universal trend throughout British India. The balance within the Bombay tenancy relationship, it should be noted, may have rested on particular conditions, notably the absence of any legal status for most tenants and the emerging difficulties of land availability throughout most of western India. Where these features were lacking, as they were in parts of northern India, the landlord's advantages and opportunities may have been decisively diminished and the opportunity for social stratification here – as our north India specialists seem to believe[145] – consequently lessened.

Identifying the rich peasantry

Until, however, we can identify some of the beneficiaries of the economic and social processes, any claims for the existence of social stratification remain, to some extent, insubstantial. On our analysis, of course, the rich peasantry is necessarily a disparate entity: indeed, that very term has long been highly misleading if it is assumed to denote a precise social class. Over western India, too, any simple classification by size of holding or

---

145 In the north, both Kessinger and Dewey for the Punjab and Stokes for the U.P. have directly questioned the thesis of social stratification. See Kessinger, 'The Peasant Farm in North India, 1848–1968', pp. 303–23; Dewey, 'Social Mobility and Social Stratification amongst the Punjab Peasantry: Some Hypotheses'; Stokes, 'The Emergence of the Rich Peasant under Colonial Rule: Some Northern Lights'.

some other asset means little.¹⁴⁶ Certainly the economic elite tended to own more resources like carts, wells and, probably, land but their position, arguably, most depended on qualitative functions; access to credit and, through modern communications, wider markets and the power to turn debt and tenancy relationships with others to their advantage. Unexpected groups, as the case of the Saswad Malis showed, might achieve this. It is easiest, therefore, to examine in turn major social groups and determine how far they each benefited from processes of change.

Firstly, then, how prominent was the alien, 'professional' moneylender? The credit relationship, as we saw, seems after 1880 a significant source of tenurial change and, even if agriculturist creditors were the main protagonists, we might expect the Marwari and Gujarati lenders, particularly in the Deccan, to have benefited heavily. Certainly nineteenth-century Bombay officialdom thought so. Evidence from the Deccan Act registrations suggested that land transactions between agriculturists and 'sowkars' sometimes even outnumbered those within agriculturist groups.¹⁴⁷ Thus the Government of India's land transfer survey of 1895 named Bombay as one province where 'the transfers to the money-lending and non-cultivating classes are serious and are progressing'¹⁴⁸ and the Bombay government itself concluded in 1899 that land transfer to alien moneylenders represented 'a political danger'.¹⁴⁹

As we saw with the Deccan Riots, however, such official fears can prove insubstantial on closer investigation. We need, too, to examine actual extent of landownership rather than volume of land transactions, for moneylenders may frequently have acquired land only temporarily, dispersing it in deals with agriculturists less open to formal notice. The first investigation of any reliability on landownership was that generated by Elgin's land transfer enquiry of 1895. Its major statistical conclusions for Bombay Presidency are presented in Table 4. Superficially, this sug-

---

146 Some attempts have been made to base categorisation of peasant status on size of holding and this may be useful for some provinces without the ecological variations which existed within Bombay Presidency. See, for example, T. J. Byres, 'Land Reform, Industrialization and the Marketed Surplus in India: an Essay on the Power of Rural Bias', in David Lehmann, ed., *Agrarian Reform and Agrarian Reformism. Studies of Peru, Chile, China and India* (London, 1974), p. 233.
147 They did, for example, every year between 1892 and 1896. *Bombay Presidency Registration Department Report, 1895–96*, para. 49.
148 IOL, Government of India Revenue Progs., October 1895, No. 73, Land Transfer Note, p. 81.
149 IOL, Bombay Confidential Revenue Progs., Vol. 5777, October 1899, No. 15, J. W. P. Muir-Mackenzie, Secretary to the Bombay Government, to the Secretary to Government of India, Dept. of Revenue and Agriculture, No. 7100–168, 7 October 1899, para. 8.

Table 4 *Estimated landownership by 'non-agriculturists': the land transfer enquiry, late 1890s*

District	Total of occupied government land investigated (acres)	In possession of 'non-agriculturist moneylenders' (acres)	Percentage held by 'non-agriculturist moneylenders'
Ahmedabad	237,833	39,633	16.7
Kaira	79,902	5,671	7.1
Panch Mahals	33,135	4,737	14.3
Broach	95,797	16,042	16.7
Surat	150,677	21,398	14.2
Thana	30,206	5,098	16.9
Total, Northern Division	627,550	92,579	14.8
Khandesh	307,402	32,875	10.7
Nasik	1,099,940	200,162	18.2
Ahmednagar	2,961,714	822,856	27.8
Poona	722,827	108,437	15.0
Sholapur	263,400	20,431	7.8
Satara	286,777	13,447	4.7
Total, Central Division	5,642,060	1,198,208	21.2
Belgaum	287,477	16,617	5.8
Bijapur	561,374	16,048	2.9
Kolaba	191,040	37,621	19.7
Ratnagiri	406,376	23,626	5.8
Kanara	193,312	12,314	6.4
Total, Southern Division	1,639,579	106,226	6.5

N.B. No figures for Dharwar District exist.
Source: IOL, Bombay Confidential Revenue Progs., Vol. 5777, October 1899, No. 15.

gested the somewhat alarming conclusion that non-agriculturists had acquired between a fifth and a third of the land in most regions of Bombay Presidency, but the impression of extensive change needs to be qualified. Firstly, these figures were for all non-agriculturist groups. Although moneylenders seem to form the major component in Gujarat and most of the Deccan, throughout the south, particularly in Satara District, most 'non-agriculturist landowners' were clearly indigenous inamdars, many of whom had probably owned their land for generations. Again, the

In possession of 'other non-agriculturist landlords' (acres)	Total held by all non-agriculturists (acres)	Percentage held by all non-agriculturists
7,440	47,073	19.8
2,377	8,048	10.1
847	5,584	16.9
9,583	25,625	26.7
2,999	24,397	16.2
11,673	16,771	55.5
34,919	127,498	20.3
27,031	59,906	19.5
46,006	246,168	22.4
214,452	1,037,308	35.0
68,085	176,522	24.4
36,796	57,227	21.7
22,512	35,959	12.5
414,882	1,613,090	28.6
61,376	77,993	27.1
50,264	66,312	11.8
48,428	86,049	45.0
132,022	155,648	38.3
44,287	56,661	29.3
336,377	442,663	27.0

overall statistics are somewhat distorted by two districts, Thana and Kolaba, where non-agriculturists apparently owned as much as 55.5 and 45 per cent of the land respectively. As in Satara, a minority of this was held by 'moneylenders' but, even so, Thana and Kolaba were special cases influenced by their close proximity to Bombay City and consequent occasional large investments in land by merchants and industrialists. Elsewhere, this investigation seemed to suggest that non-agriculturist moneylenders owned around 15 per cent of the Bombay land; more in the north and central Deccan and parts of Gujarat, less in the south.

This would appear, nonetheless, to involve some considerable advance on the Deccan Riots situation – perhaps a doubling of sowkar ownership in twenty years – because Poona and Ahmednagar were two districts where non-agriculturist holdings were now classified as particularly extensive. The land transfer enquiry, though, was far from a comprehensive survey. After 1910 the compilation of a full record of the rights in ryotwari land[150] slowly produced the first complete picture of the pattern of landownership. Its findings, as Tables 5 and 6 show, were much less disturbing than the land transfer enquiry. All non-agriculturists, not just moneylenders, were recorded as owning around 15 per cent of the land of the Deccan and there were no especially large district deviations from this norm. The completed record of rights survey of 1917 suggested that the proportion over the Presidency as a whole was even lower: just 11.5 per cent of the land, as Table 6 shows, owned by all non-agriculturists. This provided a sharp contrast from the 1890s enquiry in some individual cases, non-agriculturist holdings in Ahmednagar District, for example, apparently declining from 35 to only 13 per cent of the total. How can we explain this? It is quite possible that agriculturists, in overall terms, were gaining land from non-agriculturists during the 1910s.[151] Alternatively, and more likely, the statistics of the land transfer enquiry may have been highly misleading. Unlike the record of rights survey, the earlier enquiry only investigated sample areas – on average around a third of the total – which may or may not have been representative. Most important, however, was probably the coincidence of the land transfer enquiry with the famine era of the late 1890s. Many peasants, under pressure, may have mortgaged or even sold land which was afterwards recoverable and certainly we might expect holdings by all types of moneylender to be at their peak in the famine conditions. The Bombay government was probably right to describe, in 1911, the fears aroused by the land transfer enquiry as 'greatly exaggerated'.[152]

In this way, non-agriculturist moneylenders may have increased their proportion of the cultivated land from around 6 to about 10 per cent in the typical Bombay district in the thirty years after 1875, but they can, in no sense, be seen as the major beneficiaries of economic and social processes. As in the 1870s, their lack of familiarity with agricultural conditions and cultivation procedure probably undercut much economic advantage from the formal acquisition of land. This was particularly so in the poorer and less commercialised areas. There was simply little point in

---

150 The genesis of this will be discussed in Chapter 8.
151 The Bombay government suggested this in 1911. See IOL, Government of India Revenue Progs., April 1913, No. 1, G. Carmichael, Chief Secretary to the Bombay Government, to the Secretary to the Government of India, Dept. of Revenue and Agriculture, No. 831, 28 January 1911, para. 4.
152 Ibid.

Identifying the rich peasantry

Table 5 *Estimated landownership by 'non-agriculturists': the record of rights enquiry in the Deccan, 1910s*

District	Total area of occupied government land (acres)	Total area held by all non-agriculturists (acres)	Percentage held by all non-agriculturists
Sholapur	2,427,644	211,130	8.7
East Khandesh	2,088,000	380,475	18.2
West Khandesh	1,584,089	239,272	15.1
Ahmednagar	1,869,925	242,892	13.0
Satara	1,617,741	189,541	11.7
Poona	2,433,448	451,996	18.6
Nasik	2,362,366	451,217	19.1
Total	14,383,213	2,166,523	15.1

Source: *Bombay Presidency Land Revenue Administration Report, 1911–12*, Part 2, p. 48.

Table 6 *Estimated landownership by 'non-agriculturists': the completed record of rights enquiry, 1917*

Division	Total area of occupied government land (acres)	Total area held by all non-agriculturists (acres)	Percentage held by all non-agriculturists
Northern Division	3,193,557	720,369	22.6
Central Division	13,302,019	1,574,583	11.8
Southern Division	6,221,824	312,280	5.0
Total, whole Presidency	22,717,400	2,607,232	11.5

Source: *Report of the Bombay Provincial Banking Enquiry Committee, 1929–30*, Statement II, para. 60.

securing title to land which lacked much value. The land transfer enquiry, as Table 4 shows, revealed that only 2.9 per cent of Bijapur District was owned by non-agriculturist moneylenders: 'from its former inaccessibility and consequent backwardness, the transfer of lands has lagged in comparison to other districts'.[153] Within the Deccan, Sholapur

153 IOL, Bombay Confidential Revenue Progs., Vol. 5777, October 1899, No. 15, J. W. P. Muir-Mackenzie to the Secretary to the Government of India, Dept. of Revenue and Agriculture, No. 7100–168, 7 October 1899, para. 9.

District, again in the famine belt east, was, as Table 5 suggests, almost certainly the district with the smallest proportion of non-agriculturist ownership. The same phenomenon is observable at a lower geographical level. In Poona District one of the smallest proportions of landownership by alien moneylenders was in Indapur, the poorest and least climatically favoured of the Poona taluks.[154] Here 'sowkars have not acquired much of the land, probably owing to its general low value and it being a poor investment'.[155] Within Ahmednagar District, too, 'the percentages of land actually entered in the names of moneylenders is very small in out-of-the-way villages with poor land'.[156]

This might suggest that alien moneylenders held most of their land, however small in total, in the more advanced areas, playing a consequently important role as pioneers of commercialisation processes. Certainly a concentration phenomenon is evident: though moneylenders held little land overall, those who did often owned markedly larger than average farms. Thus in Maval taluk of Poona during the 1910s non-agriculturist moneylenders formed only 9 per cent of landowners but they held 14 per cent of the land[157] and in Sirur taluk, at the other end of the district, 8 of the 11 landowners holding over 500 acres of land were apparently non-agriculturists.[158] Yet such concentration, as here in Poona, was rarely linked to areas of most advanced agriculture and extensive commercialisation. Indeed, the district in each administrative division which was probably the most commercialised – Kaira in the north, Satara in the central and Belgaum in the south – was clearly marked, in the land transfer enquiry, by a strikingly low proportion of ownership by non-agriculturist moneylenders.[159] Again, we can also illustrate this situation within the districts. In Ahmedabad District, with a recorded total for non-agriculturist holdings of 19.8 per cent, the proportion was as low as 4 to 6 per cent in the central, cotton-growing taluk of Daskroi and as high as 32.8 per cent in

---

154 See IOL, Bombay Confidential Revenue Progs., Vol. 5777, October 1899, No. 15, Breakdown for Poona District.
155 *SBG*, NS, No. 516, Papers on the Modification of Rates in Indapur Taluk, Poona District, 1904–10, J. P. Brander, Assistant Collector, to the Collector of Poona, No. 2070, 10 November 1909, para. 13.
156 IOL, Bombay Confidential Revenue Progs., Vol. 5777, October 1899, No. 15, F. L. Charles, Collector of Ahmednagar, to J. K. Spence, Commissioner, Central Division, No. 2027, 30 March 1896.
157 *SBG*, NS, No. 565, Settlement Report on Maval Taluk, Poona District, by R. D. Bell, 26 June 1916, para. 21.
158 *SBG*, NS, No. 558, Settlement Report on Sirur Taluk, Poona District, by R. D. Bell, 25 September 1915, para. 20.
159 See Table 4. Dharwar rather than Belgaum was perhaps the most commercialised district in the southern division but this was the one district for which no statistics were collected.

the peripheral, much less commercialised, Parantij taluk.[160] Within Kaira District the areas with the highest proportions of cash-cropping were the south-eastern taluks of Borsad and Anand, the tobacco tracts, but here, too, recorded ownership by non-agriculturists was notably lower than average.[161] Examining the Deccan, there seems little evidence that aliens were ever significant purchasers of canal land. Indeed, within Purandhar taluk, where irrigation was most extensive, the record of rights survey reported only 5 per cent of the land as the possession of non-agriculturist moneylenders.[162]

In general, then, specific local patterns of landownership indicate that agriculturist groups controlled the evolution of commercial agriculture even more thoroughly than the broad structure of their landholdings would suggest. In the localities of most extensive cotton, sugar and tobacco cultivation, alien groups evidently found it extremely difficult to break into landownership. Their holdings, in practice, were determined by ease of availability and access rather than unfettered choice according to economic value. Thus, ownership by non-agriculturist moneylenders was always greatest in or near the market centre or taluk headquarters: this was the local reflection of the phenomenon whereby the districts adjoining Bombay City were those with the highest proportion of newcomers on the land. Again, inam land was a more accessible object for most alien moneylenders for 'inam land in a Government village is often owned originally by a non-agriculturist, who is more ready to part with it than a tiller of the soil would be'.[163] Any progressive trend in the ownership of inam land is very difficult to identify precisely because non-agriculturist holdings had always been so relatively extensive. Yet it seems likely that alien moneylenders were gaining somewhat on other non-agriculturist groups on the inam lands.

This, of course, has to be set against our earlier remarks on the issue of landownership, for our statistics referred only to ordinary ryotwari land. Hence the Commissioner of the Northern Division commented of the land transfer enquiry that 'if inami land had been included the results would have been still more alarming'.[164] Nevertheless, the thrust of the

---

160 IOL, Bombay Confidential Revenue Progs., Vol. 5777, October 1899, No. 15, Breakdown for Ahmedabad District.
161 It was 10% of the cultivated acreage in Borsad, only 5% in Anand. Ibid., Breakdown for Kaira District.
162 *SBG*, NS, No. 571, Settlement Report on Purandhar Taluk, Poona District, by R. D. Bell, 24 January 1916, para. 20.
163 IOL, Bombay Confidential Revenue Progs., Vol. 5777, October 1899, No. 15, M. C. Gibb, Collector of Ahmedabad, to F. S. P. Lely, Commissioner, Northern Division, No. 15/1631, 2 June 1897, para. 9.
164 Ibid., F. S. B. Lely to the Survey Commissioner, Bombay, No. 625/14, 18 February 1898, para. 4.

argument would only need to be qualified if the inam lands played a significant role in the commercialisation process. This raises an interesting and often overlooked issue: the reaction of different tenurial systems to the potential for economic change. Officialdom had a firm view that all the areas of overlord and alienated tenure tended to lag in economic performance. As early as 1880, Pedder claimed that 'on the whole, under the khoti tenure, as compared with the very poorest ryotwari or peasant proprietary districts, the trading, manufacturing and artisan proportion of the population is smaller; ... it is generally poorer, and ... comprises fewer well-to-do people'.[165] This, of course, may be simple ryotwari prejudice, though Pedder, as we saw earlier, was far from a utilitarian ideologue and his claims were supported by detailed evidence from licence tax returns.[166] Thereafter, it did seem that the areas where overlord and alienated tenure were most extensive rarely formed the commercial pacemakers. In Ahmedabad District, for example, Gogha taluk, as we have seen, was one talukdari area which conspicuously lagged. Viramgam taluk of the same district provides a marked contrast within itself. The settlement report of the 1890s on the ryotwari villages reported substantial expansion in population, cultivation and cart and well ownership[167] but in the talukdari villages numbers of carts and ploughs actually fell between 1865 and 1892.[168] This, it might be argued, was the type of geographical diversification, based on differential access to communications and ecological capability which we have seen throughout western India. Yet, ecologically, the very existence of overlord tenures here presumably rested on some traditional creation of an assured agricultural surplus. Gogha taluk, in the far south of the district, was remote, but the average twenty miles from the Viramgam talukdari villages to the nearest railhead formed no insurmountable barrier to wider production for the market.

The problem of the talukdari areas – and this was shared in khoti and inam regions – was, arguably, that no one group managed to establish a position of undisputed control over potential production for the market. The overlords were weakened, from an early stage, by excessive division of the rights. Small sharers, often absentee, were as a result little able to organise and manipulate agricultural processes. The position of their

165 *PP*, 1881, Vol. 71, Part 2, Report of the Indian Famine Commission 1880, Appendix 3, Ch. 1, Question 10, Statement by W. G. Pedder.
166 With roughly comparable populations, Ratnagiri District produced only a third the amount of licence tax as Poona District in 1872. Ibid.
167 See *SBG*, NS, No. 242, Settlement Report on Viramgam Taluk, Ahmedabad District, by T. R. Fernandez, 13 February 1890.
168 *SBG*, NS, No. 303, Settlement Report on the Talukdari Villages of Viramgam Taluk, Ahmedabad District, by H. Holland, 24 January 1895, para. 15.

tenantry, too, eventually received some official support through such policies as the extension of the revenue survey to inam villages and the Khoti Settlement Act of 1904. Even so, amongst the tenants an assertive elite rarely emerged and cases like the Mehlol estate remained exceptions in Bombay Presidency. For all the weakness of many of the landlords, tenants were largely unable to manipulate the switch to cash rents on which tenant entrepreneurship depended. In 1889, in the talukdari areas, 'the system still universally prevailing except in a few rare instances, is that known as "Bhagbatai", under which the landlord recovers his rent in the form of a fixed proportion of the actual produce'.[169] Tenants in many areas, particularly the khoti region, were themselves weakened by pressure on rights and the limited availability of alternative land or employment. The official judgement that inamdars and tenants were a 'fair match' for one another reflected a situation, on many of the non-ryotwari lands, which inhibited the successful development of any 'rich peasantry'.

The quest, then, needs to concentrate on the ryotwari lands as well as on traditionally agriculturist groups. In general, in western India, the beneficiaries of economic and social change seem to have come from groups with an established prominent position on the village land. Often their economic position was underpinned by village office holding, possibly an important aid to factors like easier access to credit. This was true of the Kanbis of Ahmedabad District who already in 1825 'raise the richest produce' and enjoyed 'hereditary rights in their own villages, with peculiar privileges'.[170] The settlement reports of the 1880s and 1890s which catalogued the highly differential spread of commercialisation in the district recorded the Kanbis as the pioneers of the process. 'They live in comfort. Their houses are two-storeyed substantial brick edifices. If money is wanted for a local improvement, they can produce it.'[171] By the late nineteenth century, the more ambitious of the group were beginning to seek openings outside agriculture and 'a good many sons of well-to-do Kanbis learn English in the Ahmedabad Government and Mission high schools'.[172] Equally ambitious and more exclusive were their fellows in

---

169 IOL, Government of India Revenue Progs., June 1890, No. 50, A. D. Younghusband, Talukdari Settlement Officer, to the Chief Secretary to Bombay Government, No. 776, 27 November 1889, para. 2.
170 *SBG*, No. 10, Report on the Portion of the Daskroi Purgunna Situated in Ahmedabad District by Capt. J. Cruickshank, 1825, para. 56.
171 IOL, Bombay Confidential Land Revenue Progs., Vol. 5777, October 1899, No. 15, M. C. Gibb, Collector of Ahmedabad, to F. S. P. Lely, Commissioner, Northern Division, No. 15/1631, 2 June 1897, para. 5.
172 *SBG*, NS, No. 214, Settlement Report on Daskroi Taluk, Ahmedabad District, by T. R. Fernandez, 20 February 1889, para. 34.

Kaira District, the Lewa Kanbis or Patidars. It was they who controlled the expansion of tobacco cultivation in the south-east of the district, 'men of substance, who, besides cultivating, carry on a thriving business as moneylenders'.[173]

The Gujarat Patidar forms the archetypal Indian rich peasant: a traditionally powerful figure, manipulating the spread of a more market orientated agriculture over the second half of the nineteenth century and, after 1900, influential, too, in local politics and the professions. Yet it is no accident that the paradigm was most evident in Gujarat. This was where groups like the Patidars had also fared well wearing their village officer hats and, in particular, the survival of the narwa and bhagdari systems underpinned their social position. In the Deccan, however, especially in Poona and Ahmednagar Districts, the decline of the old village officer class had, as we have seen, created a more fluid situation. Most of the pioneers in the more commercialised areas still came from impeccably solid agriculturist backgrounds. It was Kunbis in Satara District who 'add field to field'. Yet their economic success was not linked so assuredly with a tradition of unchallenged social influence and a few, like the Saswad Malis, were nouveaux-riches. In this way, the rich peasantry in the Deccan was a much more disparate, less socially coherent and self-conscious entity.

In the end, the evidence of economic and, particularly, tenurial processes suggests a not unfamiliar story: during the late nineteenth century, in Bombay Presidency, a pacemaking group, typically comprised of traditional landowners, but varying greatly in size and particular composition according to local conditions, drew economically ahead from and strengthened its social control over the bulk of the peasantry. Yet, before we move on, two particular lessons, which are not often drawn, need to be stressed. Firstly, this type of stratification was not, as sometimes seems to be assumed, necessarily permanent and irredeemable. The process was accompanied by the creation of a much more fluid tenurial system in which many, perhaps most, peasants became part-owner, part-tenant and part-labourer. This was predominantly forced on the middle and poor peasants through stratification, but in itself it offered the potentiality for future change. In the late nineteenth century, as we have seen, the tenancy relationship typically favoured the landlord, but, in theory, there was no reason why this should not change. The conditions, for example, which labourers enjoyed might shift substantially according to variations in supply and demand. As we proceed in the twen-

---

173 SBG, NS, No. 337, Settlement Report on Borsad Taluk, Kaira District, by E. Maconochie, 19 January 1895, para. 40.

tieth century, the continuance of stratification and even the maintenance of now established divisions needs to be continually proved, not simply assumed.

The second lesson is very different. It concerns the developmental implications of what we have described. Agrarian social structure is sometimes seen as an obstacle to India's economic development. The rural elite, on such a thesis, is condemned as 'parasitic', siphoning off the rewards of agricultural change without, in any sense, using them productively.[174] Yet it is hard to see how, fundamentally, the Bombay rich peasantry differed from, say, the Meiji landlords in Japan. Like them, Bombay's agrarian elite were mainly established village leaders, who, amidst a more fluid tenurial situation, exploited developing commercialisation through devices like kind rents. There is no evidence that the Bombay landlord was more 'parasitic'. Indeed, widespread credit supply by the rural pacemakers may explain phenomena like the general expansion in better houses in the late nineteenth century. Again, there is no evidence that the Bombay elite were content to bask in their success rather than reinvest in developing enterprise. Groups like the Saswad Malis and the Kaira Patidars could not spend money enough on acquiring more land and, if they were hardly able to pioneer new techniques extensively, that role among the Meiji landlords may have been exaggerated.[175] The real distinction, perhaps, from the Japanese situation lies in external stimuli. If the Bombay rich peasants had enjoyed the advantages of their Japanese counterparts – the consistency of high international demand for a crop such as Japanese silk enjoyed, a government energetically developing and publicising technical improvement in agriculture, and, above all, connections with and opportunities to invest in a buoyant rural handicraft sector – they, too, might have played a publicised role in wider processes of economic development.

174 See, e.g., Barrington Moore Jr., *Social Origins of Dictatorship and Democracy* (London, 1967), pp. 354–5.
175 Moore, for instance, concedes this. Ibid., p. 281.

7
The agricultural economy, 1900–1935: the critical watershed?

If the nineteenth-century history of Bombay's agrarian economy raises complex problems of interpretation, when we turn to the period after 1900 the difficulties multiply. During the first third of the twentieth century the economic influences and fluctuations seem markedly more dramatic and violent than those experienced previously. Price trends are the most conspicuous indices of this. Levels in the Bombay village – partly in consequence of the stabilisation of the rupee in the 1890s – became increasingly shaped by international price movements, but at the very time when they were particularly volatile. Hence rapid inflation during the 1910s and then the sharp falls associated with the onset of the depression during the late 1920s afflicted the countryside and might be expected to have had substantial consequences.

This instability appears to be repeated in climatic trends. There was no further great disaster like 1899–1900, but rainfall levels in many localities still seem lower than was typical during the late nineteenth century and more uneven in distribution between years and over geographical areas. Karjat taluk in the Ahmednagar famine belt, for example, experienced average annual rainfall levels between 1901 and 1921 below even the average for the 1890s and over 20 per cent under the average of 1876–86, further complicated by 'the irregular and untimely nature of the fall'.[1] Under these circumstances occasional famine persisted and there were outbreaks affecting many areas of the Presidency in 1918–19 and again in 1920–1. Yet these occurrences did not have the deep-seated impact of the great famines of the late nineteenth century. Peasant reaction during these famines and other local 'scarcities' seemed, as we shall see, more subtle and flexible, marked by substantial migration. Hence in 1920–1 'the numbers on government works were not very large'[2] and there were few signs of serious social tensions with 'no appreciable rise in crime'.[3]

---

1 *SBG*, NS, No. 605, Settlement Report on Karjat Taluk, Ahmednagar District, by A. H. Dracup, 24 August 1922, para. 5.
2 IOL, Bombay Land Revenue Progs., Vol. 11540, August 1926, p. 449, Report on the Famine Operations in the Bombay Presidency, 1920–1, 12 August 1926, para. 7.
3 Ibid., para. 13.

Famine and scarcity, too, were no longer the major determinants of mortality in the early twentieth century. This role was now played by epidemic and, in particular, by the great influenza pandemic of 1918, which caused high death rates, especially in the Deccan. As a result, Bombay Presidency's total population, which had risen from 18.53 million in 1901 to 19.65 million in 1911, fell back to 19.29 million in 1921,[4] creating an overall increase in twenty years of just 4.1 per cent. Thereafter, however, as the impact of both epidemic and seasonal scarcity moderated, the rate of population growth became much quicker and more consistent. The Presidency's population rose by 13.4 per cent between 1921 and 1931, from 19.29 million to 21.88 million.[5]

Interpreting these multifarious trends is not easy. Contemporary officials often seemed bewildered by events and the settlement reports of the era vary widely in tone between panegyrics on expanding commercialisation and gloomy descriptions of famine, stock loss and decline. Even Harold Mann, who as Director of Agriculture was best placed to review developments, regarded the 1910s as a confusing period of extreme short-term fluctuations, though he believed that little true 'progress' was achieved.[6] These difficulties of interpretation are reflected in the modern literature. Within specialist writing on Gujarat, for example, Hardiman's belief in agrarian crisis during the early twentieth century can be contrasted sharply with Bates' claims for a resurgence of commercial agriculture among independent smallholders.[7]

The basis for these interpretative clashes is clear. On the one hand, conditions for commercial agriculture seem highly favourable down to the mid-1920s and, even following the depression, some recovery in prices is evident by the mid-1930s. Alternatively, one could argue that the famines of the late 1890s, involving heavy plough cattle losses, depressed agricultural performance into the 1920s, when the onset of the depression struck a further body-blow at agricultural diversification. Nevertheless, we should, at one level, be on firmer ground in the early twentieth century in view of the existence of much fuller agricultural statistics. George Blyn has presented these in his study of agricultural trends in India and, as is well known, they seem to indicate the onset of a crisis of per capita

---

4 *Census of India, 1931*, Vol. 8, Part 2, Bombay Presidency, Imperial Table 2.
5 Ibid.
6 Harold H. Mann, *Economic Progress of the Rural Areas of the Bombay Presidency 1911–1922* (Bombay, 1924). I am very grateful to Dr I. J. Catanach for making available to me his material on this source.
7 David Hardiman, 'The Crisis of the Lesser Patidars: Peasant Agitations in Kheda District, Gujarat, 1917–34', in D. A. Low, ed., *Congress and the Raj. Facets of the Indian Struggle, 1917–1947* (London, 1977), pp. 47–75; Bates, 'The Nature of Social Change in Rural Gujarat: the Kheda District, 1818–1918', pp. 771–821.

output during the early twentieth century. According to Blyn, in Bombay Presidency 'the disparity in trends started about midway during 1911–1921'[8] and by the 1920s per capita production of foodgrains was clearly deteriorating. Basing solid conclusions on these statistics, of course, has been widely criticised. Not only was their collection procedure highly uncertain[9] but, because of the central role of official judgements about the quality of individual seasons, a strong built-in bias was possible.[10] Yet Blyn's interpretation of Bombay trends may still be correct, for it remains likely that productivity must have had to increase substantially to maintain per capita output in view of possibly severe problems of land availability. In the late nineteenth century, as we saw, official estimates were of very little good virgin land left in western India. As a result, population growth after 1900 may well now have created perceptible, even drastic, decline in landholdings per head. The phenomenon is certainly suggested by the statistics on landholding: 'between 1886 and 1920, in the Central Division of the Bombay Presidency, the number of holdings below five acres increased from 61,319 to 358,754 and the average size of holdings fell from 31.2 to 17.8 acres'.[11]

Even this evidence, however, needs to be set against other contradictory indicators. In practice, it seems that more land was consistently being brought under cultivation. The statistics on acreage, much less tainted than those on productivity, indicate the Bombay Presidency's gross cultivated area as around 24 million acres in the first years of the twentieth century rising to typical levels of just over 27 million acres by the early 1920s and about 29 million acres by the mid-1930s.[12] Substantial expansion in land in use, then, continued as it had done since the

---

8 Blyn, *Agricultural Trends in India, 1891–1947: Output, Availability and Productivity*, p. 100. Note that Blyn always deals with Bombay and Sind combined, whilst our concern is exclusively with the Presidency proper.

9 On the vagaries, see Dewey, 'Patwari and Chaukidar: Subordinate Officials and the Reliability of India's Agricultural Statistics', in Dewey and Hopkins, eds., *The Imperial Impact*, pp. 280–314.

10 In Bombay, crop yield statistics were determined by the 'standard yield' as modified by a 'condition factor' (i.e. the quality of the season, any special circumstances, etc.). Heston has shown how the condition factor was invariably reported as below average during the first half of the twentieth century, creating the impression of declining per capita output. Yet this may be simply a by-product of growing political pressure on local administrations in the countryside, leading officials to report conditions pessimistically as a means of moderating revenue demands and hence mollifying opposition. See Heston, 'Official Yields per Acre in India, 1886–1947: Some Questions of Interpretation', pp. 303–32.

11 Quoted in *Report of the Peasant Enquiry Committee of the Maharashtra Provincial Congress Committee*, p. 17.

12 See *Annual Crop and Season Reports of the Bombay Presidency*, Appendix A. All subsequent references to crop and acreage statistics, where not explicitly stated, come from these reports.

Wingate era. Indeed, down to 1920, the rate of expansion, as I have argued elsewhere, considerably exceeded the aggregate growth in population[13] and Harold Mann accepted that the overall area of cropped land per head must have been 'decidedly increasing' in the 1910s.[14] Even if the new land was considerably less productive, it seems no difficult matter to have maintained or even increased per capita performance over the early part of the twentieth century. Yet even the evidence of cultivated acreage suggests some eventual crisis and turning-point, although pushing it at least a decade later than Blyn indicated. The expansion of cultivation was slowing by the 1920s at the very time when population was beginning to grow most rapidly and in the 1920s the 13.4 per cent population increase was matched by an average growth of around 7.1 per cent in gross cultivated area. Between 1928–9 and 1938–9, the expansion in land use was down to under 3 per cent.

By the outbreak of the Second World War, then, productivity would undoubtedly have needed consistent improvement to maintain the increasing population on the land and this might suggest that a quite sharp turning-point in per capita performance occurred during the late 1920s or early 1930s. Superficially, therefore, the history of Bombay's agricultural economy under British rule might seem a pattern familiar to many students of the Third World. Increases in production probably occurred over a long period as well as some qualitative changes associated with extending commercialisation and agricultural diversification. Yet, whether the fault remained within the absolute levels of agricultural performance or in the response of other sectors of the economy or with the conditions of international trade and European control, the process did not have the consequences which seemed to develop in mid- and late nineteenth-century Japan. Instead, quantitative growth was spent mainly on increased population. This, since mortality from famine and some diseases did start to recede in the twentieth century, did not face the natural checks, like land availability, which always constrained agricultural production. Hence, eventually, it overwhelmed the whole process of expansion.

And if such a turning-point in quantitative performance did occur during the 1920s, it was aggravated by changing patterns of international

---

13 See my 'Trends in the Agricultural Performance of an Indian Province: the Bombay Presidency, 1900–1920', in K. N. Chaudhuri and Clive Dewey, eds., *Economy and Society. Essays in Indian Economic and Social History* (New Delhi, 1979), Ch. 5. Whilst Bombay's population grew by 4.1% between 1901 and 1921, gross cultivated area increased by 13.8% between 1902–7 and 1919–24.

14 See Mann, *Economic Progress of the Rural Areas of the Bombay Presidency 1911–1922*.

demand for the products of commercial agriculture. The onset of the world-wide agricultural depression in the late 1920s brought a sharp fall in both cash-crop and foodgrain prices. As a result, price statistics suggest two periods, between 1900 and 1935, of very different economic character. The first, before the depression, is marked, as the foodgrain prices presented in Table 7 show, by substantial price rise; rapid inflation even during the First World War. Demand for export crops was especially buoyant. Prices of Gujarat cotton more than doubled between 1901 and 1917 and reached further heights in the super-inflation year 1918 and the early 1920s.[15] This created a continued, possibly accelerating, expansion of commercialisation. Total cotton cultivation in Bombay Presidency, for example, averaging just under 3½ million acres in the years 1902/3–1906/7, rose over the 4 million acres level for most of the First World War period and during the early 1920s.[16] In this way, the themes of 1900–20 might seem like those of 1880–1900 writ larger. The onset of the depression, however, marked the decisive end of a long period when the underlying trend of prices, whatever the short-term fluctuations, was upward. Prices remained relatively low during the 1930s; in 1935, for instance, typically around three-quarters the levels of July 1914.[17] The rich peasants – cash-crop producers, landlords who received their rents in kind – would appear to have been the obvious victims of price decline, thus ensuring a sea-change in social trends, too. The consequence of this, coupled with intensifying population pressure, may well have been some retreat into the foodgrains. The extent of cotton cultivation, for example, was back down to 3.6 million acres by 1938–9.[18]

In this way, we can present a consistent thesis of expansion and decline in the agricultural economy of 1900–35. Yet the pieces of this particular jigsaw do not all fit very neatly. A particular problem is the land–population ratio and its consequences. The official evidence for declining size of holdings is substantial for the whole period after 1880 but if this was a simple response to population pressure, we should not have expected its emergence until after 1920, for in overall terms the ratio between cultivated acreage and population could not have been deteriorating much before then. Either aspect of the evidence might be misleading here. Even the thesis of declining per capita production by 1935 cannot be accepted uncritically. The absence of any major famine throughout the

15 See *Index Numbers of Indian Prices, 1861–1926*, Table 5, p. 7.
16 *Annual Crop and Season Reports of the Bombay Presidency*, Statistical Table 3.
17 See *Report of the Peasant Enquiry Committee of the Maharashtra Provincial Congress Committee*, p. 14.
18 *Annual Crop and Season Report of the Bombay Presidency, 1938–39*, Statistical Table 3.

Table 7 *Foodgrain price trends, 1900–22: prices of rice in husk in two taluks*

Figures in number of seers per rupee

	Khed taluk, Ratnagiri District	Kalyan taluk, Thana District
1900	15.4	14.0
1901	20.3	18.1
1902	20.9	20.3
1903	19.8	18.75
1904	18.4	17.7
1905	17.3	18.6
1906	16.5	14.6
1907	15.1	16.8
1908	15.75	15.3
1909	17.1	17.3
1910	16.75	18.1
1911	15.4	14.1
1912	13.9	12.5
1913	14.4	12.8
1914	14.7	12.4
1915	14.6	12.4
1916	12.6	11.9
1917	12.6	10.25
1918	9.6	8.9
1919	8.7	6.5
1920	9.25	8.3
1921	10.4	9.1
1922	11.75	9.2

Sources: SBG, NS, No. 627, Settlement Report on Khed Taluk, Ratnagiri District, by R. G. Gordon, 1924, Appendix I; SBG, NS, No. 613, Settlement Report on Kalyan Taluk, Thana District, by M. J. Dikshit, 5 September 1923, Appendix I.

period between 1902 and 1935 is very striking. Some famines and scarcities, as we have noted, did occur, but even they tended to diminish in frequency during the latter part of the period. This makes it hard to understand how a peasant economy could have experienced fundamental decline in per capita production without some sign of crisis of subsistence and social anomie. Most of the basic questions about the direction of the agricultural economy in the twentieth century remain dependent on more specific analysis.

Before the depression: expansion and inflation

Viewed broadly, the twenty years following the ending of the great famine in western India mirror the period 1880–96 as an era of apparent expansion. In aggregate terms, as we have seen, the land–population ratio was probably even improving. Since the years of scarcity involved considerable temporary contraction in cultivation, this meant that times of favourable climate and harvest produced striking increases in land in use: the three good years of 1914–15 to 1916–17 witnessed an expansion in the cultivated area of over 2 million acres. This might suggest that the growth of the cultivated acreage was fundamentally a positive response to economic opportunity rather than necessarily a desperate reaction to population pressure and, if so, price movements may have provided the incentive. The value of all crops was rising rapidly in the early twentieth century. In Poona City, the price of jowar by 1913 was nearly 30 per cent higher than the average of 1893–1902[19] and similar increases were evident in rice prices in the Konkan. One official estimate was that the value of all Bombay's crops rose from Rs. 25.8 per head of the agrarian population in 1902 to Rs. 54.2 per head by 1911.[20] This reflected international price trends following the recovery from the 'Great Depression' at the end of the nineteenth century and these continued to exert a decisive impact when the conditions of war then produced a further rapid rise in the inflation rate after 1914. When these influences were compounded by a poor harvest in 1917–18, this was enough, in 1918 and 1919, to push prices to nearly double pre-war levels and to produce desperate requests from the Bombay government for powers to control prices.[21]

Inflation may have been a considerable spur to extending commercialisation, for it seems possible that prices of some cash crops, notably cotton, were now rising more quickly than those of the foodgrains. International demand for raw cotton was growing rapidly, particularly in Asia, with Japan's emergence as a major cotton goods manufacturer and exporter. Exports of Japanese cotton cloth increased by 185 per cent between 1913 and 1918 alone[22] and the bulk of her raw cotton imports now came from

---

19 *SBG*, NS, No. 577, Settlement Report on Haveli Taluk, Poona District, by R. D. Bell, 7 March 1916, Appendix N.
20 IOL, Bombay Land Revenue Progs., Vol. 10556, January 1919, No. 9, Memorandum on the Value of the Gross Agricultural Produce of British India, 5 February 1917, para. 22.
21 For an account of some of the problems, see IOL, Papers of Edwin Montagu as Secretary of State for India, Vol. 18, Lord Willingdon, Governor of Bombay, to Montagu, 5 October 1918.
22 W. W. Lockwood, *The Economic Development of Japan* (expanded ed., Princeton, 1968), p. 38.

India.[23] As one Gujarat settlement officer noted in 1912, 'the world's demand for cotton fabrics has outgrown the supply of raw cotton' and Gujarat cotton could now enjoy high prices even in years, like 1912, when there was a bumper American crop:[24] hence the more than doubling of Gujarat cotton prices between 1901 and 1917 whilst typical foodgrain prices, as Table 7 suggests, rose by around 75–80 per cent. The differential can be demonstrated most strikingly in the locality. In Bardoli taluk of Surat District, for example, cotton prices leapt from Rs. 18.2 per hundred seers in 1902 to Rs. 72.5 by 1924, at the same time as the price of the leading foodgrain, jowar, increased from Rs. 6.5 to Rs. 13.6.[25] Not surprisingly, the acreage under cotton was expanding rapidly here, from just under 26,000 acres in 1894 to over 40,000 acres by 1924.[26]

This represented, arguably, a new phenomenon. The spread of commercialisation in the late nineteenth century had depended primarily on communications improvement both in terms of railways and the local resource of carts. In the first quarter of the twentieth century, however, – as very briefly in the early 1860s – there may have been considerable windfall gains to be made out of real price increases for commercial crops. Cotton was not the only one to experience substantial expansion. The extent of sugarcane cultivation increased from an average of 55,753 acres between 1902–3 and 1904–5, to 81,441 over 1916–17 to 1918–19 and the equivalent rise in tobacco acreage was from 67,225 to 93,724.[27] The process was accompanied by intense inflation in land values, particularly in the more commercialised areas. Land prices in the east Khandesh cotton tracts, Keatinge estimated, were converted from an average of Rs. 30–50 per acre in 1888–90 up to Rs. 160–170 by 1910–14.[28] 'The great desire', the Bombay Governor commented in 1918, 'seems to be to invest in land, and there is a regular land boom on here.'[29] The Banking Enquiry of 1929–30 later estimated that by the early 1920s agricultural land cost between Rs. 50 and Rs. 400 per acre in the Bombay districts, with the mean at about Rs. 250 per acre.[30]

23 India was consistently Japan's most important raw cotton supplier between the late 1890s and the 1930s. For details of the trade, see S. J. Koh, *Stages of Industrial Development in Asia* (Philadelphia, 1966), pp. 141–5.
24 *SBG*, NS, No. 529, Settlement Report on Ankleshwar Taluk, Broach District, by H. Denning, 15 August 1912, para. 23.
25 *SBG*, NS, No. 648, Settlement Report on Bardoli Taluk, Surat District, by M. S. Jayakar, 30 June 1925, Appendix I.
26 Ibid., para. 11.
27 *Annual Crop and Season Reports of the Bombay Presidency*, Statistical Table 3.
28 Keatinge, *Agricultural Progress in Western India*, p. 45.
29 Montagu Papers, Vol. 18, Lord Willingdon to Edwin Montagu, 8 June 1918.
30 *Report of the Bombay Provincial Banking Enquiry Committee, 1929–30*, Vol. 2, p. 95, Evidence of the Provincial Co-operative Institute, Bombay.

Superficially, this is evidence of an expanding commercial economy, but how secure was its agricultural base? Technically, there is little evidence of any wider use of resources and implements. Inflation was widespread, raising costs as well as prices to be gained on the market and its effects on most cultivators were therefore complex. Ownership of many resources – carts, wells and ploughs – may not have been expanding as steadily under the inflationary conditions as they had during the 1880s and 1890s. The settlement officer of Dholka taluk in Ahmedabad District in 1920 noted that, despite 'a remarkable advance' in cotton cultivation,[31] numbers of carts and ploughs had actually fallen appreciably over the period since 1896.[32] Initial problems had probably been created by the famine which had hit this area hard, but recovery was then seriously impaired by the rising costs of materials. By 1920, 'a Redwa [ordinary heavy cart] of good type cannot be bought new under Rs. 200–300', whereas 'a Redwa could be obtained some five years ago for as little as Rs. 70–100'.[33] As a result 'the stock of carts is low and a considerable number are lying unrepaired or without wheels, owing to the high prices'.[34]

The most serious problem, however, was arguably that of plough cattle numbers. As we have seen, losses during the famine had been extensive and recovery was apparently slow and faltering. The official census of agricultural stock for the Bombay Presidency suggested that total plough cattle numbers, standing at 3.21 million in 1895–6, had not regained this level until nearly twenty years later. Thereafter, there was a further slight decline and in 1924–5 plough cattle numbers totalled just 3.12 million.[35] Undoubtedly cultivated acreage had extended considerably over this period and so it is hard to see how, in general, some deterioration in ploughing extent or technique could have been avoided. Such difficulties, of course, may have varied considerably from region to region. There is firm evidence that plough cattle remained in especially short supply in the Deccan famine belt, for numbers had declined since the onset of the famine by 33 per cent in Indapur taluk down to 1908–9[36] and by 20 per cent in Sirur taluk as late as 1913–14,[37] when most areas were on the

31 *SBG*, NS, No. 607, Settlement Report on Dholka Taluk, Ahmedabad District, by J. F. B. Hartshorne, November 1920, para. 56.
32 Ibid., para. 96.
33 Ibid., para. 44.
34 Ibid.
35 *Annual Crop and Season Reports of the Bombay Presidency, 1909–10, 1924–25*, Appendix E.
36 *SBG*, NS, No. 516, Papers on the Modification of Rates of Assessment in Indapur Taluk, Poona District, 1904–10, J. P. Brander, Assistant Collector, to the Collector of Poona, No. 2070, 10 November 1909, para. 10.
37 *SBG*, NS, No. 558, Settlement Report on Sirur Taluk, Poona District, by R. D. Bell. 25 September 1915, Appendix F.

verge of recovery. Even so, few if any localities could have been experiencing long-term improvement in cattle numbers relative to land over the period 1890–1920.

Why was this? Natural replacement was clearly a slow process and, for the peasant who had lost all his stock, purchase of cattle on the market proved an increasingly expensive operation, when by 1920 'a good pair of bullocks can cost as much as Rs. 400' at the Ahmedabad cattle market.[38] The main explanation, however, for the consistent failure of cattle numbers to rise overall may lie with problems of land availability. It was little use acquiring cattle if they could not be given the extensive grazing facilities they required: 'the very salutary old Mahomedan rule' was that 'there should be one acre of grazing land to every ten acres of cultivated land'.[39] By the early 1920s, however, as a result of the expansion of cultivation, 'the area available for common grazing ... has been reduced in almost every part of the Bombay Presidency' and the remaining lands were 'being seriously overgrazed'.[40] Significantly, a new category of 'fodder crops' had, for the first time, appeared in the Bombay agricultural statistics for 1907–8 and they expanded rapidly in area from 26,419 acres in that first year to over 110,000 acres by 1918–19.[41] This, it seemed, was not merely the product of any change in enumeration technique but the genuine recognition of a new agricultural trend and Mann commented in 1920 that 'the growing of crops specially and only for the sake of their fodder value is almost a new thing in western India'.[42] Now some of the cultivated lands had to be put to use specifically to maintain the cattle population.

To take up waste grazing land for the specific cultivation of fodder crops might, of course, lead to more effective feeding of cattle, producing animals who could work harder and more efficiently. Anderson, the Settlement Commissioner in the 1920s, strongly argued that 'one acre cultivated maintains more useful cattle than an acre waste under grass' because of the absence of any efficient organisation of grazing.[43] Nevertheless, it seems likely that the dynamic behind the expansion of

38 *SBG*, NS, No. 607, Settlement Report on Dholka Taluk, Ahmedabad District, by J. F. B. Hartshorne, November 1920, para. 47.
39 L. B. Kulkarni, *Improvement of Grazing Areas in the Bombay Presidency* (Bombay Department of Agriculture Bulletin No. 112 of 1923, Poona, 1923), p. 1.
40 Ibid.
41 Harold H. Mann, *Fodder Crops of Western India* (Bombay Department of Agriculture Bulletin No. 100 of 1920, Bombay, 1921), pp. 1–2.
42 Ibid., p. 1.
43 'Grazing is allowed at all seasons and in all weather without any control, cattle trample down and spoil more than half the total crop of grass.' IOL, Bombay Land Revenue Progs., Vol. 11678, February 1928, p. 59, Report by F. G. H. Anderson, Nos. ST358, LR420, 16 November 1926, para. 3.

fodder crops was always limited availability of grazing land. The spread of commercial crops, with their requirement of more thorough preparation of land, had presented the problem of tending 'at the same time to increase the required number of cattle and to reduce the grazing'.[44] This, in turn, raises the central question about the early twentieth-century expansion concerning how far it was constrained by difficulties of land availability and quality. The growth in cultivation, of around an eighth in twenty years, may obviously have involved increasingly marginal land. Some of the new cultivation, it was alleged as early as 1908, involved 'poor lands which will hardly ever yield a good crop' and hence 'much of the labour, which is wasted on these poor lands, might be more profitably applied in giving better cultivation to the good lands'.[45] Certainly, if poor peasants were taking up marginal land in a desperate quest to feed their families, then that might represent a less favourable development for total agricultural production than their employment as labourers on the more intensive cultivation of better quality land.

Dogmatic assertion on the broad issue is impossible, but it is possible to approach the problem through examining more specific matters associated with land availability. Was, for example, the expansion of commercial crops 'robbing' the foodgrains? In absolute overall terms, this does not appear to be the case. The foodgrain area was extending steadily, though, as we would expect from the rapid expansion of some commercial crops like cotton, at a slower than average rate. Total acreage under jowar and bajra increased by 4.7 per cent between 1902–7 and 1919–24.[46] This aggregate, however, masks considerable regional variation. In the Deccan, particularly in the eastern famine belt, foodgrain cultivation seems to have been growing much more quickly and in Poona District, for example, the acreage producing jowar and bajra rose by 18.1 per cent between 1903–4 and 1922–3, whilst the total cultivated area increased by 15.6 per cent.[47] In other areas, notably Gujarat and parts of the Southern Maratha Country, foodgrain cultivation was undoubtedly falling, particularly in the period after the outbreak of the First World War. Between 1914–15 and 1924–5 total cultivation of all foodcrops (i.e. jowar, bajra, rice and 'other cereals and pulses') declined by as much as 25.2 per cent in Broach District, by 23.9 per cent in Surat District and by 15.2 per cent in Dharwar District.[48] The phenomenon is well illustrated, at local level, by the case of Bardoli taluk in Surat. As we saw earlier, cotton cultivation

---

44 Kulkarni, *Improvement of Grazing Areas in the Bombay Presidency*, p. 1.
45 *Annual Crop and Season Report of the Bombay Presidency, 1907–08*, para. 11.
46 *Annual Crop and Season Reports of the Bombay Presidency*, Statistical Table 3.
47 Ibid.
48 Ibid.

here grew by over 54 per cent between 1894 and 1924 but over the same period the extent of land growing jowar, the predominant foodgrain, decreased by 32.3 per cent: 'the necessary effect of this has been that people have largely to import their food-stuffs from outside the taluk'.[49]

What seems to be revealed here is increasing regional specialisation in the production of crops. This may have put pressure on the very poor in a pacemaking locality like Bardoli, but inter-regional trade was in the 1910s possibly well able to cope with the food import needs of the commercial specialists. Bardoli, too, illustrates another theme of changing crop patterns. It was a frontier area where the extension of the cultivated acreage had been a particularly rapid process, for during the settlement of 1896–1925 over 14,000 acres were brought under the plough for the first time.[50] Once again, then, the expansion of commercial crops appears to have occurred most easily in localities where we know that land was not so scarce. This would have limited the direct 'robbing' of the foodgrain area. In any case, any occurrence of the latter phenomenon seems to have been mainly concentrated in localities particularly suited to the cultivation of commercial crops. The anxiety raised by some officials and subsequent historians – that, in conditions of extending commercialisation, small tenant farmers may be pressurised by landlords into growing commercial crops contrary to their own interests and the suitability of the soil – probably had limited substance in pre-1920 Bombay. The foodgrain security of the bulk of the peasantry was not apparently being gratuitously undermined by the commercialisation process.

Regional specialisation, however, can also be seen as an intensification, on a grander scale, of the spatial differentiation evident during the late nineteenth century. Cotton producers, for example, could now probably enjoy real price improvements for their commodity but cotton cultivation was becoming highly regionally concentrated. Table 8 illustrates the pattern of district variations between 1914–15 and 1924–5. This reveals a rapid expansion of cotton cultivation in Gujarat and steady advance in the Southern Maratha Country coupled with absolute decline in every Deccan district. Khandesh was still a major cotton-producing region, but in the central and east Deccan the retreat of the crop had been dramatic. In general, Gujarat seemed to be emerging strongly after 1900 as the commercial pacemaker of western India, followed by Dharwar and Belgaum Districts in the south. This may have been because, as we saw in

---

49 *SBG*, NS, No. 648, Settlement Report on Bardoli Taluk, Surat District, by M. S. Jayakar, 30 June 1925, para. 11.
50 IOL, Public and Judicial Department Papers, Vol. 1957 of 1928, No. 984, Official Report of Bombay Legislative Council Debate, 13 March 1928, Speech by F. G. H. Anderson.

Table 8 *Changes in district patterns of cotton cultivation, 1914/15–1924/5*

Figures in acres

District	Extent of cotton cultivation in 1914–15	Extent of cotton cultivation in 1924–5	Percentage change
Ahmedabad	284,343	515,429	+ 81.3
Kaira	38,336	134,771	+251.6
Panch Mahals	34,417	46,357	+ 34.7
Broach	237,832	316,796	+ 33.2
Surat	122,976	193,593	+ 57.4
West Khandesh	478,199	464,893	− 2.8
East Khandesh	939,129	924,170	− 1.6
Nasik	110,881	75,115	− 32.3
Ahmednagar	272,742	140,822	− 48.4
Poona	25,720	2,296	− 91.1
Sholapur	117,840	64,596	− 45.2
Satara	17,878	16,926	− 5.3
Belgaum	257,997	332,736	+ 29.0
Bijapur	693,558	864,712	+ 24.7
Dharwar	566,866	736,384	+ 29.9
Thana	2	0	
Kolaba	0	0	
Ratnagiri	3	1	
Kanara	0	0	
Total, whole Presidency	4,198,719	4,829,597	+ 15.0

Source: *Annual Crop and Season Reports of the Bombay Presidency, 1914–15, 1924–25*, Statistical Table 3.

Bardoli, land availability was often much less of a serious constraint than in the Deccan and the Konkan but equally it can be seen as a continuation of late nineteenth-century trends. Within the Deccan, too, the localities of extending commercial production also tended to be the pioneers of the earlier period, like the central taluks of Satara District. In Koregaon taluk, for example, 'the area of the more valuable crops seems to have increased considerably' by 1923: cash crops like groundnuts, condiments and spices and fruit and vegetables, all expanding at the previous settlement of 1891, had roughly doubled in area since then.[51]

[51] SBG, NS, No. 642, Settlement Report on Koregaon Taluk, Satara District, by M. Webb, 15 March 1923, para. 21.

This type of evidence suggests that a deep process amounting to marked regional stratification within Bombay's agricultural performance was firmly in train after 1900. Differential price movements now offered new incentives for cash-crop production, but speed and extent of recovery from the famine, ownership of agricultural resources and, particularly, availability of good new land all varied extensively from area to area and influenced the degree of response. Even if the overall movement of long-term per capita production was favourable, some localities clearly felt the need to concentrate more heavily on foodgrains; though occasionally this may have represented, where neighbouring food deficient areas existed, an effective entrepreneurial response to developments. Mann believed that the most favoured and buoyant regions in the 1910s were eastern Khandesh and the canal irrigated areas of the Deccan, whilst stagnation or even decline was the rule in parts of the Konkan and within 'a region of decay' in south-west Ahmedabad and Kaira.[52] Kopergaon taluk in Ahmednagar District seems a classic example of the former category. This was a nouveau-riche locality, its agriculture transformed by the provision of extensive new canal irrigation facilities during the 1910s. As a result, by the settlement of 1923 commercial crops, notably sugarcane, had extended markedly, land prices had moved up sharply and even numbers of plough cattle had increased substantially.[53] The most striking index of success, however, was demographic. Whilst every other taluk in Ahmednagar suffered population losses between 1911 and 1921, mainly due to the influenza epidemic, in Kopergaon population had actually risen by around 10 per cent.[54] Clearly, significant immigration into this favoured locality was occurring with outsiders from neighbouring areas seeking to establish themselves as cultivators. The opposite extreme of decline and emigration was represented by Matar taluk in the 'region of decay' in Kaira District: here population fell by as much as 29 per cent between 1891 and 1921.[55]

What were the social consequences of these patterns of change? Social stratification within the village, as well as regional differentiation, might seem an obvious theme to be highlighted, for many of the influences we identified in the previous chapter continued to be powerful, most obviously the process of extending tenancy. Money rents appear to have

52 See Mann, *Economic Progress of the Rural Areas of the Bombay Presidency 1911–1922.*
53 See *SBG*, NS, No. 608, Settlement Report on Kopergaon Taluk, Ahmednagar District, by A. H. Dracup, 3 April 1923.
54 Ibid., para. 16.
55 *SBG*, NS, No. 646, Report on the Settlement of Matar and Mehmedabad Taluks, Kaira District, by M. D. Bhat, 28 February 1931, para. 18.

shared in the inflation, so landlords in general were clearly capable of maintaining their real income. Assessments of changing levels of cash rents suggest that they roughly doubled in Alibag taluk of Kolaba District in the thirty years before 1922–3[56] and, in the cotton tracts of east Khandesh, increased by the same proportion over the shorter period of 1888/90–1910/14.[57] Rents, therefore, were rising both in the heavily populated areas and those where land was more extensively available. Most Bombay rents, however, were kind rents and, whilst it is more difficult to make precise judgements about these, inflation should have provided the landlord with a built-in increment. Merely to maintain his proportion of the harvest was, almost certainly, to secure an appreciating asset. In general, then, the thesis of social stratification might appear to rest on even stronger ground in the period 1900–20. Cash-crop producers, benefiting in most localities from differential patterns of price increase, should have enjoyed growing advantages over subsistence farmers or even those with occasional grain surpluses to sell. This might have created a novel and substantial economic foundation to stratification.

In the sense that tenurial changes appear to have been developing steadily, stratification should be seen as an on-going process. However, there is little evidence that the social power of the landlord was necessarily extending anew after 1900. In conditions where some crops like cotton were appreciating in real value, one might have expected widespread pressure on tenants to shift to cash crops but the developments in cropping patterns, particularly the maintenance, even expansion, of foodgrain production in the Deccan, suggests that such moves could not have been automatically successful. Again, concerning the commercial producer, the costs of cash-cropping were always relatively high and profits, therefore, depended on a fluctuating, narrow margin between these costs and prices. An investigation of sugarcane production on the canal lands in 1914, for example, revealed that profits could reach the 'highly remunerative' level of 'Rs. 100 to 200 or more per acre',[58] but this depended on selling the crop for Rs. 700 or 800, since average costs of cultivating an acre of sugarcane were estimated as high as Rs. 597.[59] Even if, therefore, the price showed potential to increase above that Rs. 700 or 800, a heavy proportionate rise in costs might still balance or even outweigh any benefit.

---

56 *SBG*, NS, No. 632, Settlement Report on Alibag Taluk, Kolaba District, by J. R. Hood, 30 April 1924, para. 19.
57 Keatinge, *Agricultural Progress in Western India*, p. 45.
58 J. B. Knight, *Sugar Cane, its Cultivation and Gul Manufacture* (Department of Agriculture Bulletin No. 61 of 1914, Bombay, 1914), p. 1.
59 Ibid., p. 33.

Costs like the purchase of manure (Rs. 100 of the Rs. 597) were clearly liable to rise with inflation, but the most important cost with a wide potential to fluctuate was that of additional labour. Many of the Bombay commercial crops, particularly cotton and sugarcane, were relatively labour-intensive and required the employment of non-family labour for extensive production. What, then, was happening to labour costs after 1900? Most officials argued that they were increasing substantially, even in real terms. Keatinge believed that the process had been gathering pace since the 1870s. 'During the last century', he remarked in 1912, 'the real wages of a field labourer have increased by 20 to 50 per cent, and possibly in some cases by nearly 100 per cent; and practically the whole of this increase has taken place during the last forty years.'[60] He instanced the case of Haliyal taluk in Kanara District where such wages had apparently tripled between 1900 and 1915, certainly a faster rate of increase than that of average prices.[61] Similar comments were made about wages for all types of employment. Settlement reports throughout the first quarter of the twentieth century typically agreed, like that on Kalyan taluk, Thana District, of 1923, that 'wages have gone much higher than the rise in prices would warrant' and that as a result all labourers were 'quite well off in material prosperity'.[62] Willingdon, the Bombay Governor, believed that this process played a major role in mollifying the social and political tensions of wartime: 'people are extraordinarily quiet in our Presidency at present chiefly because wages have been so high'.[63]

Deciding on the validity of such comments is clearly a difficult matter. Yet it is vital to our whole view of the evolution of rural society, for there is little doubt that peasants were resorting to all types of labour work more extensively after 1900. Mann, questioned by the 1926–8 Royal Commission on Agriculture, conceded that in western India 'the number of people who depend partially on the land and partially on labour has increased'.[64] By the 1920s, many peasant families involved themselves in a variety of 'odd-jobs', often outside agriculture and involving travel over some distance. Among the Kolis in Surat District, for example, 'steamers and sailing vessels employ a number. A considerable number finds employment on bridges and other construction works on the railways all over India. A large body works in the salt pans at Panvel and elsewhere in

---

60 G. Keatinge, *Rural Economy in the Bombay Deccan* (London, 1912), p.72.
61 Ibid., p. 70.
62 *SBG*, NS, No. 613, Settlement Report on Kalyan Taluk, Thana District, by M. J. Dikshit, 5 September 1923, para. 48.
63 Montagu Papers, Vol. 18, Lord Willingdon, Governor of Bombay, to Edwin Montagu, 5 October 1918.
64 *Report of the Royal Commission on Agriculture in India, 1926–28*, Vol. 2, Part 1, para. 3567, Evidence of Harold H. Mann.

the Presidency.'[65] The process was particularly marked in the poorest areas. For example, in Karmala taluk of Sholapur District, in the heart of the famine zone, 'emigration... takes place almost every year' and 'in an ordinary village it is almost always possible to find Marathas who have been or still periodically go to Bombay to work as coolies'.[66] It needs to be emphasised, however, that even where labour work involved emigration this was no creation of a landless proletariat, for those who spent much time away from cultivation still typically retained close contact with agricultural operations. This applied even to the Konkanis who worked in the Bombay cotton mills. In the months of March, April and May, when the pressure of agricultural work called, numbers returning by steamer from Bombay City to Ratnagiri District usually outweighed considerably numbers leaving the district for the metropolis.[67]

The process of the expansion of labouring activities might be explained in entirely contrary ways. If real wages for all types of labour work were rising, then it is hardly an unexpected phenomenon: peasants were simply allocating the family's time and effort in order to secure the maximum return. Socially, the effect of this would surely be to counteract tendencies towards stratification, since agricultural entrepreneurs would face rising labour costs whilst the poor peasants and the landless would enjoy opportunities to supplement their income significantly. Alternatively, increases in real wages might still have been occurring, even if the growth in labouring owed most to 'peasant decline'. Peasants might have found themselves forced to look for additional labour work as a consequence of falling economic and social status and then discovered that the conditions of the early twentieth century, perhaps temporarily, offered enhanced returns for this. More likely, the alleged rise in real wages, granted the high rates of overall inflation in the period, may be simple official illusion. If cultivating peasants were forced by poverty out on to the labour market, one would have expected a relative glut of available labour and the maintenance of a powerful position for employers in and outside agriculture. The expansion of labouring may well be evidence of emiseration among the poor peasants and extending social stratification.

Which is the most likely explanation? One central problem with any thesis which presents labouring work as ameliorating the general position

---

65 *SBG*, NS, No. 405, Papers on the Settlement of Jalalpur Taluk, Surat District, J. W. A. Weir, Acting Collector of Surat, to F. S. P. Lely, Commissioner, Northern Division, No. 1199, 9 May 1899, para. 6.
66 *SBG*, NS, No. 436, Settlement Report on Karmala Taluk, Sholapur District, by H. L. Painter, 30 July 1903, para. 18.
67 *SBG*, NS, No. 574, Settlement Report on Ratnagiri Taluk, Ratnagiri District, by J. A. Madan, 11 September 1914, para. 39.

of the poor peasantry concerns the breadth of opportunity for such employment. Officials spoke of whole villages 'emigrating' to remunerative work, but it is hard to believe that wage labour employment outside agriculture could really have been so extensive in an economy experiencing little structural transformation. Within agriculture the labour-intensive commercial crops were still grown on a relatively small proportion of the Presidency's land and even they, where cultivated by smallholders, would not automatically have required more than family labour.

Yet much behaviour by the Bombay peasantry after 1900 does suggest a more deliberate and flexible allocation of their effort, often involving a conscious choice towards wage labour. It is already evident that a more mobile society was emerging from the way in which resources and peoples flowed towards the favoured localities like Kopergaon taluk. This mobility was especially conspicuous amidst the famines and scarcities of the early twentieth century: indeed, migration from threatened areas seemed of the very essence of the problem, lending weight to Morris' interpretation of modern famine as the product of considered reactions to an initial climatic deficiency.[68] As we noted earlier, the new famines and scarcities involved, by no means, the calamities of the late 1890s and the cultivating peasantry seemed capable, in most areas, of producing sufficient crops to survive unaided. Thus in 1905–6 despite the widespread declaration of famine and the organisation of relief works, 'there was no such increase in the death-rate of the affected districts as is generally to be expected' and 'no cases of emaciation or starvation were reported'.[69] Instead the 'famine' had been created by decisions by some peasants, at an earlier stage of the season, to abandon some or all of their cultivation operations, in a difficult but less than disastrous season, in favour of other forms of work. In 1905–6 temporary emigration from affected areas was of 'an unprecedently large scale': an estimated 100,000 people (10 per cent of the total population) left Poona District and up to 70,000 each of the other Deccan districts.[70] Again, in the poor season of 1920–1 'emigration on a large scale was noticeable very early in the season' and, in addition, 'the numbers on Government works were not very large, as people have now learnt to migrate to great centres of labour'.[71] As a result, cultivation was, in many cases, simply not attempted rather than being tried and failing: the area sown with foodgrains across the Bombay

68 See Morris D. Morris, 'What is a Famine?', paper read to the SSRC Conference on Indian Economic and Social History, Cambridge, July 1975.
69 *Report on the Famine in the Bombay Presidency, 1905–06*, para. 32.
70 Ibid., para. 35.
71 IOL, Bombay Land Revenue Progs., Vol. 11540, August 1926, p. 449, Report on the Famine Operations in the Bombay Presidency 1920–1, 12 August 1926, para. 7.

Presidency fell by 10.1 per cent in 1920–1 compared with the previous year.[72] The settlement officer, A. H. Dracup, working in Ahmednagar District in the early 1920s, claimed to have discovered a direct correlation between population levels in the individual taluks and the previous year's rainfall. When the season was poor, around a third of the population subsequently migrated to external labour work.[73] These 'famines', therefore, were not at root serious crises of subsistence, for one would then expect desperate attempts to extend cultivation and increase production. Instead, it seemed, labour opportunities and the returns from them were such, after 1900, to form for some peasants a preferable alternative to cultivation activities about every one season in five.

Nevertheless, many factors may have influenced this peasant choice, the difficulties of obtaining credit for agricultural needs in poor years, for example. It does not automatically follow that labour work was well paid. In fact, it is hard to believe that real wages were rising very quickly amidst rapid general inflation. In Alibag taluk of Kolaba District, for example, it was claimed that wages for unskilled labour had increased substantially from 3 to 5 annas per day in 1891 to 10 annas by 1924[74] but, at the same time, rice prices in the taluk had leapt from under Rs. 5 per 100 seers in the early 1890s to around Rs. 11 and more in the early 1920s.[75] The labourer, in other words, was roughly maintaining his income in real terms. Improvements, however, may have come, with such absolutely low levels of wage, from apparently minor ameliorations in working conditions: if, for example, field labourers started to receive meals, as occasionally happened in Alibag by the 1920s.[76] Shorter working hours were another possible source of improvement. Landlords complained by the 1920s that even agricultural labourers kept 'sarkari hours', working the 11 a.m. to 6 p.m. of government office workers rather than, as formerly, from dawn until dusk.[77]

Whatever their broad significance, these developments seem to have produced substantial change within the traditional semi-feudal systems of employment of labourers. The hali system in south Gujarat was, in a formal sense, breaking down in the early twentieth century. The official explanation was the attractiveness of labouring work elsewhere: in

---

72 Ibid., para. 5.
73 *SBG*, NS, No. 605, Settlement Report on Karjat Taluk, Ahmednagar District, by A. H. Dracup, 24 August 1922, para. 8.
74 *SBG*, NS, No. 632, Settlement Report on Alibag Taluk, Kolaba District, by J. R. Hood, 30 April 1924, para. 21.
75 Ibid., Appendix I.
76 Ibid., para. 21.
77 *SBG*, NS, No. 607, Settlement Report on Dholka Taluk, Ahmedabad District, by J. F. B. Hartshorne, November 1920, para. 124.

Broach District 'the Bhils are beginning to realise the market-value of their labour and to free themselves from the semi-serfdom which the system involves'[78] and in Surat, too, 'imbued with the new sense of independence they run away to get higher wages'.[79] Yet even this evidence is not incontrovertible. The impetus for the dissolution of the formal system may have come from the employers. Some officials spoke of declining fortunes among the Anavla Brahmins, who employed many of the halis, in the aftermath of the great famine, so that 'they find it hard enough to feed themselves, let alone halis'.[80] Breman's account of the hali system, however, weds acceptance of the formal decline of the system with scepticism about any genuine shift in the rural balance of power, arguing that the landlords were intensifying their control over the village and real wages among labourers failed to increase.[81] Clearly, it may now have been in the employer's interest to allow the old formal links of dependence to dissolve. If labour could be acquired, when needed, cheaply and efficiently on the market, then it might make sense to discard the hali system with its potentially irksome rigmarole of patronage.

In overall terms, it is probably impossible to generalise about the trend in real wages in early twentieth-century western India, although we can say that officialdom almost certainly exaggerated any increase. Most arithmetical assessments do suggest that labourers were sharing in the income from price rises in the crops they helped to produce but that, in real terms, actual wage levels changed little. However, circumstances probably varied considerably according to local conditions and, in particular, to landholding and cropping patterns. Sugarcane producers on the Deccan canal lands or tobacco growers in Kaira District might be able to limit labour costs because 'improvement' here came typically from the more intensive cultivation of smallholdings and family labour might still suffice. Cotton cultivation, however, was extending steadily, sometimes on large farms in frontier zones like Bardoli, and hence it may have been making considerable demands on labour resources in areas where population pressure was not acute. McAlpin suggests that the position of labour in the cotton-growing areas may have been improving during the

---

78 SBG, NS, No. 529, Settlement Report on Ankleshwar Taluk, Broach District, by H. Denning, 15 August 1912, para. 15.
79 SBG, NS, No. 405, Papers on the Settlement of Jalalpur Taluk, Surat District, F. S. P. Lely, Commissioner, Northern Division, to Government, No. 2814, 24 July 1899, para. 4.
80 SBG, NS, No. 381, Settlement Report on Chikhli Taluk, Surat District, by E. Maconochie, 17 June 1897, para. 21.
81 See Breman, *Patronage and Exploitation. Changing Agrarian Relations in South Gujarat, India.*

late nineteenth century:[82] if so, the process was possibly more powerful after 1900. In Bardoli taluk, for example, – where population rose by just 4.5 per cent between 1891 and 1921[83] – whilst total cultivated acreage and cotton cultivation both extended rapidly, the settlement of 1925 revealed that rent levels had increased less quickly than land prices and the wages of agricultural labour most of all. 'It may be', commented one official, 'that owing to scarcity of effective labour, the cultivating tenants and labourers are now in a stronger position and are able to derive better bargains with the landowners and obtain a bigger share of the gross produce.'[84] There is certainly evidence, in the Bardoli of the 1920s, of considerable tenurial fluidity, of labourers as well as established landowners enjoying the cash to buy or lease land to grow cotton. Not only 'cultivators of every class and position' but also 'landless men and adventurers' participated in 'the mania for "getting rich by cotton" '.[85]

For wider reasons, also, social stratification in the more commercialised areas was possibly nothing like so intense in the period 1900–20 as in the last two decades of the nineteenth century, though spatial differentiation between localities was still working very strongly. Contrary to initial impressions, the nature of the different stimuli to commercialisation in the two periods favoured stratification more strongly in the first. In the 1880s and 1890s linking up quickly with expanded modern communications and new market opportunities could give the rich peasant pioneer a major, windfall advantage over the bulk of the village, unaware of the potential advantages or unable, through lack of resources like carts, to exploit them. After 1900, however, a major new stimulus to cash-crop production came from price increase, often representing improvements relative to the value of the foodgrains. This created intensified geographical stratification, because agriculture was apparently becoming more geared to the specialist production of crops suitable to local conditions. Within the commercial crop areas, however, this permitted a socially more broadly based response. Price incentives, as is now commonly recognised, were a force to which much of the Indian peasantry could quickly reply. In all, the economic expansion of the first quarter of

---

82 McAlpin, 'The Effects of Expansion of Markets on Rural Income Distribution in Nineteenth Century India', pp. 289–302.
83 *SBG*, NS, No. 648, Settlement Report on Bardoli Taluk, Surat District, by M. S. Jayakar, 30 June 1925, para. 14.
84 *SBG*, NS, No. 648, Papers on the Settlement of Bardoli Taluk, Surat District, A. M. Macmillan, former Collector of Surat, to M. S. Jayakar, 18 October 1925.
85 R. S. Broomfield and R. M. Maxwell, *Report of the Special Enquiry into the Second Revision Settlement of the Bardoli and Chorasi Talukas* (Bombay, 1929), para. 24.

The agricultural depression and its legacy

the twentieth century may have brought more widely distributed benefits within Bombay agrarian society.[86]

The agricultural depression and its legacy

Yet it might be natural to assume that any potential 'improvement' had been decisively reversed by the mid-1930s. Not only had the land–population ratio in western India apparently shifted to steady deterioration but the depression had undercut the market value of crops and the incentive for the advance of commercial production. At their lowest point in the early 1930s prices of all crops had plummeted to under half the levels of September 1929 and stood, typically, at only around two-thirds of those of 1914.[87] This collapse had come after a period, in the mid- and late 1920s, when prices, particularly of export crops, had moved downwards, the boom era firmly broken. If any decisive turning-point ever existed in the history of the Bombay countryside, this process would appear to have strong claims.

The major immediate effects of the depression on the Bombay countryside were seen exclusively in terms of commercial agriculture and the cash-crop producer. Officialdom argued that the latter experienced a massive blow, whilst the subsistence cultivator and, especially, the labourer were much less adversely affected. As early as 1922–3, a downswing in the prices of all the leading crops except cotton produced, it was alleged, 'a rise in the prosperity of the labouring classes in the rural areas' whilst most employers of labour 'suffered badly': 'this fact may lead to very marked economic changes in the near future, the exact nature of which it is difficult at present to forecast'.[88] By the late 1920s the trend seemed clear. Conditions for labourers were 'especially good' because 'while prices have greatly fallen the reduction in the rates of wages has been relatively small ... The class of the agricultural population which has suffered most is the big cultivator employing hired labour on a large scale in producing his crop ... especially ... the Deccan

---

86 For similar lines of arguments, see Dewey, 'Social Mobility and Social Stratification amongst the Punjab Peasantry: Some Hypotheses', where the interesting hypothesis is suggested that social stratification occurs most clearly in conditions of 'primary commercialisation'. I would, rather, argue that historical conditions in late nineteenth-century India favoured stratification more strongly than those between 1900 and 1920. In this way, I think, India mirrors other agrarian societies, notably Russia, as I attempt to argue in my 'The Russian Stratification Debate and India'.
87 For a discussion of this phenomenon, see *Report of the Peasant Enquiry Committee of the Maharashtra Provincial Congress Committee*, p. 14.
88 *Annual Crop and Season Report of the Bombay Presidency, 1922–23*, para. 24.

Canal Irrigator and the cultivator of commercial crops like cotton and groundnut.'[89] This seemed logical. If one must be sceptical about the extent of any rise in labourers' real wages in inflationary terms, it does seem unlikely that employers could have forced down wages as rapidly and consistently as the general fall in prices demanded. Similarly, the balance of the landlord–tenant relationship might have shifted decisively against the landlord. As one Bombay Governor argued, 'where there is a crop share system, and this prevails to a fairly considerable extent in this Presidency, the tenant class have no great difficulty, but the landlords are hard hit':[90] clearly the value of the produce the landlord received was dropping.

Further, commercial agriculture and its proponents may have been adversely affected by differential price movements. The price falls of the mid- and late 1920s affected all crops relatively evenly but there is evidence that over the 1930s cash crops were faring worse than the foodgrains. The basic story was again common to all crops – stability at very low levels from 1933, a genuine rally in around 1936, followed by further decline in the late 1930s – but the strength of these short-term trends varied considerably from crop to crop, with significant results for the overall pattern. Most commercial crops were still declining in value over the six years before the outbreak of the Second World War. The average price of Broach cotton, for example, fell by 19.6 per cent between 1932–3 and 1938–9 and that of groundnuts in Karad taluk of Satara District by a third over the same period.[91] Foodgrain prices, however, remained at similar levels in overall terms and jowar prices in Sholapur District even showed a rise of 31.3 per cent over 1932–3 to 1938–9.[92] In general, levels of internal demand, with rapidly growing population, were such as to ensure no further wild fluctuations in foodgrain prices. Export crops, however, continued to be influenced by international overproduction and the further dip in world-wide prices in the late 1930s.

All this provides a powerful case that the depression and its legacy proved a watershed in both economic and social terms. Yet the victim of change in the official literature is a stereotype of the big rent-receiving and labour-employing landlord. Actual tenurial status, as we have seen, was much less rigid and clear-cut and even the major commercial producer may have benefited from some of the roles he typically played. Leasing-in

---

[89] *Annual Crop and Season Report of the Bombay Presidency, 1930–31*, para. 44.
[90] IOL, Papers of Sir Frederick Sykes as Governor of Bombay, Vol. 3B, Sykes to Lord Willingdon, Viceroy, 20 December 1931.
[91] *Annual Crop and Season Reports of the Bombay Presidency, 1936–37*, para. 29; *1938–39*, para. 32.
[92] Ibid.

entrepreneurs, for example, probably enjoyed declining rent levels, for the evidence suggests that even cash rents fell substantially.[93] Nevertheless, the onset of lower prices almost certainly did create striking short-term effects. Cash-crop cultivators, as we might have expected, initially tried to expand production to maintain their incomes but this merely increased their labour costs – and, probably, the real wages received by labourers – whilst exacerbating the price problem by flooding the market. The phenomenon is particularly evident in cotton cultivation. The price boom in the commodity reached its peak in 1924, but then came the first abrupt fall: by 1926 prices in Bardoli, for example, were under half the level of two years previously.[94] Meanwhile, however, cotton cultivation had extended rapidly, passing the 5 million acres mark for the whole Presidency for the first time in 1925–6, an increase of 12.5 per cent since 1923–4.[95] There is little doubt that, under these conditions, labourers in commercial agriculture enjoyed a sudden improvement in their position whilst landowners, with no quick price recovery, had committed themselves further to now much less profitable enterprises. An investigation into sugarcane cultivation on the Poona canal tract revealed that the value of the crops fell short of total costs in many cases in 1925–6 and 1926–7.[96] In general, it seemed, 'the cane-growers on the canals are running their businesses at a loss nowadays. This is because the prices of gul have fallen very low, while the costs have remained high on account of the high price of labour.'[97]

All this evidence, however, concerns a strictly short-term phenomenon in the mid- and late 1920s. Long-run trends may have been entirely different, for already by the early 1930s, with continued depression, cultivators had made substantial adjustments. The extent of cotton cultivation, for example, had fallen to 3.8 million acres by 1932–3, a decline of nearly a quarter on the levels of seven years previously.[98] Cotton

---

93 Sykes, for example, stressed that the fall in rent levels was general to all types: 'where cash rents prevail... tenants are unable to pay them in full'. IOL, Sykes Papers, Vol. 3B, Sykes to Willingdon, 20 December 1931. Possibly, however, the situation was that the tenants consciously reacted to falling crop values by speedy resistance to rent payment.
94 Broomfield and Maxwell, *Report of the Special Enquiry into the Second Revision Settlement of the Bardoli and Chorasi Talukas*, Appendix A.
95 *Annual Crop and Season Reports of the Bombay Presidency, 1923–24, 1925–26*, Statistical Table 3.
96 P. C. Patil, *Studies in the Cost of Production of Crops in the Deccan No. 1. Crops in the Neighbourhood of Poona* (Department of Agriculture Bulletin No. 149 of 1927, Poona, 1928), p. 6.
97 Ibid., p. 8.
98 *Annual Crop and Season Report of the Bombay Presidency, 1932–33*, Statistical Table 3. All subsequent statistics in this paragraph are taken from this table in the appropriate annual report.

growers by now, with less demand for labour, may well have forced down wages successfully and, even if they had not, the opportunities for earning high wages in the cotton fields had clearly become more limited. Yet the long-term reaction to the depression in western India seems to have been subtle. We might have expected a substantial retreat into foodgrains and subsistence agriculture, particularly if population was now beginning to press heavily on land availability. However, there is no evidence of any such occurrence. The foodgrains continued to benefit from the expansion of total cultivation but not to increase their share of the land. In 1922–3 all the cereals and pulses combined covered 18.55 million acres out of Bombay's gross cultivated area of 27.74 million acres, or 66.87 per cent. By 1938–9 they mustered 19.42 out of 29.74 million acres, 65.30 per cent of the total. Similarly the commercial crops, in overall terms, seem to have maintained their position, though here the aggregate reflected substantial individual variations. Cotton cultivation, as we have seen, contracted: by 1938–9 its total extent of 3.63 million acres represented a further decline since 1932–3. Other cash crops, however, had markedly extended their acreage. Sugarcane cultivation, for example, despite the alleged difficulties of the producers, continued to expand, with only a short-lived decline in the mid-1920s. Its total acreage in the Bombay Presidency, standing at 58,354 in 1922–3, had risen to 72,952 by 1938–9, an increase of 25 per cent. Tobacco experienced a boom during the 1930s. Its pre-depression peak was the total 113,965 acres of 1924–5, but by 1938–9 tobacco cultivation covered 153,189 acres, a rise of 34.4 per cent.

How do we explain this? Cotton, by far Bombay's leading crop export, was most affected by international competition and the world-wide depression, but one might not have expected any expansion in commercial crops. It may be that here calculations other than simple ones about 'profit' may have been involved. Sugarcane producers, as the investigations in the Deccan in the mid-1920s showed, persevered even though, in terms of monetary calculation, they seemed in the short run to be losing cash by growing the crop. The follower of Chayanov might explain this in terms of the necessity of using family labour to the full. On the other hand, highly 'rational' calculations about future prospects may have been involved, in view of the growing protection granted to the Indian sugar industry during the inter-war period. In addition, crops like sugar may have been of simple, subsistence value as an efficient and high-yielding foodstuff. A report in 1914 had noted that 'sugarcane is to be classed as a food-crop, since from 7,000 to 8,000 lbs of human food is obtained per acre and that in a highly digestible and readily assimilable form. This is far greater than can be obtained by any grain-crop.'[99]

[99] Knight, *Sugar Cane, its Cultivation and Gul Manufacture*, p. 1.

The agricultural depression and its legacy

Yet by the 1930s crops like sugarcane and tobacco, grown under the right conditions, may have been perfectly profitable in strictly capitalist terms. One major problem of many Bombay cash-crop producers in the mid-1920s was that they had acquired very high cultivation costs. It had made sense, as prices had steadily increased, to try to produce as much as possible without too meticulous investigation of outlays. Thus, on the Deccan canal lands, 'during the boom period when the prices of gul were most alluring, the tendency of sugarcane cultivators was to get the maximum outturn, spending more money on manuring and labour'.[100] It was this process, arguably, which created relative labour scarcity in some commercial areas and improved conditions for wage earners and even a year like 1924–5, when prices rose slightly after a previous fall, was now said to help labourers.[101] Under these conditions, the initial impact of the depression on the big commercial landowner was clearly disastrous, particularly if he reacted by trying to increase production. Costs came down at best only 'slightly', whilst the fall in the value of commodities was 'relatively much greater'.[102]

Nevertheless, in the long run, these problems did not need simply to be endured. The main reaction of cash-crop producers to the depression may well have been firm recognition of the necessity now to minimise costs as a priority. Labour was probably the most important and expendable of these. Hence there was no panic retreat from commercial crops into foodgrains, simply because there was no need. Instead, the 'retreat' in the long term was into methods and types of commercial agriculture which were less labour-intensive. This would explain the different trends in the cultivation of various crops. For the large cash-crop producers in Gujarat and the Southern Maratha Country there were few opportunities for such adjustment within cotton cultivation and many may therefore have switched crops. Breman gives an excellent example of the process in south Gujarat: the agrarian elite in his villages turned, particularly, to the tending of mango trees, which manifestly presented very few labour demands.[103] As a result, they were able to strengthen their hold over the village and increase the economic divide between themselves and the halis. Elsewhere, entire shifts in cropping patterns were unnecessary. Sugarcane cultivation, for example, had experienced a rapid growth in costs before the depression – these had reached as much

---

100 Patil, *Studies in the Cost of Production of Crops in the Deccan No. 1*, p. 8.
101 See *Annual Crop and Season Report of the Bombay Presidency, 1924–25*, para. 34.
102 Patil, *Studies in the Cost of Production of Crops in the Deccan No. 1*, p. 8.
103 See Breman, *Patronage and Exploitation. Changing Agrarian Relations in South Gujarat, India*.

as Rs. 1,000 per acre on some Deccan canal plots by 1925–6[104] – but retrenchment was clearly possible. Most plots on the canal lands were only 1 or 2 acres in size and more intensive work by the family might diminish costs significantly. The same logic would apply to the tobacco growing smallholders in areas like south eastern Kaira District.

This suggests that the effects of the depression may have been to reverse rather than strengthen any centripetal social tendencies within the Bombay village. By the last years of the early twentieth century boom, one gets the impression, many of the richest agricultural producers were retreating from day-to-day involvement in cultivation, preferring leisure and social position to maximum efficiency within the family budget. The depression probably drove them back to greater use of their own and family labour. But if this, for the rich peasant, was a social and cultural loss, the main economic consequence would be lower wages and loss of opportunities for those who sought labouring work. By December 1931 the Bombay government noted a firm readjustment: 'as compared with two years ago wages in rural areas have now fallen by 22 per cent'.[105] During the 1930s – a reassertion of late nineteenth-century trends and a reversal, arguably, of more recent ones – renewed social stratification may well have been at work in Bombay agrarian society.

At the same time, however, the 'crisis' of the depression can clearly be exaggerated. In wider terms, the fall in crop values obviously created deterioration in India's foreign trade position and hit the finances of provincial governments which still relied so considerably on land revenue. Yet within the local economy such structural shifts were not automatically evident. The monetary value of crops, land, rents and labour had fallen markedly but, once the short-term dislocations were overcome, by roughly proportionate amounts so that the social balance within the village was not necessarily disturbed. The changed relationship was with the external world, identified by officialdom as the general 'lack of purchasing power among the cultivators'.[106] Even here, however, the adjustment was smoothed, as the Indian economy became more autarkic in the 1930s and consumer needs were met increasingly by home industry. Hence the peasant response was typically a subtle reaction to changing circumstances, rather than a decisive shift in cropping patterns. This, too, was probably what changes in the land–population ratio demanded. There is little doubt that this was dropping steadily during the 1930s, since gross cultivated acreage now remained fairly static until a slight rise

---

104 Patil, *Studies in the Cost of Production of Crops in the Deccan No. 1*, p. 6.
105 IOL, Sykes Papers, Vol. 3B, Sykes to Willingdon, 20 December 1931, para. 6.
106 *Annual Crop and Season Report of the Bombay Presidency, 1938–39*, para. 35.

in the late 1930s took the level of 1938–9 2.1 per cent above that of 1930–1.[107] Evidence of growing problems of availability of land is further provided by new shortages of grazing land for cattle. By 1938–9 the extent of 'fodder crops' in the Bombay Presidency had risen to 2.35 million acres, an increase of 31.4 per cent on the figure for 1922–3.[108] Nevertheless, there was still no sign of any subsistence crisis as well as no 'retreat' in patterns of crop production. Historical accounts and models of agricultural performance in less developed countries in the twentieth century frequently talk of, or even assume, some decisive and self-evident turning-point where agriculture lost any capacity to stimulate or participate in economic expansion. In western India, however, it would be misleading to see the onset of decline in the land–population ratio automatically in these terms. In 1935 Bombay agriculture, as in the mid-nineteenth century, was continuing to expand production and adapt positively to external stimuli.

The problem of land ownership and organisation

This sanguine conclusion would, however, need to be heavily qualified if the problem of availability and organisation of land was really more severe than the overall trends in population and cultivated acreage imply. We need to return to the paradox posed earlier: although in aggregate terms the land–population ratio did not seem to be deteriorating seriously before the mid-1920s, still a welter of official statistics and comment indicate accelerating decline in units of landholding from the late nineteenth century. The process seemed particularly dramatic during the first two decades of the twentieth century. Between 1904–5 and 1916–17 the total number of holdings in the Presidency apparently rose from under 1½ million to over 2 million and the number of those of under 5 acres more than doubled.[109] By 1924–5, the Royal Commission on Agriculture was told, 48 per cent of Bombay holdings were smaller than 5 acres and a further 40 per cent between 5 and 25 acres in size.[110] This, in all-India terms, was hardly untypical but in many parts of western India, especially on the Deccan plateau, relatively large holdings had been

---

107 *Annual Crop and Season Reports of the Bombay Presidency, 1930–31, 1938–39*, Appendix A.
108 *Annual Crop and Season Reports of the Bombay Presidency, 1922–23, 1938–39*, Statistical Table 3.
109 IOL, Bombay Land Revenue Progs., Vol. 11330, February 1923, p. 65, Note by the Deputy Secretary to Bombay Government, Revenue Department, on 'Proposals to Check Excessive Sub-division and Fragmentation of Holdings', Attached Table.
110 *Report of the Royal Commission on Agriculture in India, 1926–28*, Vol. 2, Part 1, paras. 3504–7, Evidence of Harold H. Mann, Director of Agriculture, Bombay Presidency.

traditionally required to support a peasant family. In addition, these statistics provide a sharp contrast with earlier estimates. In 1892–3 the average size of holdings had been recorded at just over 30 acres in the Central Division and nearly 20 acres in the Southern Division.[111]

Two problems, however, need to be identified. Our exposition so far has identified a trend towards 'subdivision', a decline in size of holdings or units of ownership. This was apparently exacerbated by 'fragmentation', the tendency for those holdings to become split up into a growing number of small plots or units of cultivation. Both problems were highlighted in the writings of Keatinge and Mann, Bombay's leading agricultural and agrarian experts of the early twentieth century.[112] Mann's study of Pimpla Soudagar, a village eight miles from Poona City, revealed that, whilst 14 acres had been an average sized proprietary holding in the Wingate era, by 1914–15 5 acres and under was the norm.[113] The 156 holdings now existing in the village included only two larger than 30 acres, but they were further fragmented into as many as 718 distinct plots, some as small as 0.05 acres.[114] Mann's conclusion – that 'in the last sixty or seventy years the character of the land holdings has altogether changed'[115] – seemed self-evidently correct. From this evidence, crisis seemed to be gathering inexorably both in individual access to land and in the conditions and organisation under which it was held.

Officialdom assumed that this situation was both the outcome of rapid population growth and 'the direct result of the laws of inheritance of the country'.[116] When a peasant head of family died his land would be equally divided between sons, driving down the size of holdings in conditions of increasing population. At the same time, as Anderson commented of one Poona village, 'each of the brothers insists on having every superior and inferior patch exactly equally divided amongst them',[117] so that intensifying fragmentation also developed. Yet if even the general census figures are correct, this could hardly explain the speedy decline in the size of units of

---

111 *Annual Jamabandi Reports for the Central and Southern Divisions of the Bombay Presidency, 1892–93*, Appendix 9.
112 See, e.g., Keatinge, *Rural Economy in the Bombay Deccan, Agricultural Progress in Western India*; Harold H. Mann and N. V. Kanitkar, *Land and Labour in a Deccan Village, No. 2* (London and Bombay, 1921).
113 Harold H. Mann, *Land and Labour in a Deccan Village. No. 1* (London and Bombay, 1917), pp. 45–6.
114 Ibid., p. 47.
115 Ibid., p. 46.
116 IOL, Bombay Land Revenue Progs., Vol. 11330, February 1923, p. 65, Note by the Deputy Secretary to the Bombay Government, Revenue Department, on 'Proposals to Check Excessive Sub-division and Fragmentation of Holdings', para. 1.
117 IOL, Bombay Land Revenue Progs., Vol. 11473, June 1925, p. 441, Report by F. G. H. Anderson, Settlement Commissioner, No. SV465, 6 April 1925.

ownership and cultivation recorded in the official statistics. Were numbers of sons to be allocated land really growing so quickly? The increase in Bombay's total population noted earlier of just 4.1 per cent between 1901 and 1921 – the very period, apparently, of dramatic deterioration – hardly suggests so. Pandit's case study in south Gujarat fully supports this wider census evidence:[118] in practice, since migration, bachelorhood and absence of male heirs constantly acted to diminish competition, numbers to be newly allocated land, amidst slow overall population growth, remained relatively constant. When we add our evidence for steadily extending cultivation, then the thesis that population pressure alone created a substantial decline in size of holdings begins to look fundamentally misconceived. Yet how, then, can we explain the results of Mann's investigations, for these were undoubtedly more thorough and detailed than any previously attempted? In Pimpla Soudagar the number of cultivators had risen nearly three times between 1840–1 and 1914–15,[119] so that sub-division caused by population pressure almost certainly developed. However, we might wonder how typical Pimpla Soudagar was. It was close to Poona City and strongly influenced by the neighbourhood's commercial and industrial opportunities: in spring 1916 89 men, around a quarter of heads of household, were working in the ammunition factory at Kirkee.[120] Across the Deccan, villages such as these, like the most favoured canal villages, probably received steady infusions of immigrants, boosting their rate of population growth much above the average. To this extent, subdivision created by population pressure was possibly extending in a village like Pimpla Soudagar, but elsewhere it is hard to see how it could have been widely prevalent before the 1920s.

Only one general interpretation fits the bulk of the evidence: fragmentation of holdings, for whatever reason, probably was developing very rapidly but officials confused aspects of the process for sub-division, in reality a less powerfully progressive phenomenon. Most Bombay agriculturists' total landholdings were probably not falling but becoming split into a much larger number of small, geographically separated plots. Amidst a growingly more mobile society in which migration was common, some plots would often be held in different and occasionally distant villages. In these villages' records, such small pieces of land held by the outsider would likely be delineated as 'holdings' when clearly they were

118 Dhairyabala Pandit, 'The Myths around Sub-division and Fragmentation of Holdings: a Few Case Histories', *The Indian Economic and Social History Review*, 6, 2, June 1969, pp. 151–63.
119 Mann, *Land and Labour in a Deccan Village. No. 1*, p. 45.
120 Ibid., p. 126.

actually 'plots'. In general, the likelihood is that many 'plots' were recorded as holdings, ensuring, in the overall statistics, a steady deterioration of the average size of holding. In this way, officialdom created twin problems of sub-division and fragmentation, working in parallel. It is, however, the changing organisation of land, not questions of absolute levels of ownership, which forms the central issue which we must confront for the period before the 1920s.

The problem is a serious one because fragmentation, on this interpretation, must have been intense. In the Konkan, for example, where holdings were traditionally small and a more consistent pattern of population growth did exacerbate the problem, the most minute plots seem to have developed. In Ratnagiri District 'the intricate parcelling of the land' was such that 32 typical villages surveyed in 1917–19 'were found to contain 88,000 subdivisions'.[121] Here the process reached its logical culmination with, by 1920, 'cases in the coast talukas... where a single cocoanut tree is a hissa (plot) and is moreover jointly owned by several persons'.[122] What had caused such developments? Inheritance procedure, as in Anderson's Poona village, clearly extended the process, but we need to understand why it was so important for each brother to receive some of the best and some of the worst land. Quality of land, of course, varied greatly within most villages: in Anderson's, Kamthadi in Purandhar taluk, some lands were 'deep, rich and well watered' whilst others formed 'a high plateau, almost a mountain side, where only scanty grass will grow and rocks emerge abundantly'.[123] Even so, it is hard to explain why, as Anderson discovered, most peasants refused to accept an equivalently valued but integrated block of poorer land in place of their typically tiny allowance of the best. Anderson found that in Kamthadi 'they would not think even when they already hold 20 or 30 fields in different parts of the village of surrendering say one of their shares in the superior land soil in exchange for an area of the inferior even on payment of the panchayat assessed value of the difference'.[124]

This only becomes explicable when we recall the historical pattern of economic and social change across the Bombay Presidency. Since the later nineteenth century, a highly variable process of agricultural commercialisation had developed, creating local differentiation at every level.

---

121 IOL, Bombay Land Revenue Progs., Vol. 11193, March 1922, p. 293, Report by the Settlement Commissioner, No. SV322, 30 August 1920, para. 1.
122 IOL, Bombay Land Revenue Progs., Vol. 11330, February 1923, p. 65, Note by the Deputy Secretary to the Bombay Government, Revenue Department, on 'Proposals to Check Excessive Sub-division and Fragmentation of Holdings', para. 17.
123 IOL, Bombay Land Revenue Progs., Vol. 11473, June 1925, p. 441, Report by F. G. H. Anderson, No. SV465, 6 April 1925, para. 8.
124 Ibid., para. 5.

The problem of land ownership and organisation

Initially, at any rate, this had thrown substantial advantages to peasants enjoying favourable resource, communications and ecological conditions who could latch on to market opportunities. Ownership of fertile land in a favourable location, therefore, became viewed as vital. In Kamthadi, for example, the best land became 'tremendously cut up' because 'it seems to be the ambition of everybody in the village to own about a guntha of irrigable land'.[125] From the peasant's point of view this made firm sense. Fertile land produced not only higher output but, between 1900 and the mid-1920s, crops whose price position seemed to be improving perceptibly in real terms. The economic and social lessons of the previous thirty years, therefore, dictated that by the First World War era a peasant heir must demand his right to a share of the best land. Inheritance procedure, however, could only have been one contributor to the process. Peasants readily acquired the smallest units of the best land by purchase and leasing-in. In western India, in the early twentieth century, fragmentation of land seems to have been predominantly a by-product of agricultural commercialisation.

This thesis can be confirmed in other ways. Fragmentation, for example, seems to have increased most rapidly in the period of steadily extending commercialisation between 1900 and 1920 and the process was typically most striking in the more commercialised areas. Keatinge included 'the west Deccan and parts of Gujarat' as among the worst victims, not the famine belt east Deccan.[126] Fragmentation was most notoriously intense on the Deccan canal lands. In the famous sugarcane-producing tracts of Haveli taluk by 1916 'the land is much subdivided. In the area around Sinhgad and Khadakwasla tank few cultivators pay as much as ten rupees land revenue.'[127] In neighbouring Junnar taluk 'patasthal[128] lands sell and let in very minute lots' and an investigation of 17 acres of such land in the village of Hatban in 1916 revealed 45 distinct individual rights scattered among many more plots.[129] This, also, serves to define the real problem once more. Officials described cases like this as 'subdivision' because they could identify small individual units of ownership as well as the further 'fragmentation' into smaller plots. It is, of course, possible that extensive migration into favoured villages like Hatban created genuine 'subdivision', but more likely that many of the 45 landholders concerned also held land or lived elsewhere and that their Hatban lots were, therefore,

---

125 Ibid., para. 8.
126 Keatinge, *Agricultural Progress in Western India*, p. 70.
127 SBG, NS, No. 577, Settlement Report on Haveli Taluk, Poona District, by R. D. Bell, 7 March 1916, para. 23.
128 I.e. land irrigated by a water channel.
129 SBG, NS, No. 561, Settlement Report on Junnar Taluk, Poona District, by R. D. Bell, 12 April 1916, para. 33.

strictly 'plots' rather than small 'holdings'. Some may have been, in total, owners of large amounts of land, for the evidence suggests that the great landowners and entrepreneurs were particularly afflicted by fragmentation. Keatinge found, in a survey of 206 holdings in the Dhond Petha of Poona District, that, whilst just over 2 plots per holding was the norm, the 24 holdings of over 50 acres in size were fragmented into as many as 142 plots.[130] Plots, therefore, tended to be of similarly small size, however much land their owner held in total.

In most areas of the Bombay Presidency, then, units of land organisation had become greatly cut up by the 1920s. Yet if the process was, as we have argued, predominantly a deliberate response to commercialisation rather than the inevitable product of population pressure, was it necessarily so harmful? Clearly peasants, as Anderson discovered, preferred to hold a number of fragmented plots, giving the opportunity of more diversified production, rather than an integrated monocultural unit. Rich peasants obviously valued highly the purchase or renting of the tiniest amounts of irrigated land. The process, also, had egalitarian social implications. It permitted larger numbers of villagers to acquire a stake in commercial production, limiting the opportunity for village leaders to establish or maintain a tight monopoly. Acquisition of small amounts of land outside one's own village might even mitigate geographical stratification. In this sense, fragmentation may have played an important role in the broader distribution of the returns from commercialisation which one senses in the early twentieth century.

The obvious handicap, however, was the technical problems which some might expect with very small units of cultivation. Keatinge argued that fragmentation made 'effective tillage and irrigation impossible'.[131] Some of the plots he discovered were not only very small but often 'very awkward and unsuitable shapes'. 'Some of these narrow strips running the whole length of the field are only 22 feet wide,' so that it was very difficult to run a plough effectively along them.[132] Yet as is now well known from modern economic studies the assumption that small units of occupation are necessarily 'inefficient' shows a serious misunderstanding of peasant actions and motivations. The peasant acquiring and working a tiny plot had to make it pay and the evidence of sugarcane production before the depression – with its high costs and extensive use of labour and manure – suggests that much cash was invested in intensive cultivation. In practice, in the early twentieth century, cultivation of small plots was

130 Keatinge, *Rural Economy in the Bombay Deccan*, p. 40.
131 Keatinge, *Agricultural Progress in Western India*, p. 221.
132 Ibid., pp. 220–1.

The problem of land ownership and organisation 237

possibly one method, along with extending cultivation and growing regional specialisation, of steadily increasing absolute levels of output in Bombay agriculture.

However, the argument cannot be abandoned here. It may be that down to 1920 fragmentation presented fewer difficulties to aggregate agricultural performance than commonly assumed, but the process had progressive implications. The major problem was the exact opposite to that suggested by Keatinge. Cultivators of the tiny plots on, for example, the canal lands in fact typically went for maximum short-term efficiency. They tried to wring as much as possible from the soil to recoup often extensive levels of investment. With the inflationary prices reigning in the ten years after 1914, this also made wider economic sense, but, in the long run, the process may have created ecological problems. By the 1930s, there were signs on the Deccan canal tracts of some soil exhaustion, largely caused by over-watering to increase short-term productivity. An estimated 43,420 acres were affected to some degree in 1930.[133] With the depression and its associated retrenchment by cultivators on their costs, the rate of subsequent soil deterioration may well have slowed but there was now little opportunity of reversing previous trends in land exploitation and organisation. Population, too, was now growing more rapidly and consistently. By the mid-1930s there were genuine problems of sub-division to add to those of fragmentation on the land in many villages of western India.

In conclusion, however, the issue of land organisation needs to be seen in a wider context than yet considered. Very small units of cultivation might not seem superficially to accord to principles of capitalist agriculture but, in fact, they were the logical creation of agrarian commercialisation in an economy and society like British India. Massive demand for the best land was likely to develop within a rural economy, like Bombay's, which had experienced a locally intensive but generally highly variable process of commercialisation. The other key pillar of fragmentation was the absence of wider transformation in the Indian economy which ensured that there were very few other outlets for investment anything like so attractive as land. Land in western India, particularly as its output and the worth of its crops grew, became increasingly valuable and desired but, with land necessarily a finite object, demand on the market started to outrun supply from the late nineteenth century. Cultivating peasants reacted by trying to obtain as much land as they could, where they could, manoeuvring within the extended family, on the market and

133 *Report of the Peasant Enquiry Committee of the Maharashtra Provincial Congress Committee*, p. 83.

within the credit relationship. In the early twentieth century, social trends, lacking the clear stratifying tendencies of the late nineteenth century, were equivocal enough to give many the opportunity for such manoeuvres. This was the root reason why land organisation had become so intensely broken up by the mid-1930s.

What we have described in this chapter are perceptible changes at work in the rural economy of the twentieth century, changes which defy, however, any simple categorisation as factors for 'advance' or 'decline'. Most evident is the emergence of a more flexible economy and society. Peasants switched crops, migrated, engaged in part-time labour and acquired small units of land by purchase or lease, all on a scale unknown in the nineteenth century. We will return to consider the wider implications of these developments in the final chapter.

8
The impact of government policy, 1880–1935

In carrying forward the story of the agricultural economy's evolution down to 1935, we have, however, omitted any consideration of one possibly vital influence: the role of the state and its policies. In 1880, with the Deccan Agriculturists' Relief Act, the Bombay government had just assumed new powers to influence rural economic relationships in the four central Deccan districts to add to its traditional prominent role within the village as an arbiter of land rights and the beneficiary of land revenue. The Deccan Act, as we saw, was a complex measure and its impact may have been highly consequential, not necessarily in the manner intended. At the same time, it could hardly set a final seal on debate over agrarian reform. The range of developments we have described, particularly tenurial changes and those in the organisation of land, provided a further challenge to policy. Motives would involve not only the inevitable security fears aroused by the apparent tenurial revolution recorded in the revenue papers and census material. Even if tenurial changes were regarded in sanguine vein as the product of a more fluid system, they still carried the danger of destroying that direct administrative contact with and knowledge of the cultivating peasantry which was the ryotwari system's special claim. In a simple technical sense, too, the changes in land organisation developing in the early twentieth century made nonsense of the revenue department's intricate survey maps. So we might expect further calls to reform and legislative action, like those which had proved so successful in the 1870s.

In considering policy, our predominant concern will be with its practical impact on the ground not only because this is a history of agrarian society, not the Bombay revenue system, but also since the intricacies of debate fare better in the modern literature than most of our other themes.[1] However, we first need to outline the major issues of discussion and the evolving character of agrarian policy.

---

1 There are three major modern studies in which the Bombay debates on agrarian policy feature prominently: Kumar, *Western India in the Nineteenth Century*; Catanach, *Rural Credit in Western India*; and Dewey, 'The Official Mind and the Problem of Agricultural Indebtedness in India, 1870–1914'. I have also discussed the detail of debates more fully in my 'Agrarian Society and British Administration in Western India 1847–1920', Ch. 7.

The policy debates, 1880–1935

In one sense 1879 formed no watershed. The enactment of the Agriculturists' Relief Act did not end the conflicts between the advocates of legal reform and their opponents and the Act remained a source of controversy during its early years. Minor amendments proved necessary in 1881 and 1882 and a major strengthening of court and conciliation procedure in 1886.[2] Critics like Raymond West, who was to become the Judicial Member of the Bombay government in 1890, continued to attack the legislation as a crude attempt to manipulate 'natural' economic processes by which 'all freedom of contract is annulled'.[3] Inevitably such attitudes often struck a chord with new appointees unfamiliar with Bombay conditions. Although the Governors of the 1880s, Ferguson and Reay, had retained a benevolent neutrality in the debates, Lord Harris, who arrived as Bombay Governor in 1890, was soon declaring the Deccan Act as 'entirely opposed to my views':[4] 'Government have unmistakenly stepped in and taken from A any chance he had of recovering what he had lent to B.'[5] At the same time, official investigations into the working of the legislation came to mainly favourable conclusions. Special reports, within Bombay, by H. Woodward in 1883 and A. F. Woodburn in 1888 concluded by recommending the extension of the Deccan Act to other parts of Bombay and, when the Government of India decided to investigate matters for itself, its Commission of 1892 also supported wider legislative regulation of debt transactions.[6] As in the 1850s and 1860s the debate at government level in the decade and a half after 1879 remained finely balanced, but now the Deccan Act rather than official non-interference formed the status quo. As Harris conceded, as 'the Act has been running for ten years, and the people are accustomed to it', it seemed reasonable to support 'its continuation in the four districts'.[7]

Yet the proponents of the alternative reform programme – to extend

---

2 For example, the conciliation system had had limited success simply because parties involved did not attend hearings. The 1886 Act empowered conciliators, if they wished, to enforce attendance.
3 Raymond West, 'Agrarian Legislation for the Deccan and its Results', *Journal of the Society of Arts*, 41, 1892–3, p. 712.
4 IOL, Papers of Lord Harris as Governor of Bombay, Lord Harris to Lord Lansdowne, Viceroy of India, 26 January 1891.
5 IOL, Harris Papers, Lord Harris to Lord Cross, Secretary of State for India, 20 May 1890.
6 For these reports, see *Selections from the Records of the Government of India, Home Dept.*, No. 342, Papers Relating to the Working of the Deccan Agriculturists' Relief Act during the Years 1875–94 (hereafter Deccan Act Papers, 1875–94).
7 IOL, Harris Papers, Harris to Lansdowne, 26 January 1891.

credit availability in the village rather than regulate debt transactions – had certainly not abandoned their efforts with the enactment of the Deccan Act. Indeed, these reached their own culmination with William Wedderburn's famous scheme of 1882–4 to develop trial agricultural banks in the Deccan.[8] The successful publicity which Wedderburn attracted – the meeting of 'the leading capitalists of Poona' with the Bombay Governor in November 1882 and his declaration that he was 'favourably disposed towards the scheme';[9] John Bright and Manchester's support for the concept during 1883; and Wedderburn's favourable contacts with influential members of the Indian Council like Barrow Ellis and Evelyn Baring – all this coupled with Wedderburn's own enthusiastic idealism suggested that the proposals came very near acceptance. By this interpretation, the firm rejection of the agricultural bank plan by Kimberley, the Secretary of State, in 1884 was the defeat of a practical proposal by a hostile group in the India Office, when, as Wedderburn later complained, 'every interest and every authority in India had been brought into substantial agreement as to the merits of the scheme'.[10]

However, this interpretation, the most common in the modern literature too, overlooks the depth of opposition to the project within the Bombay government. In particular, most revenue department officials rejected the idea of widespread government involvement in rural credit banks. The three Revenue Commissioners meeting in 1881 had declared that 'we dislike intensely the idea of Government becoming a partner in a concern which to succeed must supplant the local moneylenders': the only state aid they would recommend was immunity from stamp and registration fees.[11] The proposals of 1882–4, however, demanded much more than this: government financial aid to liquidate peasant debts in an experimental area, the postponement of revenue reassessments here and the use of revenue officials to collect debts owed to the banks. With such commitment, one official argued, peasants in debt to the banks 'would look on Government officials as they looked on the moneylenders in

---

8 The details of the scheme are discussed in Kumar, *Western India in the Nineteenth Century*, pp. 228–36; Catanach, *Rural Credit in Western India*, pp. 26–30; and, most fully, in Dewey, 'The Official Mind and the Problem of Agricultural Indebtedness in India, 1870–1914', pp. 261–95. The official papers are collected in *PP*, 1887, Vol. 62, Correspondence Respecting Agricultural Banks in India.
9 *PP*, 1887, Vol. 62, Correspondence Respecting Agricultural Banks in India, Reply by Sir James Ferguson to the Poona Deputation.
10 William Wedderburn, 'Renewed Consideration of Agricultural Banks for India. When will Something Practical be Done?', *Asiatic Quarterly Review*, 3rd series, 11, January 1901, p. 7.
11 MARD, Vol. 24 of 1881, No. 1231, Joint Letter by E. P. Robertson, Arthur Crawford, G. F. Sheppard, Revenue Commissioners, to Bombay Government (undated).

1875, and as Irish tenants look on landowners and rent collectors now'[12] and another wryly commented on the scheme's financial backers: 'if these terms were acceded to, Messrs. Mandlik and Macaulay and their three colleagues could soon retire into private life as millionaires'.[13] Kimberley's rejection in 1884, in practice, saved the Bombay government the embarrassment of openly rebuffing the scheme itself and was widely welcomed within the revenue department. The Deccan Collectors, for example, including that of Poona where the experiment was proposed, were unanimous in support of Kimberley's despatch, the Collector of Nasik rejoicing that this would 'have the effect of shelving this highly utopian scheme'.[14] T. H. Stewart, the Settlement Commissioner, justly claimed to have advanced previously all the arguments contained in the Secretary of State's veto.[15]

For us, the episode is instructive because it demonstrates that the 'legal, regulatory' strand was still dominant in thinking about potential reform within the Bombay revenue department. The same men, who rejected the banks scheme, strongly supported the maintenance and extension of the Deccan Act so that even Harris conceded that 'the revenue officers, who ought to know what the people themselves feel about it ... are almost unanimous that their condition is improved'.[16] When Harris left, the pressure to extend the Act grew. In 1899 the Bombay government, on the strong recommendation of H. E. M. James, the Commissioner in Sind, proposed the introduction of the Deccan Act in Sind, but the Government of India, still equivocal about its effects, presented obstacles.[17] Nevertheless, by 1901 the Bombay government was declaring that the legal regulations introduced by the 1879 Act 'should really be the general law and only the question of expense stands in the way'.[18]

Strengthening the measures contained in the Deccan Act and expanding its geographical scope beyond the central Deccan was one clear way forward for the legal reform school. However, the logical extension of this philosophy was to go beyond regulatory legislation to the supposed real evil by a total ban on land transfer to non-agriculturist groups. By the mid-1890s, when the Government of India launched its major land

12 Ibid., E. W. Ravenscroft to Sir James Ferguson, 24 June 1881.
13 MARD, Vol. 30 of 1883, No. 395, Minute by J. Nugent, Secretary to Bombay Government, Revenue Department, 10 June 1883.
14 MARD, Vol. 30 of 1887, No. 395, Collector of Nasik to the Commissioner, Central Division, No. 834, 7 February 1885, para. 2.
15 Ibid., T. H. Stewart to the Secretary to Bombay Government, Revenue Department, No. 78, 13 January 1885, para. 2.
16 IOL, Harris Papers, Lord Harris to Lord Cross, 20 May 1890.
17 MARD, Vol. 8 of 1900, No. 1039, Part 1.
18 MARD, Vol. 14 of 1902, No. 1039, Minute by J. Monteath, 10 April 1901, para. 3.

transfer enquiry, many in Bombay had made this logical adjustment. The statistics gathered during the investigation convinced the Bombay government that land transfer now constituted a serious political danger and that 'the sooner the matter is taken in hand the better it will be'.[19] Among the prominent officials of the revenue department, Muir-Mackenzie, its Secretary, could claim that 'only one officer . . . takes the view that state interference is inadvisable'.[20] The Bombay government therefore advocated in 1899 firm legislation for eight districts of the Presidency[21] prohibiting all land sales to non-agriculturists unless specifically sanctioned by a revenue officer.[22] Only the Punjab in the late 1890s seemed similarly determined and, as Ibbetson later remarked, Bombay apparently offered at least as good a trial ground for India's first land alienation legislation.[23]

Throughout the last quarter of the nineteenth century, then, Bombay remained in the van of agrarian reform in India. Its administration now consistently favoured increasing legal regulation of village transactions as the mechanism of reform but, as the episode of 1882–4 showed, alternative programmes also had their advocates. This prominence, as we have stressed before, owed little or nothing to any stronger ideological commitment within the Bombay administration. Simply, the Deccan Riots were assumed to have demonstrated a more pressing agrarian problem in western India and, in the late 1890s, the evidence of the land transfer enquiry and the extent and severity of the famines further strengthened fears.[24] In the first years of the twentieth century, however, we seem to reach a decisive watershed in Bombay debates on agrarian policy. At all-India level, the protracted reform debates were brought to fruition with land alienation legislation in the Punjab, Bundelkhand and

19 IOL, Bombay Confidential Revenue Progs., Vol. 5777, October 1899, No. 15, J. W. P. Muir-Mackenzie, Secretary to Bombay Government to Secretary to the Government of India, Revenue Department, No. 7100–168, 7 October 1899, para. 8.
20 Ibid., para. 12.
21 These were Ahmednagar, Poona, Sholapur, Nasik, Thana, Bijapur, Khandesh and Kolaba, all districts (except for Bijapur, included presumably as a relatively poor and backward district severely hit by famine) where the land transfer enquiry had apparently revealed high proportions of landownership by non-agriculturists.
22 IOL, Bombay Confidential Revenue Progs., Vol. 5777, October 1899, No. 15.
23 Ibbetson commented that although 'the Bombay raiyat is not so important a political element as the Punjab yeoman', he was 'more apt to become the tool of the most dangerous political agitators in India' and 'the history of the Deccan Riots shows what may ensue upon his expropriation'. Sir Denzil Ibbetson to Lord Curzon, 6 August 1905, quoted in V. C. Bhutani, 'Agricultural Indebtedness and Alienation of Land', *Journal of Indian History*, 47, 2, August 1969, p. 279.
24 As Catanach puts it, the extent of famine in the late 1890s, particularly its occurrence in Gujarat, 'created an atmosphere of shock in some Bombay Government circles'. Catanach, *Rural Credit in Western India*, p. 38.

the North-West Frontier Province and with recognition of the dream of 'agricultural banks' with the enactment of the Co-operative Credit Societies Act of 1904. Bombay, however, was now isolated from these developments. After its early enthusiasms, the Bombay government issued, in February 1903, a firm declaration against land alienation legislation, simply stating, in conclusion, that land 'should be in the hands of those who can make the best use of it'.[25]

Similarly, the new initiative on co-operative credit evoked little initial response in western India. Consulted in the winter of 1901–2 about the developing all-India proposals, only one of the Deccan Collectors, Maconochie of Sholapur, showed any enthusiasm.[26] The Bombay government reported in 1902 that, if pressed, it was prepared to regard Sind as a possible trial region but, even here, the particular schemes evolved locally were dismissed as impractical and visionary.[27] Even after the passing of the Co-operative Credit Societies Act, there was no rush in the Bombay Presidency to stimulate new institutions. By the middle of March 1905, despite the appointment of James McNeill as registrar in the previous November, still no credit societies existed in Bombay.[28] By October there were eight, but they were all concentrated in Dharwar District.[29]

What had caused so major a shift in the Bombay government's attitude to agrarian reform? Clearly so abrupt an occurrence could hardly be attributed to ideological developments. One possibility is that, assuming security fears were typically the prime stimulus to action, reform suddenly came to seem as contentious and potentially dangerous as inaction amidst the increasingly more vociferous and turbulent political life of the Bombay Presidency. In 1901 the Bombay government introduced an amendment to the Land Revenue Code enabling it to grant land involving no existing private rights – mainly small holdings in tribal areas – on an inalienable tenure. This, a mere minor precursor to a full-blown land alienation act, produced a storm of opposition. There were protest petitions and political demonstrations, which undoubtedly stunned the Bombay government.[30] Immediately, in June 1901, a letter went to the Govern-

---

25 IOL, Government of India Revenue Progs., October 1905, No. 37, W. T. Morison, Secretary to Bombay Government, to the Secretary to the Government of India, Revenue Department, No. 719, 2 February 1903, para. 17.
26 MARD, Vol. 22 of 1902, No. 720, Part 5.
27 MARD, Vol. 20 of 1902, No. 720, Part 3.
28 MARD, Vol. 73 of 1905, No. 428, Part 1, Telegram from the Registrar of Co-operative Credit Societies, 14 March 1905.
29 MARD, Vol. 74 of 1905, No. 428, Part 2, Revenue Department, Resolution No. 10145, 13 December 1905.
30 See Catanach, *Rural Credit in Western India*, pp. 40–1. Catanach sees this episode as decisive in the rejection of land alienation legislation for Bombay.

ment of India postponing any action 'until the results of the measure about to be adopted and of the application in the Punjab of the Act passed for that Presidency are demonstrated'.[31]

Even so, this was still far from the curt rejection of 1903 or the stubborn suspicion of change evident throughout the revenue administration in reaction to the co-operative credit legislation of 1904. The most striking single influence in the early twentieth century almost certainly came with the publication of the 1901 Famine Commission Report. The Commission, which contained no member versed in Bombay conditions, singled out the Bombay ryotwari system for fierce condemnation. Allegedly it had proved both inflexible in its demand and too lax in its supervision of credit and tenurial change: 'the rigidity of the revenue system forced them [the peasants] into debt while the valuable property which they held made it easy to borrow'.[32] Unless there were reform, which included, apparently, the swift enactment of a land alienation act,[33] the Commission concluded that 'indebtedness in the Bombay Presidency must continue and increase'.[34]

Throughout the Bombay administration, the reaction to this criticism was immediate and intense. One of the most prominent themes of Indian policy formation is the evolution and protection of the distinct systems and approaches of the different provinces. In Bombay we have challenged traditional descriptions of the 'agrarian system' as a set of principles deeply rooted in official ideology. Yet even a technical methodology, at root empirically evolved, could produce, if individual enough, a passionate commitment over time among its administrators. By the late nineteenth century, particularly after the enactment of the Deccan Act, there was a distinct 'Bombay system', different in detail even from other ryotwari provinces and jealously guarded within the provincial revenue department. At the same time, this system was subjected to considerable sniping and criticism in all-India discussions. T.C. Hope, the protagonist of the Agriculturists' Relief Act, spoke in 1882 of 'general ignorant hostility' to Bombay practice on the Viceroy's Council from such as Crosthwaite, Ilbert and Stewart Bayley.[35] His problems were exacerbated by a visit from William Wedderburn who 'talked right and left

---

31 IOL, Bombay Revenue Progs., Vol. 6238, June 1901, No. 28, J. W. P. Muir-Mackenzie, Chief Secretary to Bombay Government, to Secretary to the Government of India, Revenue Department, No. 4180, 17 June 1901, para. 5.
32 *PP*, 1902, Vol. 70, Report of the Indian Famine Commission, 1901, para. 331.
33 Ibid., para. 343. Only one member of the Commission, F. A. Nicholson of Madras, was uncertain of the advisability of this.
34 Ibid., para. 332.
35 IOL, Sir James Ferguson Collection, Box 4, T. C. Hope to Ferguson, 23 December 1882.

against the system so freely that others have taken it up.'[36] Not surprisingly, what Ferguson, the Bombay Governor, described as Wedderburn's 'want of discretion in his political action and writing'[37] hardly influenced the revenue department in favour of the agricultural banks scheme of 1882–4. The proposal was the work of a known and vociferous critic of the Bombay revenue system, not evolved, like the Deccan Act, as an attempt to bolster it.

Nevertheless, Wedderburn's intrigues were of the very stuff of Indian official in-fighting. They reached and affected a small, inbred world. As we have remarked before, too, revenue department orthodoxy had itself stoked the flames of criticism through its emphasis on the extent and severity of indebtedness in the pressure for legislation of the 1870s. The condemnation of the Famine Commission, however, raised the attack to a new level and brought it before a much wider audience. Not surprisingly, since the criticism was fundamentally the crude one of 'amount' of indebtedness, the backlash was violent. Typical was the reaction of the Collector of Ahmednagar who remarked: 'if it has taken over 50 years for the system of the Joint Report to deprive of their lands something distinctly less than one quarter of the rayats and to embarrass something very much less than 11/20 with debt, it certainly cannot be said of that system that it is a rapidly operating cause'.[38] At the same time, harsh criticism had been linked with support for legislation on land transfer. So the Bombay government's reply of February 1903 was a firm assertion that indebtedness was not increasing rapidly coupled with the simple rejection of a land alienation act.[39]

The criticisms of the 1901 Famine Commission, therefore, inaugurated a new period when the Bombay revenue department's predominant concern was to stress the individuality of Bombay conditions and to insist that dramatic agrarian change was not occurring. This dictated, too, the reaction to the debates on co-operative credit. 'There is no probability of co-operative credit societies... being established on a considerable scale in this Presidency in the near future', wrote the Bombay government in 1902, because in western India the collectivist ethic was unknown and,

---

36 Ibid., T. C. Hope to Ferguson, 5 November 1882.
37 IOL, Ferguson Collection, Box 3, Ferguson to Lord Kimberley, Secretary of State for India, 24 December 1884. Ferguson was here discussing Wedderburn's candidature for a vacancy on the Bench of the High Court.
38 IOL, Bombay Revenue Progs., Vol. 6707, February 1903, No. 22, R. A. Lamb, Collector of Ahmednagar, to the Commissioner, Central Division, No. 1574, 29 April 1902, para. 2.
39 IOL, Government of India Revenue Progs., October 1905, No. 37, W. T. Morison, Secretary to Bombay Government, to the Secretary to the Government of India, Revenue Department, No. 719, 2 February 1903.

unlike in the north, 'village communities . . . do not now exist . . . and indeed ceased to exist before the advent of British rule'.[40] The same mood also ended any serious consideration of full-blown tenancy legislation. In the reform era of the late 1890s, the concept had been discussed and, indeed, Sandhurst, the Bombay Governor, recommended tenancy legislation, firmly protecting former proprietors from expropriation, instead of a land transfer Act in 1899.[41] In the same year, one senior revenue official, after a tour of Kolaba District where tenurial changes seemed most advanced, concluded that 'a "Tenants" Act is now urgently necessary for the Bombay Presidency': he suggested the outlawing of rents in excess of half the produce with landlords also legally compelled to pay the land revenue.[42] Whilst many in 1899 seemed sympathetic to this, orthodoxy remained that land alienation legislation should be given priority and a tenants' Act introduced later.[43] After the events of 1901, however, this was forgotten. Indeed, to advocate tenancy legislation now would be the greatest concession possible to critics that all was not well with the Bombay ryotwari system; a system which presupposed the existence of only independent proprietors. In 1903, then, a tenants' law was rejected along with the rest of the reform package.[44] The same occurred when, after leaving a decent interval, the Government of India again raised the possibility of tenancy legislation in 1913.[45]

Similar objections, however, far from applied to the Deccan Relief Act. Indeed, as a distinct Bombay measure, its extension was an obvious rebuff to the critics but one which, if they claimed to desire reform in the Presidency, they could hardly oppose. As a result the Bombay government won approval to extend the provisions regulating court procedure

---

40 MARD, Vol. 22 of 1902, No. 720, Part 5, W. T. Morison to the Secretary to the Government of India, Revenue Department, No. 5574, 12 August 1902, para. 2.
41 Sandhurst's idea was that 'every registered occupant who parts with his rights of ownership should retain the right to remain on the land as a tenant at a privileged rate of rent'. IOL, Bombay Confidential Revenue Progs., Vol. 5777, October 1899, No. 15, Minute by Lord Sandhurst, Government of Bombay, 28 July 1899.
42 MARD, Vol. 10 of 1902, No. 331, Superintendent of Land Records and Agriculture, Southern Division to the Secretary to Bombay Government, Revenue Department, 19 February 1899, para. 4.
43 Sandhurst's proposal was seen as a 'half-measure' since it would not deal with the prior root problem of land transfer (it being assumed that tenancy fundamentally arose through the dispossession of a peasant proprietor by a non-agriculturist moneylender). See IOL, Bombay Confidential Revenue Progs., Vol. 5777, October 1899, No. 15, J. W. P. Muir-Mackenzie, Secretary to Bombay Government, to the Secretary to the Government of India, Revenue Department, No. 7100–168, 7 October 1899, para. 44.
44 IOL, Government of India Revenue Progs., October 1905, No. 37, W. T. Morison, Secretary to Bombay Government, to the Secretary to the Government of India, Revenue Department, No. 719, 2 February 1903, para. 10.
45 IOL, Bombay Revenue Progs., Vol. 9619, January 1914, No. 1.

and defining rates of interest to Sind in 1901. Next, the section empowering courts to direct that decrees passed against agriculturists could be paid by instalment was granted to Khandesh in 1902 and the whole Presidency in 1903. The biggest step was the enlargement of the Act, bringing the village munsifs and the conciliation system to Khandesh in 1903 and all Bombay in 1905. By 1911 every regulatory provision of the Deccan Act, controlling the operation of transactions between moneylender and debtor, applied throughout the Bombay Presidency.[46] In this way this complex legislation had become, unchallengeably, the Bombay way in agrarian reform.

The rejection of tenancy legislation in 1913 by now rested on more than stubborn reaction to external criticism. By the second decade of the twentieth century, with no reappearance of severe famine and with the evolving records of right investigation tempering the fears aroused by the land transfer enquiry, the Bombay government could justifiably claim that a more sanguine view of the agrarian situation in western India was required. The official explanation of the records of rights statistics, as we have seen, was that they revealed a positive improvement after a prior fall: an 'increasing ability of the agricultural classes not only to hold their own, but to recover what they had lost when they sank to the condition of tenants, or lower still of labourers'.[47] Settlement officers in the 1910s looked forward in anticipation that 'the proportion of land held by non-agriculturists is likely to decrease rather than increase'.[48] In this atmosphere, local officials firmly believed that the Deccan Act was all they required to rein in the terrors of agrarian change.

Yet if worries about the effects of land transfer and indebtedness were never again as intense as during the 1890s, the problem of land organisation raised a new challenge to official policy from the 1910s. Subdivision and fragmentation of land were both described in the influential reports and publications of Keatinge and Mann as a gross evil, rapidly accelerating: could government action check their extension? The prospect was first raised by several unofficial members on the Legislative Council – firstly by Dr Paranjpye in 1916 and then by Diwan Bahadur Godbole who advocated the creation of special 'impartible' holdings to

---

46 For details of the provisions extending the Deccan Act throughout Bombay, see K. S. Gupte, *The Deccan Agriculturists' Relief Act (Act XVII of 1879) as Modified up to 31 March 1928* (Poona, 1928), p. 9.
47 IOL, Government of India Revenue Progs., April 1913, No. 1, G. Carmichael, Chief Secretary to Bombay Government, to the Secretary to the Government of India, Revenue Department, No. 831, 28 January 1911, para. 4.
48 *SBG*, NS, No. 579, Settlement Report on Matar Taluk, Kaira District, by L. V. M. Robertson, 1 July 1916, para. 12.

prevent the spread of sub-division.[49] Keatinge's response was the submission of a draft bill in 1918 described as 'purely enabling'.[50] Peasants would be permitted to apply to their Collector to have their holding registered as an 'economic holding' and an enquiry would then be held to ascertain that nobody with rights in the land dissented. Once registered, an 'economic holding' would be impartible, occupied and cultivated by only one owner and with succession to it by primogeniture.[51] Keatinge's ideas were sympathetically received by the Governor, Lord Willingdon,[52] and the Government of India, though reserving judgement, presented no formal obstacle.[53] However, when the bill was circulated for wider opinion in June 1918, it met strong opposition within the Bombay administration. Critics seized on the bill's obvious weakness: it was a purely permissive measure but, at the same time, 'how is it to be expected that, say, nine members of a joint family are to beggar themselves for the sake of the remaining one?'[54] In November 1919 the Government of India were informed that the legislation was being abandoned.[55]

At the root of this rejection was, possibly, a greater fear. Since sub-division caused by increasing population pressure was regarded as the central problem, any major solution would seem to require the politically dangerous creation of a growing proletariat. A policy review in 1923 thus concluded that any serious attack on sub-division and fragmentation would 'mean wholesale "restripping" ' and a 'complete redistribution of most of the land of the Presidency' but this would effect 'the driving away of a large mass of population from land into the ranks of labour'.[56] Any legislation on land organisation was clearly a delicate matter. A permissive measure like Keatinge's always faced the objection that it would remain a dead letter, but any more directive legislation aroused the pros-

49 See IOL, Bombay Revenue Progs., Vol. 11330, February 1923, p. 65, Note by the Deputy Secretary to Bombay Government, Revenue Department, on 'Proposals to Check Excessive Sub-division and Fragmentation of Holdings'.
50 IOL, Bombay Revenue Progs., Vol. 10333, June 1918, No. 4, Draft Bill, Statement of Objects and Reasons, para. 4.
51 Ibid., Draft Bill.
52 Writing to the Viceroy, Willingdon described Keatinge's as a bill 'which I am most anxious to discuss in my Council'. IOL, Papers of Lord Willingdon, Vol. 1, Willingdon to Lord Chelmsford, 14 February 1918.
53 IOL, Bombay Revenue Progs., Vol. 10333, June 1918, No. 4, Secretary to the Government of India, Legislative Department, to Secretary to the Bombay Government, No. 674, 6 March 1918.
54 IOL, Bombay Revenue Progs., Vol. 10558, November 1919, No. 5, Minute by Mr Justice Marten, Bombay High Court, undated.
55 Ibid.
56 IOL, Bombay Revenue Progs., Vol. 11330, February 1923, p. 65, Note by the Deputy Secretary to Bombay Government, Revenue Department, paras. 20–1.

pect of massive political opposition: even Keatinge's proposal was condemned by one Ahmedabad official as an attempt to 'revolutionise fundamentally the whole scheme of... family life'.[57] In 1927 a draft bill was circulated attempting more radical reorganisation.[58] To prevent increasing sub-division and fragmentation, it proposed to prohibit the future sale and leasing of holdings smaller than a locally determined 'standard unit' to any but the holder of geographically contiguous land. At the same time, Collectors would be enabled to carry out consolidation of land into more logical units if just two-thirds of the holders of plots and half of the owners of land agreed. Here, clearly, was a real attempt to grapple with the problem but inevitably it was accused of riding roughshod over peasant freedom and rights, for example over choice of tenant in leasing out land: 'the policy of compelling an owner to accept a tenant in whom he has no trust on the simple ground that the latter happens to be a neighbour will be resented', predicted one politician.[59] In 1929 the Bombay government withdrew the proposal, reacting to a welter of opposition.[60]

By the late 1920s, of course, we are in a different world where official policy was formed much more publicly and subject to ever widening influences. In this situation, external pressures coupled with the revenue department's defensive protection of its major principles from non-official intrusion combined to maintain an equilibrium where decisive innovation was not attempted. Major legislation on tenancy and land organisation was not seriously considered down to 1935. The policy innovations, therefore, of the second half of our period were always acts of minor administrative adjustment rather than sweeping root and branch reforms. There were the measures we mentioned previously: the new Khoti Settlement Act of 1904 and the attempt to establish some influence over the conditions of inamdars' tenants through the introduction of a revenue survey. In 1913 an amendment to the Land Revenue Code finally required all landlords to pass on any remissions or suspensions of land revenue to their tenants and inferior holders.[61] All such measures, however, were open to the basic objection that they could not be automatically enforced on the ground.

57 IOL, Bombay Revenue Progs., Vol. 10558, November 1919, No. 5, Report by R. G. Gordon, Sub-divisional Officer, Dhandhuka, Ahmedabad, No. 191, 6 September 1918.
58 IOL, Economic and Overseas Dept. Papers, Vol. 1363 of 1927, No. 5537.
59 Ibid., Minute of Dissent to Report of Select Committee on the Bill by B. V. Jadhav M.L.C., 21 May 1928.
60 Ibid., G. S. Rajadhyaksha, Deputy Secretary to Bombay Government, to the Under-Secretary of State for India, No. 378, 8 October 1929.
61 IOL, Bombay Revenue Progs., Vol. 9352, September 1913, No. 22.

Similarly, other regulations can be accused of very limited practical effect. The only restriction on land transfer remained that amendment to the Land Revenue Code – authorising the grant of land involving no private rights on an inalienable tenure – which had provoked the storm of protest in 1901. In practice the opposition proved hardly necessary for, as the central government noted in 1905, the measure was applied only 'within very narrow limits'.[62] Almost all the beneficiaries were tribal groups, notably the Bhils in Khandesh and the Panch Mahals,[63] and even this limited degree of protection was opposed by some officials during the 1910s and 1920s, when fears about land transfer receded.[64] In 1929, orders were finally issued directing that all new grants of waste land should be made on the impartible tenure, as intended in 1901.[65] Yet even this regulation strictly applied could have little influence in an economy, like Bombay's in the 1930s, where extensions of cultivation were now relatively limited. The impact, therefore, remained peripheral. By 1936–7 the restriction on land transfer applied to only 1.1 million acres of land in the Bombay Presidency (over half in the three districts of west Khandesh, Broach and the Panch Mahals) compared with over 30 million acres under the ordinary transferable ryotwari and alienated tenures.[66]

In the end, the policy debates of the period 1880–1935 came to very little in direct practical terms, beyond the extension of the Deccan Agriculturists' Relief Act. Partly, no doubt, this was because of the grave difficulties of devising effective remedies for complex problems like fragmentation of land. Again, in the twentieth century the possibility of raising political agitation and widespread opposition made inaction always an easier and safer option than the introduction of major reform. Yet the most important theme of policy, at any rate in the crucial period of the early twentieth century, was, perhaps, the need to defend the Bombay revenue system and the 'Bombay way' in agrarian reform, epitomised by the Deccan Act, from hostile criticism. To propose substantial new legislation on land transfer, tenancy or land organisation would now, as was recognised within the revenue department, be admission that some of the sacred principles of the ryotwari system had been defeated. As a result, the

62 IOL, Government of India Revenue Progs., October 1905, No. 38, J. Wilson, Secretary to the Government of India, to the Secretary to the Bombay Government, No. 1548–76–2, 10 October 1905, para. 8.
63 K. S. Gupte, *The Bombay Tenancy Problem* (Poona, 1938), p. 11.
64 For example by F. G. H. Anderson, who favoured an entirely free market in land and argued that the inalienable tenure kept potentially valuable land in the hands of bad cultivators. See IOL, Bombay Land Revenue Progs., Vol. 11473, November 1925, p. 739.
65 IOL, Bombay Land Revenue Progs., Vol. 11916, March 1931, p. 19.
66 Sivaswamy, *Legislative Protection and Relief of Agriculturist Debtors in India*, p. 75.

reform movement within the Presidency lost coherence and momentum. For us, concerned with the practical consequences of government action in the village, the positive impact of policy has to be investigated, fundamentally, in just two areas: in terms of the effects of, firstly, the unique Bombay remedy, the Deccan Act, and, secondly, the constant acts of administrative action like land revenue revision.[67]

The impact of the Deccan Agriculturists' Relief Act

Modern orthodoxy about the limitations of the imperial impact on Indian society might lead us to expect that so complex and oblique a measure as the Deccan Act might have had merely superficial effects: that the operation of the Deccan village economy and society continued regardless. It certainly comes as no great surprise to discover that, when the Commission of 1892 investigated peasants' reactions in 37 villages of Bhimthari taluk, three villages had 'never heard of the Act'.[68] Yet the simple assumption of impotence cannot be made. Indeed, the Act's immediate impact on legal proceedings seemed revolutionary. There was a 'very marked – and I may say altogether abnormal – diminution in the number of suits instituted in all the courts in the four districts'.[69] The number of new civil suits begun in 1880 was under a third of the average of the previous ten years.[70]

This, however, was merely a temporary phenomenon while moneylenders in particular waited to see how the Act would operate. There was clearly a substantial slackening of all credit and monetary transactions in the Deccan in 1880, evidenced by a decline of over 40 per cent in licence tax receipts since 1879.[71] After 1880 credit activities undoubtedly resumed and numbers of court cases also increased. Yet the latter never returned to anything like the levels of the 1870s and Woodburn could claim in 1889 that 'the effect of the introduction of the Act has been to

---

67 A third could obviously be added in the shape of the influence of the rural co-operative movement in the twentieth century. However, since the reader can be referred to I. J. Catanach's major study of the subject, *Rural Credit in Western India*, this will not be discussed directly.

68 A further five villages could make 'no comment' on the Act. MARD, Vol. 89 of 1893, No. 1038, Part 2, Report on the Deccan Agriculturists' Relief Act Commission, 11 June 1892, Appendix 1.

69 Deccan Act Papers, 1875–94, Vol. 1, No. 76, A. D. Pollen, Special Judge, to the Chief Secretary to the Bombay Government, Judicial Department, No. 386, 7 June 1880, para. 11.

70 Deccan Act Papers, 1875–94, Vol. 1, No. 132, A. D. Pollen to the Chief Secretary to the Bombay Government, No. 255, 13 March 1881, para. 4.

71 Deccan Act Papers, 1875–94, Vol. 1, No. 147, H. Woodward to the Chief Secretary to the Bombay Government, No. 30, 25 June 1883, para. 9.

diminish litigation by one half'.[72] For the supporters of the Act, this was firm evidence of its beneficial effect: since the courts now were required to 'go behind the bond', moneylenders were not so readily exploiting the court mechanism to put pressure on debtors. Use of the ex parte decree, which avoided a legal contest, was cut down drastically and permanently. By 1892, it was applied to only 6 per cent of disposed cases in the Deccan.[73] 'Formerly it was exceptional', Woodburn's report remarked, 'that the merits of a case should be enquired into. Now a fair hearing is ensured for both sides.'[74]

Undoubtedly the Deccan Act did reform some of the famed 'abuses' of the Bombay legal system, mainly through the simple device of making moneylenders more chary of approaching the courts. Whether, however, this in any way decisively altered the balance of power between moneylender and debtor remains questionable. Peasant debtors still lost most of the cases brought against them. Clearly in any formalisation by a Western legal system the claims of creditors were, at law, totally valid and procedural innovations would make little difference. West, pointing this out, claimed that under the Relief Act a tiresome and expensive performance was enacted but still 'in about six cases out of seven contested by defendants in the Dekhan courts, as elsewhere, the judgement is for the plaintiff'.[75] Statistical evidence confirms this. Over the period 1880–3 in the Deccan 69 per cent of all cases against agriculturists were decided against them, 22 per cent were compromised and in only 7 per cent did the defendant win outright.[76] The introduction of village munsifs, with the aim of making courts more accessible to peasants, largely reinforced this tendency. Many had very little work,[77] but, where they were busy, the scheme mostly benefited the prosecution of cases by 'small traders and petty shopkeepers': 'it affords no relief to the agricultural classes, for the number of suits by or against agriculturists is almost infinitesimal'.[78] In the last resort, therefore, creditors could still almost always obtain legal backing for their position.

72 Deccan Act Papers, 1875–94, Vol. 2, No. 13, A. F. Woodburn to the Secretary to the Bombay Government, Judicial Department, No. 189, 27 April 1889, para. 26.
73 *Bombay Presidency Annual Administration Report, 1892–93*, p. 224.
74 Deccan Act Papers, 1875–94, Vol. 2, No. 13, A. F. Woodburn to the Secretary to the Bombay Government, Judicial Department, No. 189, 27 April 1889, para. 23.
75 West, 'Agrarian Legislation for the Deccan and its Results', p. 710.
76 Deccan Act Papers, 1875–94, Vol. 2, No. 13, A. F. Woodburn to the Secretary to the Bombay Government, Judicial Department, No. 189, 27 April 1889, para. 23.
77 For example, 53 of the 144 munsifs received no suits at all during 1881. Deccan Act Papers, 1875–94, Vol. 1, No. 139, A. D. Pollen to the Secretary to the Bombay Government, Judicial Department, No. 60, 4 February 1882, para. 29.
78 Deccan Act Papers, 1875–94, Vol. 1, No. 132, A. D. Pollen to the Chief Secretary to the Bombay Government, No. 255, 13 March 1881, para. 25.

Yet, as well as reforming court procedure, the Deccan Relief Act also aimed at a greater formalisation of the whole village credit network. Debtors were to receive documentation concerning transactions and the hopelessly indebted given the right to declare themselves bankrupt. These regulations achieved almost nothing. The section entitling peasants to receipts, pass-books and statements of account was found to be unworkable as early as 1882:[79] moneylenders showed no inclination to provide written accounts and the peasants 'are mostly ignorant of the important provisions made ... for their advantage'.[80] The insolvency provisions, the one measure unanimously approved and advocated, by Wedderburn and his followers as well as the legal regulatory school, remained a dead letter. During 1880 only three declarations of insolvency were made throughout the Deccan and Woodburn had to concede in 1889 that 'the insolvency provisions have rarely been resorted to by debtors'.[81] Some officials were mystified by these failures, particularly debtors' neglect, even once the regulations were more widely popularised, to insist on documentation. One settlement officer, working in Sholapur district in 1903, complained of 'the notorious evasion of the provisions of the Deccan Agriculturists' Relief Act by the very persons whom the law seeks to protect'.[82] Debtors, however, had to accept the long-enshrined conditions of the whole credit system. Since access to credit was so vital for all agricultural operations and expenditures for all peasant groups, nobody dared to endanger his position by challenging traditional arrangements, whatever the extent of legal backing. In particular, even contemplation of any declaration of bankruptcy would destroy most peasants' economic security. As the Special Judge appointed under the Deccan Act soon noted: 'the natives of the mofussil evince the strongest possible repugnance to being considered "nadar" or insolvent, ... They consider that to be declared insolvent would hopelessly damage their credit through all their future.'[83]

In all these ways, the ambitions of the Relief Act fell far short of achievement. Economic and social reality, not legislative diktat, shaped the character of the credit system. Even so, the Act's measures had clearly required some adjustment at local level. Moneylenders did not resort to

---

79 Deccan Act Papers, 1875–94, Vol. 1, No. 139, A. D. Pollen to the Secretary to the Bombay Government, Judicial Department, No. 60, 4 February 1882, para. 51.
80 Ibid.
81 Deccan Act Papers, 1875–94, Vol. 2, No. 13, A. F. Woodburn to the Secretary to the Bombay Government, Judicial Department, No. 189, 27 April 1889, para. 7.
82 *SBG*, NS, No. 443, Settlement Report on Sholapur Taluk, Sholapur District, by H. L. Painter, 27 July 1903, para. 31.
83 Deccan Act Papers, 1875–94, Vol. 1, No. 76, A. D. Pollen to the Chief Secretary to the Bombay Government, Judicial Department, No. 386, 7 June 1880, para. 28.

the courts so regularly or with quite such assurance of easy success. West argued that, since the legislation 'strongly invited the judges of first instance to make out a case quand meme for every agriculturist litigant', a peasant would, therefore, 'sometimes deny or question claims which he knows to be perfectly just'.[84] By 1938 one Poona lawyer was claiming that, throughout the Presidency, 'the act has . . . the effect of making the agriculturists dishonest in their dealings with moneylenders'.[85] The implication that peasants could use the Act to escape from debt obligations greatly exaggerates its impact, but even so some subtle alterations within the structure of credit and debt may have occurred. The major effect, wherever the Deccan Act became applied, was that all creditors had to take greater care over their lending operations. Woodward noted in 1883 that credit transactions on personal security had become confined 'within very narrow limits. Normal relations, interrupted during the famine years, have not been resumed.'[86] Woodburn's report of 1889, too, concluded that the Relief Act 'has made it much more difficult for a rayat to borrow, as there is so little of his property that can be attached'.[87]

Overall, it is likely that the new legal regulations introduced by the Deccan Act – investigating cases more thoroughly as well as exempting land from attachment unless pledged – produced some contraction, or halt, to the expansion of credit availability. Most official reports concluded that the legislation had 'reduced indebtedness'. The Commission of 1892's investigations in Bhimthari taluk revealed that, of 27 villages expressing a clear opinion, 17 believed that their debts had decreased since 1879.[88] Earlier, Woodburn, making a rough arithmetical assessment of the total debts of 371 peasants in 12 villages of Ahmednagar, claimed that the volume had been reduced from Rs. 1,32,421 in 1875 to Rs. 91,836 in 1888.[89] This, of course, to supporters of the Act was 'success': Woodburn described the 'great reduction in the burden of debt' as 'eminently satisfactory'.[90] Yet one must wonder about the wider economic consequences of this. Restrictions in credit availability, if they

---

84 West, 'Agrarian Legislation for the Deccan and its Results', p. 713.
85 Gupte, *The Bombay Tenancy Problem*, p. 4.
86 Deccan Act Papers, 1875–94, Vol. 1, No. 147, H. Woodward to the Chief Secretary to the Bombay Government, No. 30, 25 June 1883, para. 9.
87 Deccan Act Papers, 1875–94, Vol. 2, No. 13, A. F. Woodburn to the Secretary to the Bombay Government, Judicial Department, No. 189, 27 April 1889, para. 18.
88 MARD, Vol. 89 of 1893, No. 1038, Part 2, Report of the Deccan Agriculturists' Relief Act Commission, 11 June 1892, Appendix 1.
89 Deccan Act Papers, 1875–94, Vol. 2, No. 13, A. F. Woodburn to the Secretary to the Bombay Government, Judicial Department, No. 189, 27 April 1889, para. 20.
90 Ibid.

were occurring, may have had a general inhibitive impact on the expansion of 1880–96 in the Deccan.

In practice, however, as we have already seen, there is little evidence that, say, the Saswad Malis or the groundnut producers of central Satara experienced any special new problems of access to credit: quite the opposite. It seems clear that any contractions of credit availability hit almost exclusively the poorer peasants. In 1883 'solvent rayats can still borrow as before... but there is a consensus of opinion as to the credit of poor rayats having been too much curtailed'.[91] The fundamental issue, though, concerns the consequences of this. Despite official comments and estimates, actual diminutions in volume of lending probably occurred little, for poor peasants had to have credit, particularly when the famine of the late 1890s struck. How would they obtain it? If other forms of personal security were no longer sufficient, land became the sole reliable source of credit. Writing in the *Maharashtra Mitra* in December 1879, one correspondent described as a moneylender of Wai, Satara District, commented that to poor peasants moneylenders 'will refuse to lend money, except under the circumstances when these ryots are willing specifically to pledge their immoveable property'.[92] In 1883 Woodward confirmed that in many cases 'the poorer rayat is reduced to the alternative of resigning his field altogether or of foregoing the customary loan'.[93]

The Deccan Act, then, in marked contrast to its intentions, probably increased the tendency of the credit system to create land transfer and tenurial change, a mechanism which, as we have argued, had actually functioned little in the Deccan village down to the 1870s. The 'poklist' sale often entered in the train of the Relief Act. As early as 1881 Wedderburn was told by one Ahmednagar agriculturist that 'I know many cases where raiyats have sold their lands where they would before have mortgaged them.'[94] In addition, the very provision banning land from attachment for debt, by calling attention to land's potential as security, may also have extended the practice: for the moneylender to obtain title to the land first would at least prevent any trouble from that regulation.

---

91 Deccan Act Papers, 1875–94, Vol. 1, No. 168, Appendix to a Note on the Working of the Deccan Agriculturists' Relief Act by C. Gonne, Chief Secretary to the Bombay Government, Judicial Department, August 1883.
92 Report on the Vernacular Newspapers in the Bombay Presidency, week ending 13 December 1879, *Maharashtra Mitra*, 11 December 1879.
93 Deccan Act Papers, 1875–94, Vol. 1, No. 147, H. Woodward to the Chief Secretary to the Bombay Government, No. 30, 25 June 1883, para. 23.
94 MARD, Vol. 213 of 1881, No. 174, Report on a Meeting Held at Akola by William Wedderburn, District Judge, 26 January 1881, Statement by Hari Baji, Brahmin cultivator.

The impact of the Deccan Agriculturists' Relief Act   257

In 1886 one Ahmednagar settlement officer spoke of 'what I gather is growing to be the custom among sawcars, viz. to make deeds of sale take the place of mortgage deeds'.[95] By the 1890s extending commercialisation and the rising market value of land seem to have provided further incentive for the process. West, examining the Special Judge's report for 1890, inferred that 'about 25 per cent of the borrowers of sums even below Rs. 20, are compelled to give the security of mortgage or sale, 50 per cent have to give such security for loans below Rs. 100'.[96] The 1901 Famine Commission, with the statistics from the land transfer enquiry before them, concluded that 'transfers of property, both by sale and mortgage, have become more frequent in districts to which the Relief Acts apply'.[97]

If, in this way, the operation of the Deccan Act boosted social stratification and tenurial change, it was, of course, merely aiding the thrust of a wider economic and social movement in the late nineteenth century. Land transfer and extensions in tenancy were developing, too, in regions as yet free from the Relief Act's influence. Yet the legislation may have played a special constructive role in the central Deccan, granted the relative absence of a peasant creditor elite which the 1875 riots had revealed here. The operation of the Act almost certainly gave advantages to agriculturist moneylenders over the Marwaris and Gujaratis. The root feature of the Deccan Act – that creditors needed to take greater care over their activities – made local knowledge about peasants' means and position much more important and the contractions in credit supply seem to have come mainly from the alien shopkeepers and traders, who needed now to be especially wary. The 'smaller sowkars' of Otur, Junnar taluk, complained in 1883 that 'the larger sawkars have now ceased lending us money, and consequently we have ceased lending money to the agriculturists'.[98] Woodburn's report in 1889 confirmed that 'there had been a great decrease in moneylenders trading on borrowed capital among Marwaris, Gujaratis and Brahmins'.[99] Investigating 14 Ahmednagar villages over the period 1885–8, Woodburn further discovered that nearly half the moneylenders acquiring land were Kunbis or Malis, indisputably agriculturists: they also formed much the largest single caste category of lenders in the fourteen villages.[100] The groups most favoured, under the

---

95 *SBG*, NS, No. 184, Settlement Report on the Deshi Villages of Akola Taluk, Ahmednagar District, by W. M. Fletcher, No. 115, 24 January 1886, para. 27.
96 West, 'Agrarian Legislation for the Deccan and its Results', p. 726.
97 *PP*, 1902, Vol. 70, Report of the Indian Famine Commission, 1901, para. 335.
98 Deccan Act Papers, 1875–94, Vol.1, No.164, Petition by the Smaller Sowkars of Otur, Junnar Taluk, 28 July 1883.
99 Deccan Act Papers, 1875–94, Vol.2, No.13, A. F. Woodburn to the Secretary to the Bombay Government, Judicial Department, No. 189, 27 April 1889, para. 18.
100 Ibid., para. 22.

Deccan Act conditions of less assured credit supply, were those like the Saswad Malis or some Satara Kunbis who borrowed overwhelmingly among themselves. Any restrictions created by the legislation were here safely by-passed. Again, the Relief Act was in these ways mainly strengthening existing tendencies rather than inaugurating change. Yet in the central Deccan it gave a substantial boost to the emergent rich peasantry of the late nineteenth century and underpinned their economic and social status as creditors, at once entrepreneurs and village leaders.

The Deccan Act gave further opportunities to village social elites. It generated new local bureaucratic systems which largely came under their sway. All land transactions, for example, had now to be registered at village level, a 'decidedly unpopular' measure.[101] The vast majority of those originally charged with this duty were Kulkarni village clerks, who, it was claimed, were often 'under the thumb of local sowkars' and 'imbued with local prejudices'.[102] During 1883 a serious attempt was made to appoint new, more 'independent' registrars.[103] Yet the problem always remained that the village registrars, with their considerable powers in recording land transactions and peasant tenurial status, were likely to reflect and pursue the interests of the dominant groups and individuals. So, too, were the 'conciliators', appointed to effect amicable agreements between creditors and debtors. Despite an initial inability to compel attendance, in the early 1880s an 'enormous amount of work... has unexpectedly devolved upon the newly-appointed Conciliators'.[104] The early success, however, was entirely transitory: by 1887 numbers of applications for the intervention of conciliators had fallen to under half the 1881 figure.[105] The problem may have been the attitudes of and influence wielded by conciliators. Officialdom believed that the system depended on their zeal and efficiency: there were 'good, bad and indifferent conciliators'.[106] Yet it is hard not to believe that the actions of conciliators, members of the village elite, typically reflected the economic

---

101 MARD, Vol. 214 of 1881, No. 1028, Memorandum in A. D. Pollen to the Chief Secretary to the Bombay Government, Judicial Department, No. 350, 13 April 1881.
102 Ibid.
103 MARD, Vol. 15 of 1884, No. 316, Bombay Government Resolution, Judicial Department, No. 2285, 25 March 1884.
104 Deccan Act Papers, 1875–94, Vol. 1, No. 132, A. D. Pollen to the Chief Secretary to the Bombay Government, No. 255, 13 March 1881, para. 26.
105 Deccan Act Papers, 1875–94, Vol. 2, No. 42, W. H. Crowe, District Judge, Satara, to the Secretary to the Bombay Government, Judicial Department, No. 3009, 21 November 1888.
106 Deccan Act Papers, 1875–94, Vol. 1, No. 132, A. D. Pollen to the Chief Secretary to the Bombay Government, No. 255, 13 March 1881, para. 31.

interests of their group and that the limited use of the system rested on recognition of this fact by the remainder of the peasantry.

The Deccan Act's influence for social stratification was, however, most evident in those provisions which required a conscious initiative from agriculturists. Here the benefits were conspicuously seized only by an assertive minority in the more commercialised areas. This was true, for example, of the redemption facilities, allowing peasants to sue for recovery of mortgaged agricultural lands. In 1882 there were 2,438 suits for redemption of land but over half of these were in Satara, undoubtedly the wealthiest of the four districts.[107] Within Poona District, where Wedderburn's committee of the early 1880s suggested Purandhar taluk for a credit bank experiment as the most advanced area of the district, 'the Purandhar agriculturists have availed themselves of the facilities for the redemption of their debts... to a much larger extent than any of the other Poona talukas'.[108] In poorer areas, such as the region where the Deccan Riots had occurred, use of the redemption facilities was minimal. The investigations by the Commission of 1892 in 37 villages of Bhimthari taluk revealed only six individuals who specifically claimed to have taken advantage of them.[109]

In all, the major theme of the working of the Deccan Relief Act is a story of adaptation and adjustment. Most villagers sought to re-create as smoothly as possible the conditions which had existed within the credit and debt relationship before 1879. Yet the adjustment involved some change and consequences, most notably the necessity to look to wider forms of security to obtain credit, particularly by poorer peasants. In turn, this probably acted to quicken the pace of tenurial change and the relatively high proportions of land transfer recorded for the Deccan by the late 1890s enquiry may have owed something to the operation of the Relief Act as well as to the temporary impact of intense famine. In sum, the Deccan Act, wherever it was extended, was a contributor, although possibly minor, to processes of social stratification. Here, we might assume, is a prize example of legislation devised by British officialdom producing the exactly contrary effects on the ground within Indian society to that intended. Yet this would be to accept all official attitudes at face value. Some may have fully perceived the effects of the Act and

---

107 Deccan Act Papers, 1875–94, Vol. 1, No. 168, Appendix to Note on the Working of the Deccan Agriculturists' Relief Act by C. Gonne, Chief Secretary to the Bombay Government, Judicial Department, August 1883.
108 IOL, Government of India Revenue Progs., November 1886, No. 4, Viziarangam Mudliar to Sir William Wedderburn, 21 October 1885, para. 6.
109 MARD, Vol. 89 of 1893, No. 1038, Part 2, Report of the Deccan Agriculturists' Relief Act Commission, 11 June 1892, Appendix 1.

valued it in these terms. For example, J. G. Moore, Collector of Poona in 1879, had accurately predicted that 'the tendency of the bill when it becomes law will be to prevent people with bad credit from borrowing'[110] and he wrote approvingly of the process: 'the land which these needy cultivators forfeit will then pass into the hands of men who can afford to pay the Government assessment and to employ labour in cultivation'.[111] During the 1890s Moore, as Revenue Member of the Council, would argue strongly for the Deccan Act's retention and extension. For him, at least, the legislation's stimulus to social stratification within the peasantry was an added point in its favour.

The impact of administrative action

This, also, serves to define more strictly the mixture of official practice and assumptions which made up the 'Bombay system'. To extend our argument that its roots lay in empiricism, there was rarely any special ideological commitment to the small peasant. The predominant concern was to preserve the overall social structure intact. Hence the reform movement which created the Deccan Act was built more on distaste for the Gujarati and Marwari moneylenders as socially disruptive interlopers than on particular regard for the Kunbi landowner. West specifically stressed that between debtor and creditor in the Deccan 'there is no great choice in point of moral loveliness'.[112] For a poor peasantry, then, to lose economic and social status at the expense of more wealthy and powerful caste fellows would not disturb the equanimity of most Bombay officials, so long as there was no evident security problem, for, socially, the pattern of landownership would remain the same. Indeed, the existence of a rich peasantry was the surest evidence, for the wider all-India debates, of the capacity of the Bombay system to create wealth on the land. When, during the 1920s, the Settlement Commissioner, F. G. H. Anderson, argued for a completely free market in land and denounced land alienation legislation as 'a crime against the community', he did so as the major contemporary spokesman for the Bombay system.[113] For Anderson, the weak cultivator was, in any case, bound to go to the wall. Conditions of growing land

---

110 MARD, Vol. 11 of 1879, No. 422, J. G. Moore, Collector of Poona, to E. P. Robertson, Commissioner, Central Division, No. 4628, 2 September 1879, para. 2.
111 Ibid., para. 6.
112 West, *The Law and the Land in India*, para. 8.
113 Anderson's views on land transfer are presented in his *Facts and Fallacies about the Bombay Land Revenue System Critically Expounded* (Poona, 1929), pp. 64–5. As the title implies, the study as a whole was a passionate defence of Bombay practice, written at the time of the famous Bardoli campaign against one land revenue reassessment.

shortage perhaps highlighted his argument that 'it is the supreme interest of all the community that the best use be made of every acre'.[114]

This strong, but often overlooked, social Darwinistic streak in Bombay thinking further explains, of course, the reluctance to go beyond the regulatory measures of the Deccan Relief Act. It also influenced the temper of administrative action. One major example of this was the Bombay authorities' use of the takavi system of loans to cultivators. Following further legislation in 1883 and 1884, the scheme expanded rapidly and by the year 1891–2 grants totalled over 12 lakhs of rupees.[115] However, the onset of the great famines of the late 1890s and the persistence of occasional scarcity in the early twentieth century provided a further challenge to the takavi ideal. Hitherto confined to land improvement operations, could the system now be extended to offer cheap credit to the bulk of the peasantry in conditions of hardship? Certainly takavi was advanced in considerable quantity in the poor years: a large total of over 200 lakhs of rupees between September 1899 and October 1902,[116] and, later, 56 lakhs during the scarcity season of 1911–12.[117] The Bombay authorities, however, kept strictly to the rubric of 'land improvement'. In the scarcity year 1905–6, for example, around half of the advances were made for 'the construction and repair of wells and the balance for other objects amongst which repairs to field embankments or "tals" bulked largely';[118] hardly concerns of most peasants in a poor season. Officials defended such grants to rich peasants on the grounds that their activities generated employment and the state 'was thus spared the necessity of providing labour for large numbers of people'.[119] Yet the policy and its support for peasant entrepreneurship was based on much more than economic calculation. The Collector of Ahmedabad completed his review of the extensive takavi loans granted in the scarcity year 1911–12 with the boast: 'not a pie was given for "maintenance"; . . . This year's takavi was takavi in the true sense of the word, viz, a loan which supports the industrious, not the idle, rayat in . . . the time of trouble.'[120]

However, British governments' major administrative concern and much the most consequential, for the agrarian economy, of all official

114 Ibid., p. 65.
115 MARD, Vol. 258 of 1892, No. 749, Brief Memorandum on the Material Condition of the People of the Bombay Presidency, 1881–91, para. 25.
116 *Report on the Famine in the Bombay Presidency, 1899–1902*, Vol. 2, Appendix 40.
117 IOL, Bombay Land Revenue Progs., Vol. 9352, September 1913, No. 526, Report on the Famine in the Bombay Presidency, 1911–12, para. 34.
118 *Report on the Famine in the Bombay Presidency, 1905–06*, para. 29.
119 Ibid., para. 14.
120 IOL, Bombay Land Revenue Progs., Vol. 9352, September 1913, No. 526, Report on the Famine in the Bombay Presidency, 1911–12, para. 35.

actions remained, always, the determining of the land revenue. Whilst historians have frequently discussed, as in our first chapter, the social and tenurial implications of settlement policy, it is curious that the economic impact of the land revenue burden, although an issue of intense contemporary controversy, has rarely been examined in the modern literature. And yet this is a vital issue. Was the state, over the last century of British rule, increasing, maintaining or losing its share of the output of agriculture? And, if there was change, who within the village were the victims or beneficiaries? It might seem that, in the Bombay Presidency, reassessments about every thirty years, with often substantial increases in rates, offered considerable opportunities to the state to siphon off, from all landowners, a growing proportion of their returns. Such a process may have weakened the whole economic position of the Indian rich peasantry, with fatal developmental consequences. In Poona District, for example, the revision settlements of 1869–72 increased the total revenue demand by just over 54 per cent in the five sub-divisions affected:[121] was agricultural output, as late nineteenth-century critics charged, really expanding at equivalent rates?

Yet such critics often overlooked problems involved in this situation. Firstly, land in cultivation was extending steadily, producing upward movement in overall revenue returns even at constant rates. Increases in charges per acre were, therefore, nothing like so great as the total rise in land revenue initially suggested. More fundamentally, any system of land taxation undoubtedly requires constant attention to maintain the state's share of agricultural output. Regular survey of land is necessary if the authorities are to keep abreast of improvements to farms and holdings, affecting their productive capacity. In addition, periodic reassessments, as in India, will be needed just to maintain the real income from land revenue, where prices were rising.[122] In western India, as we have seen, the whole period from 1850 down to 1920 was fundamentally one of inflation. There was, also, no new resurvey in the Bombay Presidency after the mid-nineteenth century. The revision settlements which began in 1867 and again in 1897 were reassessments based, with some local adjustments, on the Goldsmid and Wingate surveys. For all these reasons, even where arithmetical increases in rates seemed large, we can make no

---

121 *BG, Vol. 18, Poona, Part 2*, p. 118.
122 These problems are highlighted in T. C. Smith's account of Tokugawa Japan. Here is another economy where the land revenue burden is traditionally alleged to have been very heavy, but Smith shows that lapses in resurveys and reassessments must have made it difficult for the state to preserve its proportion of the harvest. See T. C. Smith, 'The Land Tax in the Tokugawa Period', *Journal of Asian Studies*, 18, 1, November 1958, pp. 3–19.

automatic assumptions about the changing real burden of land revenue on the Bombay landowner. Further, the state in western India never attempted nor was able to control or even gauge the process. There was never any scientific attempt to obtain a particular proportion of agricultural output, or, at any time, consciously to increase or lower it. Reassessments remained entirely empirically based throughout the nineteenth century. During the 1910s and 1920s, some officials, notably F. G. H. Anderson, tried to give the system surer theoretical foundation by arguing for the Ricardian principle of 'rental value' as the means for calculating assessments.[123] In 1927 the Bombay government announced 'rental value' as 'the sole basis for fixing the assessment'.[124] In practice, however, such metaphysical declarations could have meant little to the revenue officer in the field. 'Rental value' as the declared means of settling land revenue, in any case, did not survive long: it was firmly denounced in the report of the enquiry into the Bardoli agitation of 1928, a document which served to undermine completely Bombay settlement practice.[125] The lack of any coherent theory or practical measure meant that the land revenue burden in any one locality always depended to a large extent on entirely random factors. Some settlement officers invariably recommended higher enhancements than others. In particular, much depended on the short-term economic climate when the revision settlement occurred. Areas revised in the immediate aftermath of the 1860s cotton boom experienced, as we have seen, large increases in rates. Those whose settlements happened to fall due following the great famines of the late nineteenth century were treated much more lightly, whatever their economic potential, so some areas undoubtedly, at times, faced 'unfairly' large enhancements whilst others, more able to pay, enjoyed relatively small rises.

This pattern was clearly repeated at the lowest level, for village influence and power, even in a ryotwari system, shaped the structure of individual assessment. Yet, forgetting this, what can we say about the basic trends in the land revenue burden in the Bombay Presidency? In absolute terms, the amount raised in land revenue was rising more

123 Anderson's *Facts and Fallacies about the Bombay Land Revenue System Critically Expounded* was largely a catechism of the rent theory of land revenue, in which the 'rental value' of land strongly featured as the measure of the surplus increment which the state could justifiably take.
124 IOL, Bombay Land Revenue Progs., Vol. 11615, May 1927, p. 301, Bombay Government Revenue Department Resolution No. 1790/24, 13 May 1927, para. 3.
125 See R. S. Broomfield and R. M. Maxwell, *Report of the Special Enquiry into the Second Revision Settlement of the Bardoli and Chorasi Talukas* (Bombay, 1929). In practice, too, the success of the Bardoli campaign ended the possibility of new formal reassessments.

rapidly in Bombay than in most other Indian provinces over the second half of the nineteenth century. Total receipts increased from Rs. 21.5 lakhs in 1856–7 to Rs. 29.5 lakhs in 1870–1 (a 37 per cent rise in fourteen years) and then to Rs. 34.5 lakhs by 1890–1.[126] This growth of 60.5 per cent in just over forty years compared with an all-India equivalent of only 31.8 per cent.[127] We need to remember, however, that cultivation was extending steadily in many parts of western India throughout this period, particularly in the early years when the land revenue increase seemed so substantial. The close correlation between expansion in the cultivated acreage and the very large rises in overall revenue is apparent from Table 9, which details the revenue enhancements of some of the revision settlements of the late nineteenth century. The new settlements raised very much more from the land, more than double the amount in Poona District, for example. Yet in Poona, the land area yielding revenue had increased by over 70 per cent. In districts like Ahmedabad and Kolaba, where the expansion in cultivation was much less, the overall rise in revenue yield was proportionately lower. As Table 9 demonstrates, the revision settlements, in general, increased rates per acre by around a fifth or a quarter.

In turn, this needs to be compared with the level of price rises, remembering, too, that the currency of settlements was, in practice, often longer than the intended thirty years. It might certainly be maintained from Table 2 (which understates the long-term level of inflation by starting in the middle of the cotton boom) that general price levels were, at least, between 10 and 20 per cent higher by 1900 compared with the 1860s. In overall terms, then, it seems very likely that the real per acre burden of the land revenue remained fairly static over the second half of the nineteenth century.

After 1900, however, prices began to rise very much more quickly. In the first two decades of the twentieth century, reassessments needed to produce enhancements in rates at considerably higher than previous levels to maintain the real value of land revenue receipts. Yet, now in the aftermath of the great famines and later with the onset of serious political agitation against land revenue payment with Gandhi's Kaira satyagraha of 1918, the tone had to be much more sceptical of increases. M. S. Jayakar's recommendation in 1925 for rises of around 30 per cent in land revenue rates in Bardoli taluk of Surat District[128] eventually provoked

---

126 IOL, Government of India Revenue Progs., May 1890, No. 15, Financial Statement for 1888–9, para. 37.
127 Ibid.
128 See *SBG*, NS, No. 648, Settlement Report on Bardoli Taluk, Surat District, by M. S. Jayakar, 30 June 1925.

Table 9 *Increases in land revenue in districts of Bombay during the late nineteenth century*

		Ahmedabad District	Ahmednagar District	Kolaba District	Poona District
Original settlement (after 1837)	Area surveyed (acres)	858,034	2,130,937	714,653	1,174,194
	Total assessment levied (Rs.)	8,70,112	8,95,337	10,44,053	6,07,118
	Average charge per acre (Rs.)	1.01	0.42	1.46	0.52
Revision settlement (after 1867)	Area surveyed (acres)	864,054	2,816,788	717,360	2,046,900
	Total assessment levied (Rs.)	11,24,885	15,10,563	13,07,064	12,43,791
	Average charge per acre (Rs.)	1.30	0.54	1.82	0.61
Percentage change in average charge per acre		28.71	28.57	24.66	17.31

Sources: *BG, Vol. 4B, Ahmedabad; Vol. 11B, Kolaba; Vol. 17B, Ahmednagar; Vol. 18B, Poona* (all Bombay, 1904), all Table 15.

the massive protest campaign of 1928: but, on the data which Jayakar used, it must have involved acceptance of declining real charges for even jowar prices had more than doubled since the previous settlement.[129] Between 1900 and 1920, it seems certain, the real per acre burden of land revenue in the Bombay Presidency was declining perceptibly. At one level, this is self-evident from our previous discussions. All the data from settlement reports and papers suggests that land prices and rent charges relative to revenue assessments were now rising rapidly throughout western India. Previously, we have seen this predominantly as an index of fast growing demand for land. Equally, it may indicate the scale of decline in the real value of land revenue in times of inflation.

The onset of the agricultural depression, however, would seem to create entirely different tendencies. The Bardoli satyagraha, for example, has to be seen against a background of the collapse of cotton prices which made Jayakar's proposals very much more burdensome than they seemed in 1925. Yet the political success of the Bardoli campaign together with the denunciation of reassessment procedure in the enquiry report de-

129 See ibid., Appendix I.

stroyed the possibility of any new revision settlements. The battle to determine land revenue charges now shifted to the localities and it is clear that administrative mechanisms were widely used to moderate rates at this level. Heston shows how pessimistic reporting of the 'condition factors' which determined crop yield estimates reduced the land revenue demand throughout western India.[130] Absolute receipts declined by over 28 per cent between 1922–3 and 1937–8.[131] The tendency of depression to increase the real value of fixed-rate taxes like land revenue was, in this way, heavily moderated.

Land revenue is clearly always a complex means of raising taxation. It is impossible to be dogmatic about its changing burden because for the individual landowner his own payments, whatever the wider trends, inevitably seemed much more oppressive in the early years of a settlement than at a time when the rates had remained fixed for nearly thirty years. Again, estimating realistically the proportion of output taken by land revenue remains a forlorn task, although officials from Wingate to Anderson made guesses. What we can say is that between 1850 and 1939 the state in Bombay Presidency was almost certainly not raising that proportion. Indeed, if the real value of the land revenue remained mainly static – falling even between 1900 and 1920 – the state was clearly not participating in the benefits from any real agricultural expansion and improvement. In western India, during the last century of British rule, it seems, any real increases in agricultural wealth were kept firmly in private hands. Yet the beneficiaries of the action of land revenue were those with specific status. Throughout western India, it was registered owners who normally paid the land revenue. At the very time, too, when, in real terms, its burden was being most moderated – during the early twentieth century – the practice of tenancy was extending steadily. On much of the land, the settlement was no longer effectively ryotwari, for the actual cultivator made no land revenue payment. The failure of land revenue to rise in real terms was, therefore, a boost to those who predominantly played the role of landlord and hence acted typically as a force for social stratification.[132]

---

130 Heston, 'Official Yields per Acre in India, 1886–1947: Some Questions of Interpretation', p. 327.
131 Ibid., n.
132 Anderson made the point with regard to settlement policy: 'we often think that we cannot put on settlement rates, even approaching the theoretical standard, in case we should be hitting "cultivators" too hard. The facts of the area in question often will show that practically the whole land or 80 per cent of it... is leased by landlords, where therefore cultivators are entirely uninterested in the assessment, but would have to pay their rent all the same, whatever the assessment was.' IOL, Bombay Revenue Progs., Vol. 11402, September 1924, p. 741, Note by F. G. H. Anderson, 29 August 1923, para. 3.

The influence of official policy

The effects of official policy and action on the Bombay village between 1880 and 1935 were not dramatic. The dynamic of policy debate steadily lost momentum in the twentieth century in reaction to the twin threats of external criticism and growing political activity and protest. The predominant impact of what remained and of inaction itself was to strengthen stratification and enhance the position of the richer peasants and landlords. Yet, during the later nineteenth century, far more potent economic and social forces had, in any case, dictated this trend. This story – fundamentally of the failure of imperial government policy to influence decisively the indigenous economy and society – is, of course, hardly unfamiliar. It may be that, given the complexity of agrarian issues, the Bombay government was right to settle for strictly limited ambitions in the twentieth century. The introduction later of the 1948 Tenancy Act proved far from a complete success,[133] and any major measure on fragmentation, considering the official misconceptions about it, could have caused havoc to the vital question of access to land.

Yet agrarian policy in the Bombay Presidency, even by the standards of imperial action, conspicuously failed to understand much less ameliorate the economic and social problems of the countryside. It was a supreme irony that legislation was enacted in 1879 to remedy an imaginary social revolution, whilst after 1879, when tenurial change was much more substantial and consequential, nothing further was attempted. The legislation of 1879, too, had actually acted to sponsor that tenurial change and, in a minor way, exacerbated the economic problem of credit supply. It may well be, as Moore's position suggests, that Bombay officialdom was consciously creating conditions for the rich peasant to prosper. Even so, the social and tenurial developments of the early twentieth century made nonsense of administrative records and land surveys and destroyed the ryotwari system's proud boast of direct contact with every cultivator. The Bombay government had lost effective local knowledge of the situation on the land well before 1935. As the Governor sadly conceded of settlement practice, in the aftermath of the Bardoli enquiry report's criticisms: 'it is at present rather doubtful . . . whether the groundwork of all our settlements, that is, the original survey classification, will stand the superstructure which we have been trying to raise on it'.[134] By then officialdom's power to influence developments within the agrarian economy and society was perceptibly less than in Wingate's day.

133 Some of the problems are discussed in Dandekar and Khudanpur, *The Working of the Bombay Tenancy Act, 1948. A Report of Investigation.*
134 IOL, Papers of Sir Frederick Sykes as Governor of Bombay, Vol. 1, Sykes to Lord Irwin, Viceroy, 26 April 1929, para. 2.

9

The peasant and politics in the early twentieth century

Economic and social change has featured prominently in our analysis of western India's agrarian history: the differential spread of commercialisation in peasant agriculture, consequent geographical and social stratification and the intensification of demand for and pressure on land. Would this, in turn, produce a political response from the peasantry? The question is important because the history of Indian politics in Bombay Presidency, as elsewhere, would seem to demand a more intricate understanding of the rural situation from at least the First World War era. With Gandhi's emergence, action and protest apparently became much more broadly based. Some of the most persuasive accounts of politics, it is true, see the dynamic of local action as mainly faction-based competition for office and pickings, resulting in wide-ranging cross-class, cross-caste alliances.[1] In theory, this interpretation firmly denies the existence of mass objectives or grievances as an impetus behind politics, but it is striking how, in practice, even here, rural economic change is frequently introduced as a decisive determinant of political developments.[2] Further, it could not be denied that agrarian protest campaigns, whatever their derivation, proved an important propagandist and practical weapon against the British Raj in the twentieth century: the Bardoli campaign of 1928, as we have noted, ended revision land revenue settlements in the Bombay Presidency. So the character of peasant political behaviour in the western India of 1900–35 demands analysis and may, in turn, throw further light on economic and social processes.

Yet investigations of peasant politics must reveal much more than just the role of the rural dimension and simple linkages with urban and national movements. The main concern is to locate the source of activity

---

1 This is, in general, the interpretation of the 'Cambridge school' of political historians. For a range of these arguments, see John Gallagher, Gordon Johnson and Anil Seal, eds., *Locality, Province and Nation. Essays on Indian Politics 1870–1940* (Cambridge, 1973).
2 To take just one example: in Gallagher's account of Bengal in the 1930s, a major element in Congress's failure to rouse the east is the development of a powerful rich peasantry within the tenantry, resentful of the Hindu landlords who had traditionally controlled Congress here. See John Gallagher, 'Congress in Decline: Bengal, 1930 to 1939', ibid., pp. 269–325, especially pp. 278–9.

within particular peasant groups and in response to particular pressures and developments. In India modern peasant politics has clearly always proved a specialist affair, lacking any trace of the mass rejection of the political and social system which some would detect in China and South-East Asia. So action needs to be isolated and defined. A traditional method is to base examination on a three-fold division of the peasantry into 'rich', 'middle' and 'poor' peasants,[3] and for western India this can be accepted as a useful working model. The Bombay rich peasantry, for example, on our account the major beneficiaries of commercialisation processes, may have sought wider fields to conquer in the early twentieth century. Rural campaigns in the Presidency may, therefore, be predominantly exercises in assertiveness and muscle-flexing by the rich peasants, seeking to find their feet in the wider world of protest and politics. Several accounts certainly see the rich peasantry as the natural allies of centre politicians and 'the nationalist movement' in the western Indian countryside.[4] The traditional objection to this, of course, is to emphasise the rich peasant's habitual conservatism. Why should he wish to heighten the tempo of political activity, when this might eventually provoke wider discussion of the economic and social structure and father agitations to endanger his own position? The rich peasantry might need to experience some check or problem, however temporary, to drive them initially into political assertiveness. Yet the developments of the early twentieth century in western India – the increase in wage labour costs; the general tempering of the previously strong stratification tendencies – might have provided exactly the stimulus required. The heyday of the rich peasant had ended[5] and, however unchallenged his economic and social authority remained, resentment that increasing differentiation would no longer automatically occur may have been rife.

However, in most accounts, the eventual political response to economic and social stratification is a Leninist reaction by the impoverished mass of the poor peasantry. Whilst there was no mass uprising in western India, it is clearly possible that twentieth-century political action was rooted in deprivation. The possibility is strengthened

3 Stokes, in an important recent survey, describes this division as 'axiomatic'. See Eric Stokes, 'The Return of the Peasant to South Asian History', in his *The Peasant and the Raj. Studies in Agrarian Society and Peasant Rebellion in Colonial India* (Cambridge, 1978), p. 281.
4 See, for example, Judith M. Brown, *Gandhi's Rise to Power. Indian Politics 1915–1922* (Cambridge, 1972), Ch. 3, and S. J. M. Epstein, 'Bombay Peasants and Indian Nationalism – Politics in the Bombay Countryside 1919–1939', Cambridge seminar paper.
5 This was probably an all-India phenomenon. Stokes comments: 'there seems general agreement that the "golden age" of the rich peasant in India spanned the period 1860–1900'. Stokes, 'The Return of the Peasant to South Asian History', p. 275.

because the most publicised peasant political action of the nineteenth century – the Deccan Riots – was conspicuously a 'poor peasant' outburst, the fruit of economic depression in a region of marginal, famine-risk agriculture. However, major reservations about poor peasant politics concern both their opportunity and capacity to sustain action and their close social links with and economic dependence on the rich peasantry. The outbreak of the Deccan Riots was provoked by the absence of an indigenous moneylending elite and social and personal antagonism between the bulk of the villagers and aliens, but most Bombay poor peasants in the early twentieth century encountered an economic and creditor elite of closely similar social and cultural background. Yet these checks to action are not likely to subdue intense deprivation. If, for example, competition for scarce land by the 1920s and 1930s was such as to threaten the poor peasant's whole economic security, then some backlash might be expected.

Wolf and Alavi, in stressing the natural limitations on rich and poor peasant political action, pioneer the concept of the middle peasantry as the potentially radical force within the countryside.[6] The middle peasant is, perhaps, a much less easily identifiable individual than his 'rich' and 'poor' counterpart, but in early twentieth-century Bombay he can be defined, typically, as the part-owner, part-tenant, part-labourer: the peasant who owned some land of his own, but needed either to rent further land or to work for cash wages to supplement family income. Wolf's argument that the middle peasantry forms at once 'the main bearers of peasant tradition' and 'the most vulnerable to economic changes wrought by commercialism'[7] is clearly possibly applicable to this group. Indeed, in India, recently, historians, as Stokes put it, 'fall over themselves to discover the radical middle peasant at the centre of disturbance'[8] and, for western India, Hardiman originally categorised agitation in Gujarat as the pursuit of his interests.[9] Whilst the problem of defining the middle peasantry needs constant attention, it is possible that a group which, on our earlier definition, had become used to allocating its time and effort flexibly according to the market returns from landownership,

---

[6] Eric R. Wolf, 'On Peasant Rebellions', in Teodor Shanin, ed., *Peasants and Peasant Societies. Selected Readings* (London, 1971), pp. 264–74; Eric R. Wolf, *Peasant Wars of the Twentieth Century* (London, 1971); Hamza Alavi, 'Peasants and Revolution', in Kathleen Gough and Hari P. Sharma, eds., *Imperialism and Revolution in South Asia* (New York and London, 1973), pp. 291–337.

[7] Eric R. Wolf, 'On Peasant Rebellions', in Shanin, ed., *Peasants and Peasant Societies*, pp. 269–70.

[8] Stokes, 'The Return of the Peasant to South Asian History', p. 282.

[9] David Hardiman, 'Politicisation and Agitation among Dominant Peasants in Early Twentieth Century India: Some Notes', *Economic and Political Weekly*, 9, 28 February 1976, pp. 367–70.

renting or labour would feel most threatened by the contractions of the agricultural depression and the threat of land shortage in the late 1920s and 1930s.

However, our threefold categorisation of peasant politics does not require us to categorise just one as the archetypal form. Even Wolf's emphasis on the role of the middle peasantry is an attempted explanation of peasant *radicalism*: it does not deny the existence of other types of agrarian action with very different political connotations.[10] In what follows, we will examine the possibility of political activity by each of these three peasant groupings in western India, before trying to define the character of peasant politics and its significance more strictly.

'Poor peasant politics': tenant protest in the Konkan

In one sense, the typicality of peasant politics was always 'poor peasant politics'. The most frequent action by the Bombay peasant remained, in echo of 1875, the 'grain riot', the outburst of protest against temporarily harsh economic conditions. In particular, the onset of famine high prices in 1896 produced a rash of demonstrations against moneylenders and grain dealers. On 8 November, for example, a crowd of up to 2,000 gathered in Sholapur town to protest against the high price of jowar and a riot soon developed in which around Rs. 10,000 worth of property was plundered and one man killed.[11] At Karad in Satara District, two days later, some 700 bags of grain were stolen in two hours of rioting.[12] The late 1890s also witnessed a marked increase in banditry, with its roots in dispossession, the final act of defiant poor peasant protest. Gangs of Koli dacoits ravaged Thana District, 'looted the Marwari class unmercifully and have indulged in the most unnecessary acts of brutal violence',[13] mainly the by now traditional cutting off of noses and ears.

Grain riots and banditry remained endemic in western Indian agrarian society. To this extent the traditional constraints on poor peasant politics – the links with and dependence on the rich peasantry, the lack of resources to make effective protest – were not sufficient to prevent action wherever economic conditions deteriorated considerably. Even so, such protests

---

10 Also, Wolf himself, it would appear, is reticent about applying his thesis to societies such as India. His ideas arise from quite specific case studies, notably those in *Peasant Wars of the Twentieth Century*.
11 MARD, Vol. 205 of 1897, No. 567.
12 IOL, Bombay Revenue Progs., Famine, Vol. 5087, November 1896, No. 9048.
13 MARD, Vol. 88 of 1899, No. 630, Administration Report for Thana District 1897–8, para. 13.

themselves demonstrated the technical difficulties which poor peasant politics always encountered. The grain riot was by definition ephemeral, incapable of being sustained beyond the immediate grievance to serve wider objectives. Banditry, too, in an economy of settled cultivation, remained inevitably peripheral to the commonplace pursuit of poor peasant interests. In twentieth-century western India, effective political action by or on behalf of the poor peasantry, in view of the previous processes of tenurial change, needed to be consciously 'tenant politics'. Whilst many whom we should define as 'poor peasants' still owned some land, it was predominantly rent and interest levels and the general shape of economic relationships with landowning and commercial elites which most affected the poor peasantry. In contrast, agitation against land revenue, for example, realised little for such groups.

Yet tenant politics in Bombay Presidency faced other obstacles beyond the characteristic impediments to poor peasant action described by Wolf and Alavi. If the poor peasant interest was typically the tenant interest, the Bombay tenantry as a whole was, after 1900, a much more heterogeneous entity containing those who, wearing different hats, enjoyed contrasting status and political objectives. The simple problem of definition went further. Even the poor peasant tenant, if he was a recently fallen former proprietor, would not recognise himself as a member of a common, united tenantry: his aspirations, if not his root economic interests, were often those of the landowners, for the fiction remained, if only in the peasant's mind, that the land was still ultimately recoverable. Creating effective tenant movements was, therefore, a highly difficult business in most Bombay localities. In addition, links with the rich peasantry, as Wolf and Alavi argue, mollified tensions. In most parts of western India landlords were closely allied socially and culturally with their tenants: in such areas officialdom believed that, as in Ahmedabad District, 'the relations between landlords and tenants were generally remarkably amicable'.[14] Conflict seems to have been most overt, as we have seen, on the inam lands. Extensive buying-in of inamdar rights often created interloper landlords, unfamiliar with traditional arrangements and more prepared than their equivalents on ryotwari lands to ride roughshod over tenant interests. In Gujarat 'where the inamdar regards his village less as a mere investment',[15] tenant grievances were said to be far fewer. Yet, except perhaps in Satara District where some third of the land was alienated and where the inamdar–tenant contest seemed par-

14 *Bombay Presidency Annual Land Revenue Administration Report, 1906–07*, Part 2, p. 71, Report by J. De C. Atkins, Commissioner, Northern Division.
15 Ibid., p. 68, Report by Otto Rothfeld, Assistant Collector of Sholapur.

'Poor peasant politics': tenant protest in the Konkan 273

ticularly acute,[16] protests by inam tenants could hardly act as a rallying-call to the tenantry as a whole. The inam lands had always been a special case and, as we argued earlier, there were possible economic advantages as well as social disadvantages to be won from an absentee and interloper landlord. Conflict on the inam lands was often a rational response to opportunities for manoeuvre rather than the reaction to oppression.

The potentiality for tenant action was always greatest in the Konkan. Here the tenantry, at any rate in the khoti areas, was a much more socially cohesive entity, whose tenurial status was both widely evident and traditionally recognised. Further, Konkan tenants may have acquired demonstrably acute grievances, as the region's economic performance stagnated and the problem of population pressure on land became ever more severe. Harris believed as early as 1890 that Ratnagiri District was 'ripe for a serious agitation, which may need the employment of troops if it breaks out'.[17] This assumption, however, was born of Harris' strong distaste for the khots and the khoti system[18] and subsequent official orthodoxy argued that the Khoti Settlement Act of 1904 created a relatively stable social and political climate in the region. Whilst the process of commutation to cash rents introduced under the legislation at first undoubtedly produced tensions and fierce jockeying for position,[19] by around 1910 it was felt that 'the strained relations between khots and their tenants are somewhat improving as both sides are getting familiar with the new position'.[20]

In theory, the commutation of the early twentieth century should have brought significant economic gains to khoti tenants in the subsequent inflationary climate, if they were able to maintain the fixed cash levels static. One settlement report in the 1920s, on Khed taluk, argued that this was so and tenants had therefore 'gained much... from the fact that the cash rates were fixed mainly in 1909–10 on the basis of the average prices of 12 years previous to 1904, when the prices of the produce were far less than they are at present'.[21] In many areas of the Konkan, however, as we

16 Satara was often used as the yardstick by which other districts were judged. For example, in Sholapur, Rothfeld commented, 'the relations of inamdars to their tenants are more satisfactory than they were in the Satara District and less satisfactory than they are in Gujarat'. Ibid.
17 IOL, Harris Papers, Lord Harris to Lord Cross, Secretary of State, 25 May 1890.
18 He wrote in the same letter: 'by an "Idiot Bill"... we gave them [the Khots] the most extravagant powers over the unfortunate Rayat'. Ibid.
19 *Bombay Presidency Annual Land Revenue Administration Report, 1908–09*, Part 2, p. 69, Report by A. F. Maconochie, Collector of Ratnagiri.
20 *Bombay Presidency Annual Land Revenue Administration Report, 1910–11*, Part 2, p. 47, Report by the Commissioner, Southern Division.
21 *SBG*, NS, No. 627, Settlement Report on Khed Taluk, Ratnagiri District, by R. G. Gordon, undated, para. 61.

have seen before, intense population pressure on land and potential competition for tenancies permitted landlords, by the 1920s, to impose new demands and probably drive up the rents even of those paying in cash. Khed taluk, where the settlement officer found, in view of prevailing conditions, that 'I have not had many complaints against khots by tenants',[22] was not typically Konkani: an inland area where, in the east, dry-crop conditions existed and where population pressure was less severe. In the coastal taluks, there is more extensive evidence of tenant discontent by the 1920s and 1930s. Here, for example, agitation by khoti tenants was rife throughout 1925. On 21 June, five hundred cultivators attended a protest meeting at Pimpli Budruk in Chiplun taluk[23] and, after discussion in the Legislative Council, the local Revenue Commissioner, L. J. Mountford, was appointed to institute a special enquiry. Mountford, whilst noting that commutation should have benefited the tenantry, acknowledged the existence of a number of grievances: unfair distinctions between the status and obligations of different tenants, demands by khots for additional levies and forced labour, and the general supremacy of the khots, underpinned by their monopoly of village office and the poor communications prevalent throughout the Konkan.[24]

Similar tensions developed in the ryotwari areas of the Konkan where population pressure was greatest. This was so in the coastal regions of Kolaba District where the large majority of cultivators were tenants,[25] who had, moreover, enjoyed no official intervention to ameliorate their conditions like the Khoti Settlement Act. As a result, in Pen taluk for example, 'practically all rents are paid in kind' and landlords could obtain 'as great a share of the crop (usually from one half to three quarters) as the tenant could be induced to pay'.[26] In addition to these high proportions, 'the practice of demanding straw as well as grain has extended, with the effect of raising rents in practice'.[27] In Pen these circumstances exploded into trouble in 1920. Then the tenants began a widespread strike, refusing to pay high rents and in preference boycotting the cultivation of many holdings. For two years the strike 'affected practically all the salt rice villages in the Taluka' and 'hundreds of acres remained uncultivated'.[28]

22 Ibid., para. 16.
23 IOL, Bombay Land Revenue Progs., Vol. 11540, July 1926, p. 405.
24 Ibid., Report on Ratnagiri District by L. J. Mountford, No. KHT 20, 26 January 1926.
25 In Pen taluk, for example, at the 1921 Census tenants numbered 34,957 and labourers 3,387 out of a total agricultural population of 40,920. *SBG*, NS, No. 624, Settlement Report on Pen Taluk, Kolaba District, by J. R. Hood, 31 October 1923, para. 16.
26 Ibid., para. 21.
27 Ibid.
28 Ibid., para. 23.

Then in 1922 a compromise was evolved so that, when the taluk's assessment was revised in 1923, cultivation was back to normal. The agitations in Pen taluk in 1920–2 and in Ratnagiri District in 1925 form two clear examples of a positive tenant response to harsh economic conditions. Tensions, in the most heavily populated regions of the Konkan, remained endemic throughout the 1920s and 1930s: in this sense, militant poor peasant politics could clearly develop in western India, if access to crucial resources like land deteriorated in an area lacking the sedative influence of close social links between the agrarian elite and the poor. At the same time, the episodes discussed illustrate the severe limitations of poor peasant politics. Firstly, even in the Konkan the poor peasantry invariably required external aid to conceptualise their grievances and organise action. The beginnings of effective protest against the khots' exactions in the 1920s undoubtedly depended in part on the growing interest and involvement of politicians with wider power and influence. Some Legislative Council members, notably S. K. Bole, now highlighted the grievances of khoti tenants and, again, newspapers like the *Navayug* took up their cause.[29]

In addition, although Konkani tenant politics was always, in the interests it represented, 'poor peasant politics', it seems likely that the main protagonists were typically assertive groups within the poor peasantry with potentially upward social momentum. The Pen protest of 1920–2 was the work of Agri peasants on the highly fertile salt rice lands near the sea. In 1892 a settlement officer had written of this area: 'the Agris' field practice is of the highest order. Full advantage is taken of all positions favourable to cultivation, and embanking with field to field irrigation is carried out in high perfection.'[30] Such peasants would, perhaps, feel most threatened by deteriorating economic conditions and landlord exaction which ended the hope of upward mobility. Yet such groups, too, might be more capable of coming to terms with the agrarian elite, if their individual interests could be aided. The 1922 compromise in Pen taluk may have been of this order: for most tenants 'the general result has been a slight reduction in rent . . ., though not more than the landlords could well afford'.[31] Equally, however, the nature of this agreement may simply reveal the impotence of poor peasant politics. In the end, no such action could ever be more than spasmodic: the most effective

---

29 For the range of debate by the mid-1920s, see IOL, Bombay Land Revenue Progs., Vol. 11540, July 1926, p. 405.
30 *SBG*, NS, No. 291, Settlement Report on Pen Taluk, Kolaba District, by W. S. Turnbull, 22 December 1892, para. 25.
31 *SBG*, NS, No. 624, Settlement Report on Pen Taluk, Kolaba District, by J. R. Hood, 31 October 1923, para. 23.

weapon open to the tenantry was a boycott of cultivation like that organised in Pen but, even so, the poor peasant's desperate need for land for subsistence was likely to break him before the landlord suffered severely from loss of rent. The inability to sustain political action was the fundamental barrier to the poor peasantry's effective emergence on the political stage.

'Rich peasant politics' in the Deccan and Gujarat

At the same time, rich peasant conservatism would appear a likely, prominent theme of rural politics in the western India of 1900. The rich peasant's comparative identity of interest with a government which, during the nineteenth century, had benevolently presided over his prosperity was clear. The Bombay government had done nothing to inhibit the activities of the rural elite, not even by extending legal protection to tenants. The sole point at which official processes might challenge the economic position of the rich peasantry lay with the mechanism of land revenue reassessment, but the failure to raise the value of land revenue in real terms destroyed any spontaneous sense of outrage here. Opposition to land revenue remained the best potential rallying call for the landowning elite, but its use would, in likelihood, be a mere vehicle for other aspirations or grievances.

Further, in the Deccan, the organisation of the rich peasantry as an effective political force was inhibited, like poor peasant politics, by a fundamental problem of identification and recognition. As we have seen, the decline of many of the traditional village office-holders in the region provided the opportunity for the emergence of a highly diverse agrarian elite, some of whom were nouveau-riche. This limited the scope for geographically wide-ranging connections and alliances. For all their common role in the process of agricultural commercialisation, groups like the Saswad Malis and the Kunbis of central Satara had little to unite them politically. There could be, among such men, far from any class recognition of a rich peasant interest.

And yet it is possible to argue that the desire of such groups to assert themselves in their own locality lay behind one emergent political phenomenon of the late nineteenth century. The initial development of a rural anti-Brahmin movement in the Deccan can be seen as the work of commercially successful groups who, nevertheless, lacked traditional social status. The first formal recorded meeting of an anti-Brahmin association was at Poona in 1873 when an organisation with Jotirao

Phooley as President and Treasurer was inaugurated,[32] but it was in the 1880s that the movement first gathered momentum. Then evidence began to grow of increasing antagonism, in some localities, between Brahmins and non-Brahmins, in particular in Junnar taluk of Poona, Phooley's homeland. Typical was a major dispute between the hill Kolis of Ambegaon Petha in the west of the taluk and their Brahmins. To avoid the high charges made by the latter for conducting marriages and religious ceremonies, the Kolis began to constitute their own village Panches, bypassing the Brahmin's traditional function.[33] Such actions were strongly supported by Phooley's Satya Shodak Samaj and its newspaper, the *Din Bandhu*. The Samaj held large gatherings in Junnar town, attended, it was claimed, by as many as three thousand people.[34] By 1890 the activities of the Satya Shodak Samaj had spread south into Satara District.[35] Here and in the Southern Maratha Districts an organised anti-Brahminism had already developed during the 1880s. In 1886, for example, an organisation called the Lingayat Educational Association of Dharwar had strongly petitioned the Governor, Lord Reay, calling on him 'to secure a fair share of Government employment to the non-Brahminical classes'.[36]

The linkages between this phenomenon and the economic changes of the expansionary era between 1880 and 1896 seem evident. Phooley was himself from the prospering Junnar Mali community and his description of himself to the 1883–4 Education Commission as a 'merchant and cultivator'[37] is highly revealing about his own view of his economic status. More striking, the Ambegaon Kolis formed one of our earlier prime examples of nouveau-riche agricultural entrepreneurs. The lucrative myrabolam trade had brought them 'a steadily increasing revenue'[38] so that by the time of their disputes with the Brahmins 'there is amongst

32 Dhananjay Keer, *Mahatma Jotirao Phooley. Father of our Social Revolution* (Bombay, 1964), p. 126.
33 MARD, Vol. 345 of 1888, No. 1813, H. F. Silcock, Assistant Collector, Poona, to A. Keyser, Collector of Poona, No. 593, 11 May 1888.
34 Keer, *Mahatma Jotirao Phooley*, p. 197.
35 In September 1890 the *Din Bandhu* reported clashes between followers of the Samaj and the Brahmin community at Gondavla in the Man taluk of Satara. Report on the Vernacular Newspapers in the Bombay Presidency, week ending 27 September 1890, para. 37, *Din Bandhu*, 21 September 1890.
36 School of Oriental and African Studies, London, Papers of Lord Reay, Box 1, Memorial by the Lingayat Educational Association, Dharwar, June 1886, para. 4.
37 *Indian Education Commission, 1883–84, Bombay, Vol. 2* (Bombay, 1884), Jotirao Phooley, 'A Statement for the Information of the Education Commission'.
38 MARD, Vol. 345 of 1888, No. 1813, H. F. Silcock to A. Keyser, No. 593, 11 May 1888, para. 16.

them no serious distress: for hill men of their class they are moderately well off: the population has increased: the number of cattle owned by them has largely grown'.[39] In the south as a whole, the coincidence of the increase in anti-Brahmin sentiments among such as the Lingayats of Dharwar with the marked economic expansion which followed the Southern Maratha Railway's construction seems suggestive.

We can tentatively argue, then, that the rural anti-Brahmin movement of the 1880s and 1890s owed its origins to groups, economically successful in the conditions of the time, but anxious to achieve a social position, a formal recognition of their standing in the locality they hitherto entirely lacked. Yet, even so, the process merely illustrated the severe limitations of Deccan 'rich peasant politics' as already outlined. Despite the superficial impression of unity and organisation presented by Phooley and the Satya Shodak Samaj, Bombay anti-Brahminism before 1900 was in no sense an integrated 'movement'. Groups like the Ambegaon Kolis were concerned exclusively with the power structure in their own locality. This made rural anti-Brahmin protests sporadic, dislocated and always narrow in outlook and impact. Before 1900 this type of action had already, probably, reached its maximum effect: the death of Phooley in 1890 deprived the protest movements of wider leadership and moral suasion and the Satya Shodak Samaj was already experiencing setbacks in the 1890s.[40]

Thereafter, what developed as the mainstream 'Bombay anti-Brahmin movement' was something entirely different, deriving its initial momentum from the campaign waged by the Maharaja of Kolhapur in the early twentieth century.[41] Vestiges of the old agitations remained in the south. In Satara District in 1921, for example, anti-Brahmin sentiment became so intense that 'burning of houses, cutting down crops and destroying wells are repeatedly taking place' and the Collector was pressed to intervene to arbitrate.[42] Remembering the prevalence of alienated land in Satara, it is tempting to argue that this represented the resentments of nouveau-riche rich peasants against the traditional inamdar elite who had not necessarily participated so effectively in agricultural commercialisation.

---

39 Ibid., Bombay Government Revenue Department, Resolution No. 4471, 7 July 1888, para. 2.
40 During the 1890s, wrote A. B. Latthe later, 'the flood tides of reaction were sweeping everything before them, and the Satya Shodak remained in solitary villages objects of ridicule and contempt'. A. B. Latthe, *Memoirs of His Highness Shri Shahu Chhatrapati, Maharaja of Kolhapur* (Bombay, 1924), Vol. 2, p. 373.
41 On this see Ian Copland, 'The Maharaja of Kolhapur and the Non-Brahmin Movement, 1902–10', *Modern Asian Studies*, 7, 2, April 1973, pp. 209–25.
42 Report on the Vernacular Newspapers in the Bombay Presidency, week ending 19 February 1921, *Lokasangraha*, 14 February 1921.

Equally, however, the continued existence of anti-Brahmin agitation in Dharwar and Belgaum Districts as well as in Satara may have owed most, as the Brahmin press and the Congress establishment argued, to contiguity to Kolhapur and its activists. In any case, rural protest was now greatly outweighed by impeccably urban and constitutional politics. The leaders of the 'Deccan Ryots Association' who met Montagu and Chelmsford in December 1917 were men like A. B. Latthe, a Jain lawyer from Belgaum, V. R. Kothari, another Jain graduate from Poona, and B. V. Jadhav, a prominent official from Kolhapur state: their demands, in contrast to Phooley's range of social interests, were entirely political, in particular the reservation of half the public service appointments given to Indians for 'the backward classes'.[43]

If rich peasant assertiveness was to provide the theme for a more effective politics, then a rural elite was required with traditional authority allied to economic position and a broad sense of group social identity. Only this would transcend the inevitably limited and localised ambitions of such as the Ambegaon Kolis. The mixture did patently exist in one region of the Bombay Presidency: Gujarat. Here, as we have seen, a powerful rich peasantry, the Kanbi Patidars of Ahmedabad, Kaira and Surat Districts, dominated rural society by the early twentieth century. The Patidars had weathered the early development of British revenue policy with minimal challenges to their authority – indeed the survival of the narwa system was formal recognition of rights beyond those of ordinary ryotwari proprietors – and, from the late nineteenth century, pioneered and controlled the expansion of cotton and tobacco cultivation and marketing. Further, the Patidars were an 'intensely hierarchical and competitive' people.[44] The striking and complex social distinctions, which we noted in our review of land tenure in the mid-nineteenth century, prospered amidst conditions of extending commercialisation. In Kaira District, for example, in the 1890s the real elite 'kulin' or aristocratic group was made up of the inhabitants of only six villages.[45] The highly competitive marriage market and the fierce social climbing and rivalry among the Patidars, however bewildering to official observers, had important organisational consequences. The system created extensive, geographically wide-ranging alliances and provided a structure of

---

43 For the Association's address and a list of its deputation, see IOL, Papers of Edwin Montagu as Secretary of State for India, Vol. 37; for general comment on its demands, see Report on the Vernacular Newspapers in the Bombay Presidency, week ending 29 December 1917.
44 Pocock, *Kanbi and Patidar. A Study of the Patidar Community of Gujarat*, p. 1.
45 SBG, NS, No. 296, Papers on the Settlement of Anand Taluk, Kaira District, J. De C. Atkins, Collector of Kaira, to G. B. Reid, Commissioner, Northern Division, No. 436, 8 February 1895, para. 3.

leadership, authority and hierarchy which the Deccan rich peasantry conspicuously lacked and which might prove invaluable in political action.

The growing political consciousness of the Patidar group therefore seems axiomatic as the agitational base of Indian politics began to broaden. As early as 1895 vociferous protests in Bardoli taluk of Surat District against the revision revenue settlement foreshadowed the major campaign of 1928.[46] Kaira District saw a flurry of Home Rule League activity during the last years of the First World War.[47] The most striking action, however, and the leading agrarian political movement of the whole period before 1920 in Bombay Presidency came with Gandhi's famous campaign against the payment of land revenue in Kaira District in 1918. The causes, as traditionally described, are undoubtedly accurate: a commercialised, relatively educated peasantry with a temporary but important grievance against the authorities in the coincidence of bad harvests, sharp rises in living costs and a full revenue demand.[48]

Yet the social basis of the movement requires further emphasis. It was exclusively a Patidar campaign, based on the three central and eastern taluks of Nadiad, Anand and Borsad, all strongholds of the Lewa Kanbis.[49] The first agitations against payment of the year's land revenue began in Nadiad town in November 1917[50] and Gandhi inaugurated his satyagraha here in March 1918. Nadiad was, par excellence, the Patidar town. It had developed from an old narwa village – Pedder had called the six leading Nadiad narwadars 'the richest and most influential men of the district'[51] – and within the population the Lewa Kanbis had always outnumbered the second largest caste group by more than double.[52] The great narwa villages, the organisational heart of the Patidari community

46 See *SBG*, NS, No. 359, Papers on the Settlement of Bardoli Taluk, Surat District.
47 *Source Material for a History of the Freedom Movement in India* (Bombay, 1963), Vol. 2, 1885–1920, pp. 729–30.
48 As Judith Brown puts it: 'in Gujarat the emergence of a prosperous peasant elite and the spread of vernacular education created a potential peasant leadership. A set of bad harvests and high prices during the war gave these peasants a grievance against the administration.' Judith M. Brown, 'Gandhi in India, 1915–1920: his Emergence as a Leader and the Transformation of Politics' (University of Cambridge, unpublished Ph.D. thesis, 1968), p. 481. See also the same author's *Gandhi's Rise to Power. Indian Politics 1915–1922*.
49 At the time of the *Gazetteer* enquiries, of approximately 130,000 Lewa Kanbis in Kaira, some 39,070 lived in Anand taluk, 31,871 in Borsad and 31,739 in Nadiad taluk. *BG*, Vol. 3, *Kaira and Panch Mahals*, pp. 154–5, 161, 164.
50 See *Source Material for a History of the Freedom Movement in India*, Vol. 2, pp. 730–1.
51 *SBG*, NS, No. 114, Correspondence Relating to the Introduction of the Revenue Survey in Kaira District, W. J. Pedder to C. J. Prescott, No. 11, 21 March 1862, Appendix 1.
52 See *BG*, Vol. 3, *Kaira and Panch Mahals*, pp. 175–7.

– Nadiad, Od, Ajarpura, Kasar and Chikhodra – were all thereafter in the forefront of the movement. Further, Gandhi himself soon recognised, indeed came to encourage, the campaign's exclusive caste complexion. At Uttarsanda in Nadiad taluk, a village strongly committed to the cause, he told his audience: 'Uttarsanda is all Patidars and, if this fight is to be won, it is only your community that will do so.'[53] Elsewhere, Gandhi gave powerful voice to Patidari aspirations: 'the Patidars are a venturesome community; they are Kshatriyas . . . they should not go back on their plighted word, should not betray their Kshatriya blood but fight on to the last'.[54]

Economically, an equally prominent theme united the areas and villages where the movement was strongest: they were all among the most commercialised and agriculturally productive of the district. Navagam village in Nadiad taluk, for example, proved a major centre of intense opposition to land revenue payment. Gandhi spoke here in April and June 1918 and it was Navagam men who committed the famous theft of onions from an attached field, which led to the imprisonment of Mohanlal Pandya and his five accomplices. Navagam was situated in the heart of the Khari, the great rice-growing depression in the valley of the Shedhi river: output was traditionally high through the area but the settlement officer of 1894 had hailed Navagam as 'the finest of the Khari villages'.[55] Further south, the centres of tobacco cultivation were prominent in the campaign. In Borsad taluk, Sunav and Palaj in the north-west were two of seven villages described in 1895 as the economic elite of the taluk: 'all . . . are equally favourably situated with regard to markets, and produce first class tobacco'.[56] In Palaj the movement was particularly strong. The local Mamlatdar spent four days here trying, without success, to persuade cultivators to pay up[57] and Gandhi spoke to large crowds at both villages on 22 April.[58] However, away from the localities of Patidar proprietorship and cash-crop cultivation in the south and east of Kaira, the roots of the satyagraha became progressively more shallow. In the west of the district the villages of Matar and Mehmedabad taluks – millet-growing, populated substantially by Koli rather than Kanbi cultivators, occasionally susceptible to climatic failure – were very little

53 *The Collected Works of Mahatma Gandhi*, Vol. 14 (New Delhi, 1958), No. 210, Speech to the Villagers of Uttarsanda.
54 Ibid., No. 234, Speech to the Villagers of Od.
55 SBG, NS, No. 295, Settlement Report on Nadiad Taluk, Kaira District, by T. R. Fernandez, 17 April 1894, para. 66.
56 SBG, NS, No. 337, Settlement Report on Borsad Taluk, Kaira District, by E. Maconochie, 19 January 1895, para. 44.
57 See *The Collected Works of Mahatma Gandhi*, Vol. 14, No. 245.
58 See ibid. and No. 246.

involved. The exceptions, like Khandali village in Matar, where incidents prompted one of Gandhi's few excursions to the west, were normally highly untypical of the area. Khandali was 'more fertile' and 'the general condition of the people is better' than in most of Matar and 'the number of Patidar cultivators is large'.[59] Even where the campaign was not strong, its character as a rich Patidar movement was highly conspicuous.

If, then, we can talk of 'rich peasant politics' in western India, the Kaira satyagraha seems a clear example. It was also typically rich peasant politics in another sense: its failure to arouse the mass of the peasantry. The compromise of June 1918 which ended the campaign was hailed as victory by the jubilant authorities. Willingdon, the Bombay Governor, believed that Gandhi 'has taken a bad fall over the Kaira business' and insisted that 'we gave no concession'.[60] The Collector of Kaira called the satyagraha 'for all practical purposes a failure'.[61] The statistics of revenue collection bore out the substance of this argument: 98.8 per cent of the net revenue demand, under 1 per cent less than the average for Gujarat, was collected in Kaira in 1917–18.[62] This, of course, totally ignored over Rs. 3 lakhs of suspensions and remissions granted to Kaira cultivators, almost entirely the results of concessions to the campaign: remissions and suspensions in other Gujarat districts were trifling.[63] Yet in Kaira in 1917–18 the authorities collected Rs. 22 lakhs from an original demand of just under Rs. 25½ lakhs,[64] by no means for this period an untypically or alarmingly low proportion.

These statistics suggest that, when it came to political behaviour, the economic and social elite of village society could never simply manipulate the actions of the bulk of the peasantry. In Kaira some of the most successful, well-organised and socially assertive peasants in the Bombay Presidency led by a Mahatma could not arouse generalised opposition. The limitations of rich peasant power were here exposed. It may be, of course, that the time for wider action was simply not ripe: if real wages for labour work were improving during the 1910s, this may have taken the edge from middle and poor peasant grievances and fears about inflation and poor seasons. Yet in Kaira as in the bulk of the Bombay Presidency

---

59 *SBG*, NS, No. 646, Report on the Current Settlements in Matar and Mehmedabad Taluks, M. D. Bhat, 28 February 1931, para. 33.
60 IOL, Willingdon Papers, Vol. 1, Willingdon to Lord Chelmsford, Viceroy, 11/12 June 1918.
61 *Bombay Presidency Annual Land Revenue Administration Report, 1917–18*, Part 2, p. 2, Report by Mr Ker, Collector of Kaira.
62 Ibid., Part 1, p. 9.
63 The next-largest total of remissions and suspensions was Rs. 59,595 in Ahmedabad District, under a fifth of the Kaira figure. Ibid., Part 2, p. 1.
64 Ibid.

the interest of most of the peasantry, who owned only small amounts of land, seemed inevitably limited in a campaign against land revenue payment. The rich peasants could simply not offer enough to their fellows to participate in action organised by and for the rural elite. Like tenant protest, rich peasant politics in the Bombay Presidency seemed inevitably constrained by strict technical limitations.

The Bardoli campaign of 1928: the middle peasant in politics?

The Kaira campaign, then, was justifiably hailed as a victory for officialdom. Ten years later, however, a protest movement against land revenue payment in Gujarat achieved very different results. In 1928 a new land revenue settlement came into force in Bardoli taluk of Surat District and it created immediate widespread and concerted refusal to pay the new charges. The Bombay government, despite the confiscation and sale of some protesters' lands, proved unable to break the movement and in August 1928 it compromised by offering an independent enquiry to re-examine the settlement.[65] When the enquiry reported in 1929 its two members, R. S. Broomfield and R. M. Maxwell, to official surprise, conceded the validity of the protest, slashing the land revenue enhancement in Bardoli from between 20 and 25 per cent to just 6 per cent.[66] The affair had widespread repercussions. Broomfield and Maxwell not only criticised the details of the Bardoli settlement but also castigated the whole methodology used to calculate land revenue reassessments in Bombay Presidency. In practice, too, as the Bombay government conceded, 'the effect of Bardoli has been to show that, if a taluka can organise itself sufficiently, it can resist the introduction of a revised settlement as at present prepared'.[67] Bardoli hence ended formal land revenue revisions

---

65 The official discussions during the course of the Bardoli campaign can be read in IOL, Public and Judicial Department Papers, Vol. 1957 of 1928, No. 984. See, also, IOL, Papers of Lord Halifax as Viceroy of India, Vols. 22 and 23, and IOL, Papers of Lord Birkenhead as Secretary of State for India, Vols. 13, 16. For a contemporary account from the protesters' point of view see Mahadev Desai, *The Story of Bardoli* (Ahmedabad, 1929). The Bardoli campaign is discussed in the modern literature in Ghanshyam Shah, 'Traditional Society and Political Mobilization: The Experience of Bardoli Satyagraha, 1920–1928', *Contributions to Indian Sociology*, NS, 8, 1974, pp. 89–107; Anil Bhatt, 'Caste and Political Mobilization in a Gujarat District', in Rajni Kothari, ed., *Caste in Indian Politics* (New Delhi, 1970), pp. 299–337. The campaign is set in its wider political connotation in Judith M. Brown, *Gandhi and Civil Disobedience. The Mahatma in Indian Politics, 1928–34* (Cambridge, 1977), pp. 29–33.
66 Broomfield and Maxwell, *Report of the Special Enquiry into the Second Revision Settlement of the Bardoli and Chorasi Talukas*.
67 IOL, Halifax Papers, Vol. 23, Sir Frederick Sykes, Governor of Bombay, to Lord Irwin, Viceroy, 1 March 1929, para. 17.

for the remainder of the British period in western India and, by so doing, closed the door on expansion of the provincial government's traditional major source of finance.[68] If the Bombay revenue department had won in 1918, by 1929 it had been humbled.

But why was the Bardoli campaign so strikingly more successful than the Kaira satyagraha? The explanation may be simply the increasing weakness of governmental authority by 1928–9 and the necessity now to come to terms with concerted action to maintain the balancing act of imperial rule: the Bombay government was highly embarrassed by resignations from the Legislative Council over the Bardoli affair and subsequent refusals by members to appoint a committee to meet the Simon Commission. Again, the effective organisation of the local Congress network can be emphasised. Most accounts stress the role of traditional caste leadership in rousing and channelling protest.[69] Yet, however efficient and well timed the Bardoli campaign's assault on official policy, the roots of its success seemed to lie in the breadth of opposition to the payment of the new revenue rates. By late July 1928 when the government was forced into the concession of an independent enquiry, only 1,830 out of a total 14,855 assessees had paid their land revenue.[70] This was despite increasingly rigorous action against defaulters involving declarations of forfeiture against lands totalling 65,000 acres in area.[71] It seems unlikely that simply organisational manipulation could have created such widespread action when Gandhi's exhortations had failed to arouse more than a minority of Kaira landowners. In addition, Bardoli's was arguably a more socially diverse community, lacking the cement which the Kanbi Patidars' supremacy achieved in north Gujarat. In Bardoli, Patidar groups were strongly represented on the land but so were other caste groups who, as Mahadev Desai commented, might have undermined the campaign: incomer Vanias, the 'meekest of the meek' Raniparaj, the 'proud and defiant' Anavla Brahmins, and the 'mild Parsis' as well as a Muslim community.[72]

68 For details on the failure of Bombay land revenue collections to increase in the inter-war period, see Heston, 'Official Yields per Acre in India, 1886–1947: Some Questions of Interpretation', p. 327n.
69 Shah, 'Traditional Society and Political Mobilization', for example, sees the campaign's leaders mobilising 'hitherto apolitical masses' through the aid of 'linkmen', 'local leaders of different castes and communities'. Bhatt, 'Caste and Political Mobilization in a Gujarat District', stresses the role of the caste association, the Patidar Yuvak Mandal, which 'plunged into the movement and assumed its leadership'.
70 IOL, Political and Judicial Department, Papers, Vol. 1957 of 1928, No. 984, House of Commons Statement by Earl Winterton, Under-Secretary of State for India, 23 July 1928.
71 Ibid.
72 The phrases are Mahadev Desai's. See Desai, *The Story of Bardoli*, pp. 54–5.

It seems reasonable, therefore, to seek the roots of the Bardoli campaign's success in the particular conditions of the local society and economy in 1928. Here Hardiman's argument that the middle peasant formed the main protagonist of rural politics in Gujarat stands as a powerful interpretation. Hardiman claims that in nineteenth-century Gujarat 'the rising class of rich capitalist peasants had consolidated their power largely at the expense of the middle peasantry'.[73] In harsher economic conditions in the twentieth century, the middle peasants came under strong pressure and 'their long-felt discontent burst forth in fanatical Gandhian movements'.[74] There are undoubtedly difficulties in accepting this interpretation wholeheartedly: it does not seem to fit the Kaira campaign and the assumption that economic performance was deteriorating in the twentieth century rests apparently on short-term judgements about the quality of individual seasons rather than wider analysis of long-term agricultural trends. Nevertheless, the heart of Hardiman's explanation may clearly account for the Bardoli campaign. This may have been an area where the middle peasantry was powerful and widespread and, therefore, if aroused by an immediate grievance, capable of creating broadly based political action. Bardoli was a land-extensive district where the frontiers of cultivation were steadily advancing. These are the very conditions which seem most to favour the application of Chayanov's model of peasant social mobility,[75] a model which in turn carries the implication that the bulk of the peasantry forms an intact, unified 'middle peasant' entity.

Attributing motivation to any middle peasant protest in Bardoli in 1928 is, also, no difficult matter, even if our general account of economic change must lead us to reject Hardiman's thesis of emiseration. At one level, the Bardoli campaign was caused quite simply by the agricultural depression. In 1924 when the settlement officer, M. S. Jayakar, had formulated his new revenue assessment proposals for Bardoli, agricultural prices, and particularly those of the leading local commercial crop, cotton, were at their peak. In 1925 cotton prices crashed and by 1928, as the increased revenue charges fell due, the price in the taluk was just half the level of four years previously, Rs. 16 per maund compared with Rs. 32 in

73 Hardiman, 'Politicisation and Agitation among Dominant Peasants in Early Twentieth Century India: Some Notes', p. 369.
74 Ibid.
75 In the sense that availability of land is the crucial factor which enables those on the upward section of Chayanov's family life-cycle to expand operations. Note, too, that attempted applications of Chayanov within India have been in relation to the relatively land-extensive Punjab: see Kessinger, 'The Peasant Farm in North India, 1848–1968', and Dewey, 'Social Mobility and Social Stratification amongst the Punjab Peasantry: Some Hypotheses'.

1924.[76] On Wolf's interpretation, the middle peasant's radicalism arises from constant challenges to his security which threaten to tumble him into the poor peasantry as well as the firm access to land which secures him the technical ability to act. In Bardoli the price collapse may have threatened a mass of middle peasants with retreat from a lucrative commercial agriculture into a desperate quest for renewed security in subsistence farming.

But was the bulk of the Bardoli population a landowning middle peasantry? In fact, the taluk's social structure seemed marked by deep traditional divisions in formal status. Not only had the hali system been extensive here, but also other wage labouring groups were numerous. In 1895 the estimated proportion of agricultural wage labourers to total population was as high as 32.34 per cent.[77] Since nearly a quarter of Bardoli's people were employed outside agriculture, there were, therefore, almost as many labourers as holders of land, 45.56 per cent of the total population in 1895.[78] This suggested, by western Indian standards, a sharply polarised society. The settlement officer of 1895 pointed out how in a typical north Gujarat taluk, Nadiad of Kaira District, the proportion of labourers to landholders was as low as one to seven: the simple lesson was that 'wealth is more equally distributed in the north than in the south'.[79] In practice, it was arguably this extensive availability of landless labour which had permitted the rapid expansion of a relatively labour-intensive crop, cotton, in Bardoli during the late nineteenth century. The proportion of a quickly extending cultivated acreage growing cotton rose from 17.45 per cent at the time of the *Gazetteer* enquiries of 1873–4[80] to 24.78 per cent by 1895.[81] At the same time, the cotton producers were benefiting from favourable price differentials. Between 1865 and 1895 Bardoli cotton rose in price by nearly 30 per cent, whilst its major competitors for the land of the taluk, wheat, rice and jowar achieved increases of only 21.4 per cent, 18.4 per cent and 14.5 per cent respectively.[82] All these, however, are circumstances which normally suggest, not the maintenance of an extensive and stable middle peasantry but the familiar story of a rich peasant elite riding to economic and social

76 Broomfield and Maxwell, *Report of the Special Enquiry into the Second Revision Settlement of the Bardoli and Chorasi Talukas*, Appendix A.
77 SBG, NS, No. 359, Settlement Report on Bardoli Taluk, Surat District, by T. R. Fernandez, 2 December 1895, para. 20.
78 Ibid.: by 'holders of land' here are meant tenants as well as cultivating proprietors.
79 Ibid.
80 BG, Vol. 2, *Surat and Broach*, p. 280.
81 SBG, NS, No. 359, Settlement Report on Bardoli Taluk, Surat District, by T. R. Fernandez, 2 December 1895, para. 16.
82 Ibid., para. 39.

power on the back of commercialisation. Stratification as a hypothesis about how change occurs in Indian agrarian society surely works best, of all Bombay localities, for late nineteenth-century Bardoli, where cultivating proprietors of cash crops formed a minority of the population, but enjoyed significant real price improvements and a large supply of wage labour which, in view of its abundance, could hardly have been expensive.

This would return us to a rich peasant orientated explanation of the Bardoli campaign and from what we have said of Gujarat's economy and society such an interpretation would always appear most realistic. Yet, how then can we explain, without resort to intense levels of organisational manipulation, how the bulk of the population, many of them traditionally landless, could be aroused for a campaign in the interests of proprietors? The solution, arguably, lies in the economic trends of the decade immediately before the onset of the price collapse. After 1890 the price rise in Bardoli cotton accelerated, roughly doubling by 1910 and reaching new levels twice as high again in the best years between 1917 and 1924, when Japanese demand for Indian cotton was so intense.[83] Consequently, cotton cultivation extended even more rapidly. By 1923–4 just over 40,000 acres in Bardoli taluk were growing cotton, nearly a third of the total cultivated area and representing an increase of over 50 per cent on the acreage of the mid-1890s.[84] In these boom conditions, the rate of expansion of cotton cultivation may have begun to press on even Bardoli's available labour resources by the early 1920s. Labourers, in turn – in Bardoli as throughout western India the complaint about allegedly rising labour costs was widely voiced – may have been able to force improvements in their wages and conditions, creating some more broadly based distribution in the returns from cotton cultivation. If so, a far wider stake in commercial agriculture could have come into existence.

This sounds highly speculative, but there is supporting evidence of more fluid tenurial patterns in the early twentieth century. The 1895 settlement had presented Bardoli agrarian society as predominantly composed of proprietary cultivators and the agricultural labourers who presumably worked for them: types of arrangement where peasants held land as tenants or crop-sharers involved at very most around 20 per cent

---

[83] For a full list of cotton prices in Bardoli from 1895 to 1928, see Broomfield and Maxwell, *Report of the Special Enquiry into the Second Revision Settlement of the Bardoli and Chorasi Talukas*, Appendix A.
[84] *SBG*, NS, No. 647, Settlement Report on Bardoli Taluk, Surat District, by M. S. Jayakar, 30 June 1925, para. 11.

of the population.[85] By 1926, however, the Bombay Settlement Commissioner considered that in Bardoli 'at least half the total area is held by landlords who do not cultivate it'.[86] What, then, was happening? One explanation of this large increase in tenancy is that peasants in the boom conditions were now renting the lands of others to expand cash-cropping activities and even, possibly, that labouring groups were also leasing in to secure direct involvement in cotton cultivation. The report of the enquiry into the Bardoli campaign certainly believed that participation in commercial agriculture had become socially extensive: 'in the mania for "getting rich by cotton", not only landless men and adventurers but cultivators of every class and position, bound themselves to pay rents'.[87] In the conditions of the early 1920s this may have made sound economic sense, for it seems likely that landlords were unable to force rents up to match the pace of the cotton price rise. The former Collector of Surat, at any rate, immediately inferred this when sent the data from the new settlement of 1925. Pointing to 'enormous' increases in sale prices of land, but only moderate rises in leasing rates, he suggested that 'the cultivating tenants and labourers were now able to derive better bargains with the landowners'.[88]

In this way, the commitment to and interest in commercial agriculture in the Bardoli of 1925 was arguably widespread throughout the community, for the taluk had travelled beyond the situation where production for the market could be closely manipulated by a narrow rich peasant elite. The collapse in cotton prices was, therefore, a disaster for many groups and individuals: 'heavy losses, debts and sometimes insolvency followed, the natural results of speculation which fails to come off'.[89] Since the authorities hit upon this moment to come cap in hand to the village, opposition to the land revenue reassessment proved an ideal focus for grievances. Increases in revenue rates directly affected only proprietors: but the tenants, too, faced the prospect that landlords, who in the boom conditions appear to have acquiesced in some decline in the real value of rents, would now be determined to pass on a share of the enhancement. In this way a broad social basis for the campaign was created.

85 *SBG*, NS, No. 359, Settlement Report on Bardoli Taluk, Surat District, by T. R. Fernandez, 2 December 1895, Appendix D.
86 *SBG*, NS, No. 647, Papers on the Settlement of Bardoli Taluk, Surat District, Report by F. G. H. Anderson, No. ST490, April/May 1926, para. 21.
87 Broomfield and Maxwell, *Report of the Special Enquiry into the Second Revision Settlement of the Bardoli and Chorasi Talukas*, para. 24.
88 *SBG*, NS, No. 647, Papers on the Settlement of Bardoli Taluk, Surat District, A. M. Macmillan to M. S. Jayakar, 18 October 1925.
89 Broomfield and Maxwell, *Report of the Special Enquiry into the Second Revision Settlement of the Bardoli and Chorasi Talukas*, para. 24.

Yet do these findings suggest that the Bardoli movement should be seen as 'middle peasant politics'? If we are right about the shifting balance of economic power in Bardoli, a 'middle peasantry' was being born out of the cotton speculation of the early twentieth century, in the sense that opportunities from agricultural commercialisation were, arguably, now being distributed more widely and evenly. Yet this is not the classic middle peasantry whose role the Wolf thesis has amplified, for Wolf, like the Marxists, appears to regard classification within the peasantry in terms of precise economic function and landholding status rather than the shifting, fluid categories shaped by economic change. Wolf's is a subsistence middle peasantry – 'a peasant population which has secure access to land of its own and cultivates it with family labour'[90] – whereas such a group in Bardoli in the 1920s would be 'a middle peasantry of commercialisation'. Orthodoxy, however, may be correct in its failure to discern the latter as a distinct social entity. Its existence depended on economic forces, subject to variation and flux: if we are right that commercial producers may have been able to retrench in the 1930s by economising on labour costs, then any centripetal social tendencies creating the appearance of a middle peasantry in Bardoli may have been decisively reversed in the decade after 1928.

In the terms, therefore, of the established debates, the Bardoli campaign cannot justifiably be regarded as 'middle peasant politics'.[91] The point was not that one dominant peasant class had acted but that a powerful and widespread economic grievance had united fundamentally distinct social groups. Bardoli was highly untypical in the western India of the 1920s: extensively commercialised, with a widespread commitment to cash-crop agriculture, the depression here was a blow at the aspirations of many peasant groups, 'poor peasants' as much as 'rich peasants'. The coincidence of a land revenue enhancement was, therefore, able to create here the most effective peasant politics of the whole British period in western India.

The peasantry and politics in the Bombay Presidency

The history of western India confirms Wolf and Alavi's reservations about the technical weaknesses of exclusively rich peasant and poor peasant based politics. However, the Bardoli campaign suggests that the achievement of broadly based unity among peasant groups, typically in

---

90 Wolf, 'On Peasant Rebellions', in Shanin, ed., *Peasants and Peasant Societies. Selected Readings*, p. 269.
91 For a fuller examination of and debate about this argument, see my dispute with Hardiman in the *Journal of Peasant Studies*, 7, 3, 1980; 8, 3, 1981; 9, 4, 1982.

reaction to a common grievance, may be a more useful explanation for effective peasant politics than the quest for a particular active social group. The same may apply to other areas of Asia: the position of the Chinese Communist Party in the Yenan period seems to rest on policies, such as debt reduction, with widely based appeal in contrast to the more sectional and arguably less successful radicalism of the days immediately before the Long March.[92] In conclusion, we might approach the problem of the unusual character of the Bardoli campaign within Bombay peasant politics from a different angle. Why were there not many more examples of an assertive and broadly based peasant politics in western India?

The explanation may be in one sense because of the degree of development of the Bombay agrarian economy, in another because of its relative backwardness. In western India by the early twentieth century, economic and social stratification had occurred enough to diversify interests and create different impulses towards political action. Tenurial change, for example, by converting many poor peasants into, predominantly, tenants rather than landowners, limited the social appeal of crusades against land revenue payment. Hence distinctive 'rich peasant' and 'poor peasant' movements, like those discussed earlier, came into being. Since, too, there is no evidence of any widespread sudden downturn in agricultural performance or any creation of a mass proletariat, the grievances which underlay such campaigns were necessarily limited and localised. The Yenan region of China, however, and, indeed, the whole of northern Shensi province had not, by the 1930s and 1940s, reached this stage of development. The peasantry here in an area, uncommercialised, of arid soil and small-scale, subsistence agriculture and lacking any development of tenurial diversity, could more easily coalesce in united action.[93] In this sense, it was the relative absence of economic and social stratification which might create the opportunity for a more turbulent and effective peasant politics.

Yet in western India the one significant united action had occurred in Bardoli, by Indian standards a highly advanced, heavily commercialised locality. Bardoli, then, illustrates alternative conditions for an assertive peasant politics. Where the commitment to commercial agriculture was socially extensive, then fluctuations in demand conditions could create grievances across different social groups. To be more specific, if the bulk

---

92 On the policies of the Yenan period, see Mark Selden, *The Yenan Way in Revolutionary China* (Cambridge, Mass., 1971).
93 For Wolf, of course, the Yenan peasantry contains many elements of his 'tactically mobile' middle peasantry. See *Peasant Wars of the Twentieth Century*, Ch. 3 and Conclusion. Yet the ability to act as a relatively united force may still disguise substantial distinctions of economic and social status.

of the Bombay peasantry had been engaged in cash-crop production in the 1920s, then the price falls of the agricultural depression might have reaped a whirlwind of mass peasant protest. Bardoli, however, was matched by only a few localities in western India in degree of dependence on an export crop and the breadth of involvement in its cultivation. As it was, the economic and social history of the Bombay countryside, marked in the British period by the highly differential spread of commercialisation and both geographical and social stratification, was a mixture little conducive to the development of effective mass peasant politics.

10

Conclusions: the problem of differential commercialisation

How did Indian agrarian society react to the imperial impact? Why did peasant agriculture fail to be transformed or itself transform the Indian economy? In reviewing these broad issues, the case study of the Bombay Presidency provides answers, which seem superficially familiar. Economic expansion and social change occurred, but they seem to have been accommodated to indigenous systems and structures, altering but not revolutionising them. One fundamental early example of the process at work was the failure of the ryotwari system, despite the intrusive implications of its administrative arrangements, decisively to reshape the existing social structure of most western Indian regions. Ravinder Kumar's conclusion – that 'by the opening decades of the twentieth century... the villages of Maharashtra were dominated by a nouveau-riche class of Kunbis'[1] – can be accepted for the central Deccan districts around Poona upon which he focuses: but throughout most of Bombay Presidency long-established office holding elites had survived and even flourished, emerging seemingly more virile from resistance to the administrative drive of the Wingate–Goldsmid settlements.

It was true that the differential impact of British revenue policy had some significant economic consequences. Like the contrast between Jat enterprise and Rajput languor in north India,[2] the distinction drawn between the thrusting, adaptive agriculture permitted by ryotwari arrangements and the economically restrictive atmosphere of talukdari and khoti society perhaps rests on more than simple official prejudice. Agriculture in the khoti south Konkan, for example, despite its intensive methods of cultivation and traditionally high levels of output, failed to diversify and develop a pioneering, commercialised component. Equally, however, serious economic problems can be blamed, particularly the lack of modern communications in the region. The most conspicuous effect of the interaction between British revenue arrangements and village society remained the completion of the district and village officers' decline in the central Deccan. In turn, this situation provided the real roots of the riots

---

1 Kumar, *Western India in the Nineteenth Century*, p. 330.
2 For a recent discussion, see Eric Stokes, 'Dynamism and Enervation in North Indian Agriculture: the Historical Dimension', Ch. 10 of his *The Peasant and the Raj*.

Conclusions: the problem of differential commercialisation 293

of 1875. Yet by 1900, as Kumar argues, the power vacuum had been smoothly filled by emergent peasant groups from the regionally dominant caste and by the early twentieth century, the legacy of the revenue impact was one of subtle shifts in local conditions, not a revolutionary change in agrarian society.

Similar remarks seem apposite about the dominant qualitative development within agriculture, the spread of commercialisation. Between the mid-nineteenth century and the onset of the agricultural depression in the 1920s, cultivation of commercial crops, particularly cotton, and the marketing of foodgrains both extended steadily. The consequence was notable differentiation, repeated down to the smallest physical unit, since commercialisation depended on favourable ecological and particularly transport conditions. However, broad regional variations are also clearly discernible. The areas where the process was most buoyant and wide-ranging were in much of Gujarat, in parts of Belgaum and Dharwar districts in the south and, within the Deccan, in the canal zones and the districts of Khandesh and Satara. These regions remained the pacemakers of agricultural performance, for cumulative advantages became most evident after 1900 when growing regional specialisation in crop production coincided with conspicuous price improvement in real terms among the commercial crops. Initially, commercialisation seemed to favour, as traditional orthodoxy has it, rich peasant elites. Social stratification is evidenced, not just by logical deduction from the character of primary commercialisation, but also by tenurial change and the rise of leasing arrangements which, typically resting on kind rents, suggested a balance of power and advantage lying with landlords.

To this extent, then, the agrarian economy and society changed, but the consequences clearly remained limited. Many poor peasants undoubtedly 'lost their lands' in a formal sense, but tenurial change, as is now widely recognised, did not necessarily produce shifts in real social relationships or methods of economic operation within agriculture. The cultivator of the soil usually remained the same and, indeed, local recognition of innovation might be entirely absent where rent payment merely replaced debt obligation. Also, in so far as the first stage of commercialisation was pioneered by the minority with the resources – carts, implements and access to credit – to latch on to market opportunities, in many localities it developed as a stealthy process, peripheral to the central concerns of the village economy. After 1900 commercialisation, in the more advanced regions and localities, had eaten into the heart of the village economy and its operation, but, coincidentally, the distribution of rewards from commercial agriculture and economic opportunity in general seems to have become more broadly based. Extensive labour requirements for cash-crop cultivation forced up some labourers' wages and in the villages which produced widely for external markets opportunities for emigration and supplementing incomes outside agriculture became more known and utilised. The rule may be, then, that more extensive agricultural commercialisation tempers social stratification and this may apply in other parts of Asia as in India.[3] In turn, the process arguably had vital implications for the preservation of a peasant society. In Bardoli taluk by the early 1920s many groups, including some traditionally landless, had acquired a valuable stake in cotton cultivation. Alternatively, in the radically different case of the east Deccan village examined by Mann, Jategaon Budruk, the estimated average income per family rose from Rs. 166 in 1917 to Rs. 203 in 1925–6, despite a fall in receipts from agriculture, because opportunities for external employment widened.[4] These developments provided peasants with an economic base to preserve their position within the village intact: the forces of commodity production were here not destroying the peasantry but being successfully ridden by them precisely to maintain the social status quo.

In this way, the process of agrarian change in the Bombay Presidency differed radically from the classic Marxian 'agrarian transition', involving the complementary formation of a class of rising rural capitalists and a landless proletariat. Instead, it accords more closely with Goodman and Redclift's 'second formulation', drawn from Latin American cases, where 'peasant producers may be incorporated into commodity production and exchange but retain ownership in the means of production and control of the immediate production process'.[5] However, since this development is

[3] Recent work on China seems to lend weight to the thesis. Studies by Jack Potter and Ramon Myers argue that social stratification in Chinese agrarian society was by no means as intense as is often assumed. It is striking, however, that the localities investigated – in Potter's case a village in the Hong Kong territories; in Myers' the regions of the north, subject in the 1930s to the economic stimulus of Japanese occupation – were hardly typical of the country as a whole. These were the Chinese Bardolis: places where the commercialisation of agrarian society had, by the Second World War era, gone comparatively very far. An interpretation which sees broadly based commercialisation as productive of centripetal social tendencies would marry these findings with the more traditional accounts of agrarian change in the bulk of China. See Jack M. Potter, *Capitalism and the Chinese Peasant: Social and Economic Change in a Hong Kong Village* (Berkeley, 1968); Ramon H. Myers, 'Socio-economic Change in Villages of Manchuria during the Ch'ing and Republican Periods: Some Preliminary Findings', *Modern Asian Studies*, 10, 4, October 1976, pp. 591–620; Ramon H. Myers, *The Chinese Peasant Economy: Agricultural Development in Hopei and Shantung, 1890–1949* (Cambridge, Mass., 1970).
[4] *Report of the Royal Commission on Agriculture in India, 1926–28*, Vol. 2, Part 1, Evidence Taken in the Bombay Presidency, Evidence of Harold H. Mann, Appendix A, pp. 16(i)–16(ii).
[5] David Goodman and Michael Redclift, *From Peasant to Proletarian. Capitalist Development and Agrarian Transitions* (Oxford, 1981), p. 215.

Conclusions: the problem of differential commercialisation 295

held to have involved fundamental differentiation and the subordination of mass peasant interests, even it does not seem to accord with the western Indian case. One important contrast with Goodman and Redclift's Latin America seems to be the absence of any decisive state structural role. The impotence of British official policy and action to shape events on the ground has become an almost wearisome orthodoxy, but clearly the practical impact of the Deccan Agriculturists' Relief Act underpins the case for western India. The Bombay administration may have favoured the rich peasantry but it totally lacked the authority, as well as the inclination, consciously to carve out hegemony for one social group. In practice, for all the economic thrust of social stratification during the late nineteenth century, the power of rural elites in western India was not limitless. Their social position rested on local consent and, as the Kaira satyagraha revealed, the middle and poor peasants were not a colourless mass who could be simply and automatically aroused in support of rich peasant aspirations. No political movement, therefore, developed – for the reasons we saw in the previous chapter – with the aim and ability to reshape rural society and the village was hence able to absorb and assimilate all forms of flux and change. Commercialisation strengthened rather than weakened peasant society. In western India, in sum, it was commodity production which came to terms with the village structure, not vice versa.

Our evidence for the growth of a more flexible economy by the early twentieth century sets in context the question of trends in productivity and output. Where labour work and migration were available, stable living standards could exist, even if per capita agricultural output was declining. Yet in most localities there is no reason to assume any such decline, at any rate before the 1930s. The famines, certainly the most devastating outbreak of 1899–1900, seem attributable to climatic cataclysm rather than evidence of widespread pressure on mass consumption and land in cultivation was extending alongside population growth until the interwar period. Having said this, however, technical improvement in agriculture was clearly limited. Numbers of plough cattle, for example, seem to have risen very little – may well have fallen significantly relative to land under tillage – between 1850 and the 1930s, and the famines of the late 1890s were in many areas a drastic blow to livestock resources. From the 1910s, too, attempted improvement in stock numbers faced the problem of increasingly restricted availability of grazing land. It may be, as some officials always argued, that cattle numbers were fundamentally sufficient for requirements, but Wingate's comments of the 1840s – that many cultivators were forced to hire cattle to plough effectively – must have remained valid for many areas in the 1930s. In likelihood cattle numbers were probably typically enough to maintain an equilibrium of performance, without impinging on land availability for grazing too heavily: but not sufficient for any significant improvement in ploughing technique.

Again, irrigation facilities in western India remained sparse. By 1925–6 around 1.1 million acres in the Bombay Presidency boasted some form of irrigation, just under a quarter million acres by the canal schemes, but the total gross cropped area by then stood at nearly 28 million acres.[6] It was significant, too, that irrigation facilities were fairly evenly distributed within western India so that the highest individual proportion was in Poona District where just 10.8 per cent of the land was irrigated in 1925–6.[7] Economic change had created regional disparity, with some districts much more involved in commercial agriculture than others, but there were still no extensive districts of high farming, such as the Punjab canal colonies, in the Bombay Presidency.

In the absence, then, of significant technical improvement, agriculture had necessarily worked hard in order to increase its output. There were clearly some economic gains from developments like the greater regional specialisation in crop production which was particularly evident during the early twentieth century, but fundamentally increased output came from greater use of two inputs: land and labour. The cultivated acreage was extending steadily in western India throughout the period between the mid-nineteenth century and the 1930s and, at the same time, land was being worked more intensively, particularly in the land-scarce Konkan. Fragmentation, driving down the size of the unit of cultivation, both permitted and was a symptom of the process. Clearly, though, labour and, especially, land were not limitless resources and this might suggest that agricultural expansion in western India was strictly a 'once and for all' process. We have not encountered, certainly before the inter-war period, any serious constraints on development and diversification imposed by land and labour availability, other than some tendency for cash-crop cultivation to spread most rapidly in areas, like Bardoli, where virgin land was most widely and readily accessible. However, by the 1930s major problems over the cost and supply of both labour and land may have struck. In so far as labour, at any rate on larger holdings, was hired on the market as well as supplied by the peasant family, costs for labour-intensive cultivation, as we have argued, may have become inhibitive in the conditions of the agricultural depression. More important, the expansion of the cultivated area seemed, at last, to be overtaken by the rate of population growth after 1930. By the end of our period the dynamic of

6 IOL, Bombay Land Revenue Progs., Vol. 11916, July 1931, p. 85.
7 Ibid.

Conclusions: the problem of differential commercialisation 297

agricultural expansion in western India seemed to have been decisively weakened.

All the features we have described may, to many, suggest Geertz's familiar model of 'agricultural involution'. His account of Java shows how rapidly increasing population was soaked up on the land by ever more intensive and productive cultivation, a process, however, which was 'ultimately self-defeating' in developmental terms.[8] The social symptoms were exactly those we have outlined in western India in the twentieth century: 'tenure systems grew more intricate; tenancy relationships more complicated; co-operative labour arrangements more complex – all in an effort to provide everyone with some niche, however small, in the overall system'.[9] Yet, in economic terms, Bombay agriculture could not have become 'involuted' until the eve of the Second World War. Whilst population in Java was increasing by as much as 2 per cent per annum between 1830 and 1900, pressing heavily on means of subsistence,[10] in western India aggregate rates of population growth were slow until the 1920s. This not only postpones the onset of Bombay involution, but raises questions about the value of the concept for pointing differences in agricultural development. In a sense, involution is the end of all agricultural development, unless the sector is transformed by exogenous events, for, although the Malthusian link between economic crisis and mortality may not hold, the obverse – that increased productivity will result in more mouths to feed – usually exists to some degree. Again, all agricultural expansion is a 'once and for all' process, in the sense that it presses on the finite resource of land. However, if involution is a constant threat to all agricultural systems, its force as a road-block to development may still not be decisive, at any rate in the early decades after its onset. By most tests, Russian agriculture had become involuted during the late Tsarist period. Absolute numbers on the land were rising rapidly and, whilst cultivated acreage was extending, gains in output were mainly dependent on the more intensive working of ever smaller units of land: between the emancipation of 1861 and 1917, land was moving steadily from gentry landowners to the peasantry and the tendency towards fragmentation of holding and complication of tenure was, in Russia, greatly strengthened by the village commune's redistributory activities.[11] Yet industrial development – some would even say a 'take off'

---

8 Clifford Geertz, *Agricultural Involution. The Process of Ecological Change in Indonesia* (Berkeley and Los Angeles, 1968).
9 Ibid., p. 82.
10 Ibid., p. 69.
11 The best source on the late Tsarist countryside remains Geroid T. Robinson, *Rural Russia under the Old Regime: a History of the Landlord–Peasant World and Prologue to the Peasant Revolution of 1917* (London, 1932).

or the beginnings of modern economic growth – occurred synonymously. Here involution was counteracted by the state, allied with foreign capital, mobilising effective resources for development.

This directs our attention, in the western Indian case, to the failure of responses external to agriculture. In the vital comparison between nineteenth-century Japan and India, any distinctions based on simple agricultural expansion may not be as sharp as previously assumed, certainly if we contrast Japan with the pacemaking regions like the Punjab and Gujarat. However, what seems conspicuous in the Japanese rural economy is the buoyancy of the local handicraft sector and its linkages with rich peasant enterprise and resources.[12] Whilst Indian handicrafts were, as is now widely recognised, by no means the simple overwhelmed victims of Western capitalism, the sector in India may crucially have been deprived of a major role as a source of growth and development. Capital, therefore, generated in India by agricultural improvement may have lost the institutional opportunity for diversification into industrial investment, which an expanding handicraft sector might present, remaining tied up within agriculture and rural moneylending. This is, as yet, mere hypothesis. Yet it perhaps suggests the sort of area where work on the Indian rural economy should now concentrate.

Nevertheless, there was still in western India a fundamental agricultural problem evident by the early twentieth century. This was the tendency for land to be held in ever smaller units of cultivation, for all rights and assets in the village to be intensely sub-divided. By the 1920s in the khoti areas of Ratnagiri District 'there may be as many as 80 khoti sub-sharers' with a stake in one village.[13] This, again, is a typical phenomenon of peasant society in many parts of modern Asia, exemplified by Potter's Hong Kong village[14] as well as Geertz's Java. For Geertz, this is part and parcel of involution, the 'late Gothic' social manifestations of economic trends, but, to repeat, the case of western India does not seem to fit this interpretation since rights were becoming deeply sub-divided in the early twentieth century before the involution associated with population pressure had set in.

What, then, created the growing social intricacy in the Bombay village? Arguably, the dominant phenomenon we have described – agricultural commercialisation with a highly differential local impact – was the cause.

12 See, e.g., T. C. Smith, 'Landlords and Rural Capitalists in the Modernisation of Japan', *Journal of Economic History*, 16, 2, June 1956, pp. 165–81.
13 IOL, Bombay Land Revenue Progs., Vol. 11540, July 1926, p. 405, Report on Ratnagiri District by L. J. Mountford, No. KHT20, 26 January 1926, para. 7.
14 Here Potter notes the tendency for the size of units of cultivation to drop. *Capitalism and the Chinese Peasant*, p. 64.

Conclusions: the problem of differential commercialisation

Commercialisation during the late nineteenth and early twentieth centuries created new forces on the Bombay land but, as we have continually emphasised, reacting effectively to them depended on special circumstances: holding land suitable in terms of ecological conditions for commercial crops and enjoying transport facilities to achieve access to wider markets, for example. Extent of adaptation to commercialised agriculture, therefore, varied widely from region to region, from locality to locality, from village to village and even from holding to holding. The effects of this 'differential commercialisation' were especially deep-seated and complex because the initial beneficiaries of the process were not sufficiently powerful to maintain their grip unchallenged. In the Bombay village a range of groups were increasingly able to establish some stake in commercialisation and poor peasants maintained their position in the village rather than running proletarianised from it. As a result, as differential commercialisation developed, many individuals competed fiercely to obtain the most productive land and the most valuable rights. In most villages, land was intensely fragmented, because purchasers, renters and heirs struggled to win the smallest portion of the best soil. This was why it was typically the most productive land, particularly where it co-existed alongside much more arid, infertile soil – as, classically, on the Deccan canal lands – rather than land in the areas of greatest population density, which became most fragmented. Many within the Bombay village by the early twentieth century were capable of adapting to commercialisation and yet only differential commercialisation occurred. The social response therefore was, through the operation of the land and rental market and inheritance procedure, to permit many to acquire some tiny involvement in the process.

Yet in developmental terms the implications of this, as Geertz argues of his involution, were dangerous. The Deccan peasant who obtained a small strip of irrigated land clearly needed to work it for its maximum short-term output with the crops which were currently most highly priced on the market: the cash they realised was, in cases, vital insurance against proletarianisation. This, in turn, may have had ecological consequences, inimical to long-term productivity. More important, the process involved the dissipation of the possible developmental advantages of commercial agriculture. Instead of creating concentrations of capital resources, the income from cash-crop cultivation was used, in a type of conservative egalitarianism, predominantly to preserve the village social system. The lesson is that commercialisation and, in general perhaps, major changes in the demand for and the value of commodities needed, historically, to occur on a wide front within rural economies if they were not to become counter-productive. Differential commercialisation, such

Conclusions: the problem of differential commercialisation

as occurred in the Bombay Presidency in the British period, produces quantitative expansion but not qualitative change within peasant economies.

GLOSSARY

badhekari	Tenant-at-will in the Konkan.
bajra	Millet foodgrain.
balutedar	Village artisan or provider of village services.
bania	Merchant, trader or moneylender.
bhagdar	Member of joint proprietary system in Broach District.
dacoit	Bandit.
deshmukh	Civil Governor or Collector under the Mahratta government.
dharekari	Cultivator in the Konkan with hereditary rights to the land.
huks	Emoluments paid to village officers.
inam	Grant of land revenue free or of control over land revenue.
jagir	Grant of an estate revenue free.
jowar	Millet foodgrain.
khot	Holder of office and superior proprietary rights in the Konkan.
kulkarni	Village accountant.
lakh	One hundred thousand. Hence numbers of Indian units, such as money, are given in the text in this form: Rs. 3,00,000.
mirasdar	Cultivator with hereditary rights to the land.
mofussil	Interior of the country.
munsif(f)	Small court.
narwadar	Member of joint proprietary system in Kaira District.
panchayat	Village or caste council.
patel	Village headman.
redwa	Heavy cart.
ryot, rayat, raiyat	Peasant.
sanad	Document specifying grant.
satyagraha	Gandhian protest campaign. (lit. 'truth force')
sowkar, sawcar	Moneylender.
surinjam	Grant of an estate revenue free, usually for military service.
tacavi, takavi	Loan made by government for agricultural purposes.
tal	Embankment.
taluk	Administrative unit, comprising a sub-division of a district.
talukdar	Superior landlord, in western India mainly found in Ahmedabad District.
thakur	Superior landlord in Broach District.

thal	Hereditary family estate.
tunkha	Government share of the land revenue.
upri	Newcomer with no proprietary claim to the soil.
watan	Village office.
watandar	Village office-holder.

BIBLIOGRAPHY

1. Unpublished material

A. Private papers of agricultural and revenue officials

Indian Papers of Sir James Caird (IOL Home Miscellaneous Series, No. 796).
Papers of Sir William Lee-Warner (IOL Political and Secret Department Library).
Papers of Sir George Wingate (Sudan Archive, School of Oriental Studies, Durham University).

B. Private papers of Governors, Viceroys and Secretaries of State

(All the following, except where stated, are available in the manuscript collection of the India Office Library.)
Letter books of Sir James Ferguson.
Sir James Ferguson Collection.
Indian Papers of Lord Halifax.
Papers of Lord Harris.
Papers of Edwin Montagu.
Papers of Lord Reay (Library of the School of Oriental and African Studies, London).
Papers of Sir Frederick Sykes.
Papers of Sir Richard Temple.
Papers of Lord Willingdon.
Papers of Sir Philip Wodehouse.

C. Records of the Bombay Government

Revenue Department Papers, Maharashtra State Archives, Bombay.

D. Proceedings of the Bombay Government and the Government of India

IOL, Judicial, Land Revenue, Legislative and Revenue Proceedings.

E. Newspaper reports

Weekly Reports on the Vernacular Newspapers in the Bombay Presidency from 1877.

F. Unpublished theses

Brown, Judith M. 'Gandhi in India, 1915–1920: his Emergence as a Leader and the Transformation of Politics' (University of Cambridge, Ph.D. thesis, 1968).
Charlesworth, Neil. 'Agrarian Society and British Administration in Western India, 1847–1920' (University of Cambridge, Ph.D. thesis, 1974).
Dewey, Clive. 'The Official Mind and the Problem of Agricultural Indebtedness in India, 1870–1914' (University of Cambridge, Ph.D. thesis, 1973).

G. Unpublished papers

Dewey, Clive. 'Social Mobility and Social Stratification amongst the Punjab Peasantry: Some Hypotheses', Institute of Commonwealth Studies, London, 2 March 1976.
Epstein, S. J. M. 'Bombay Peasants and Indian Nationalism – Politics in the Bombay Countryside, 1919–1939', Cambridge Seminar Paper.
Morris, Morris D. 'What is a Famine?', SSRC Conference on Indian Economic and Social History, July 1975.
Stokes, Eric. 'The Emergence of the Rich Peasant under Colonial Rule: Some Northern Lights', SSRC Seminar on Social Stratification, April 1977.

2. Official publications

A. Censuses

Census of the Bombay Presidency, 1872.
Census of India, 1881.
Census of India, 1891.
Census of India, 1901.
Census of India, 1911.
Census of India, 1921.
Census of India, 1931.
Census of India, 1951.

B. Major official reports and commissions

Report of the Royal Commission on Agriculture in India, 1926–28. Vol. 2, Evidence Taken in the Bombay Presidency. Vol. 14, Appendix to the Report.
Report of the Bombay Provincial Banking Enquiry Committee, 1929–30.
Report of the Deccan Riots Commission, 1876–78.
Indian Education Commission, 1883–84. Vol. 2, Evidence Taken before the Bombay Provincial Committee and memorials addressed to the Commission.

Report on the Famine in the Bombay Presidency, 1899–1902. Vol. 1, *Report.* Vol. 2, *Appendices.*
The Indian Statutory Commission, 1930. Vol. 7. Memorandum Submitted by the Government of Bombay.

C. Parliamentary papers

1852–3. Vol. 69. Statistical Papers Relating to India.
1852–3. Vol. 75. Official Correspondence on the System of Revenue Survey and Assessment in the Bombay Presidency.
1862. Vol. 55. Return of Cotton Goods Exported to and Imported from India.
1862. Vol. 55. Statistics on Indian Cotton.
1866. Vol. 52. W. H. Sykes, Report on the Land Tenures of the Deccan.
1878. Vol. 58. Report of the Deccan Riots Commission (Report only).
1880. Vol. 52. Report of the Indian Famine Commission, 1880. Report, Parts 1 and 2.
1880. Vol. 53. Report on the Condition of India by James Caird.
1881. Vol. 71. Report of the Indian Famine Commission, 1880. Part 1: Report, Part 3 and Appendices 1 and 2. Part 2: Appendix 3. Part 3: Appendices 4 and 5.
1887. Vol. 62. Correspondence Respecting Agricultural Banks in India.
1902. Vol. 70. Report of the Indian Famine Commission, 1901.

D. Publications of the Government of India

Index Numbers of Indian Prices, 1861–1926 (Department of Commercial Intelligence and Statistics, New Delhi, 1928).
Selections from the Records of the Government of India, Home Department, No. 342. Papers Relating to the Working of the Deccan Agriculturists' Relief Act during the Years 1875–94 (Calcutta, 1897).

E. Publications of the Government of Bombay

(i) Printed annual reports
Annual Administration Reports from 1860.
Annual Crop and Season Reports from 1903–4.
Annual Jamabandi Reports from 1881–2.
Annual Land Revenue Administration Reports from 1903–4.
Annual Police Reports from 1872.
Annual Registration Department Reports from 1876.
Annual Reports on Income Tax Operations from 1886.

(ii) Gazetteers of the Bombay Presidency
From 1877. Date and place of publication specified.

(iii) Assorted selections from the Bombay Government records

SBG, No. 1, Report by Captain Wingate on the Survey and Assessment of Khandesh, 1852.

SBG, No. 2, Report by Captain Wingate on Introducing a Survey and Revision of Assessment in the Ratnagiri Collectorate, 1852.

SBG, No. 4, R. N. Gooddine, Report on the Village Communities of the Deccan, 1852.

SBG, No. 10, Report on the Portion of the Daskroi Purgunna Situated in Ahmedabad Collectorate by Captain J. Cruikshank, 1853.

SBG, NS, No. 2, Report on the Southern Districts of the Surat Collectorate by A. F. Bellasis, 1854.

SBG, NS, No. 29, Correspondence Regarding the Concealment by the Hereditary Officers and Others of the Revenue Records of the Former Government and the Remedial Measures in Progress, 1856.

SBG, NS, No. 30, Selection of Papers Explanatory of the Origin of the Inam Commission and its Progress, 1856.

SBG, NS, No. 31, Correspondence Exhibiting the Results of the Scrutiny by the Inam Commission of the Lists of Deccan Surinjams, 1856.

SBG, NS, No. 106, An Account of the Talukdars in the Ahmedabad Zillah by J. B. Peile, 1867.

SBG, NS, No. 114, Correspondence Relating to the Introduction of the Revenue Survey Assessment in the Kaira Collectorate, 1869.

SBG, NS, No. 134, Selections, with Notes, from the Records of Government Regarding the Khoti Tenure by E. T. Candy, 1873.

SBG, NS, No. 157, Papers and Proceedings Connected with the Passing of the Deccan Agriculturists' Relief Act.

SBG, NS, No. 180, Correspondence Relating to the Conditions on which Certain Estates are Held in the Salsette Taluk of the Thana Collectorate, 1886.

SBG, NS, No. 446, Papers Regarding the Proprietary Rights of Khots in the Ratnagiri District, 1907.

SBG, NS, No. 524, Character of Land Tenures and the System of Survey and Settlement in the Bombay Presidency, 1914.

SBG, NS, No. 646, Report on the Current Settlements in Matar and Mehmedabad Taluks, Kaira District, by M. D. Bhat, 1932.

(iv) Settlement and assessment reports

The reports run into several hundreds and are specified in the references.

(v) Miscellaneous official publication

Broomfield, R. S. and Maxwell, R. M. *Report of the Special Enquiry into the Second Revision Settlement of the Bardoli and Chorasi Talukas* (Bombay, 1929).

Bibliography

3. Other published works

A. Non-official reports and source material

The Indian Central Cotton Committee, *General Report on Eight Enquiries into the Finance and Marketing of Cultivators' Cotton, 1925–28.*

Report of the Peasant Enquiry Committee of the Maharashtra Provincial Congress Committee (Poona, 1936).

Source Material for a History of the Freedom Movement in India (Bombay, 1963), Vol. 2, 1885–1920.

B. Published books and articles

Alavi, Hamza. 'Peasants and Revolution', in Kathleen Gough and Hari P. Sharma, eds., *Imperialism and Revolution in South Asia* (New York and London, 1973), pp. 291–337.

Anderson, F. G. H. *Facts and Fallacies about the Bombay Land Revenue System Critically Expounded* (Poona, 1929).

Attwood, D. W. 'Why Some of the Poor Get Richer: Economic Change and Mobility in Rural Western India', *Current Anthropology*, 20, 3, September 1979, pp. 495–508.

Baden-Powell, B. H. *The Land Systems of British India* (Oxford, 1892).

Ballhatchet, Kenneth. *Social Policy and Social Change in Western India, 1817–1830* (London, 1957).

Bates, Crispin N. 'The Nature of Social Change in Rural Gujarat: the Kheda District, 1818–1918', *Modern Asian Studies*, 15, 4, October 1981, pp. 771–821.

Bhatia, B. M. *Famines in India. A Study in Some Aspects of the Economic History of India, 1860–1965* (London, 1967).

Bhatt, Anil. 'Caste and Political Mobilization in a Gujarat District', in Rajni Kothari, ed., *Caste in Indian Politics* (New Delhi, 1970), pp. 299–337.

Bhutani, V. C. 'Agricultural Indebtedness and Alienation of Land', *Journal of Indian History*, 47, 1, April 1969, pp. 65–89, and 47, 2, August 1969, pp. 261–83.

Blyn, George. *Agricultural Trends in India, 1891–1947: Output, Availability and Productivity* (Philadelphia, 1966).

Boserup, Esther. *The Conditions of Agricultural Growth: the Economics of Agrarian Change under Population Pressure* (London, 1965).

Braudel, Fernand. *Afterthoughts on Material Civilization and Capitalism* (Baltimore and London, 1977).

Capitalism and Material Life, 1400–1800 (London, 1973).

Breman, Jan. *Patronage and Exploitation. Changing Agrarian Relations in South Gujarat, India* (Berkeley, Los Angeles and London, 1974).

Broadbridge, Seymour. 'Economic and Social Trends in Tokugawa Japan', *Modern Asian Studies*, 8, 3, July 1974, pp. 347–72.

Brown, Judith M. *Gandhi and Civil Disobedience. The Mahatma in Indian Politics, 1928–34* (Cambridge, 1977).
  Gandhi's Rise to Power. Indian Politics, 1915–1922 (Cambridge, 1972).
Byres, T. J. 'Land Reform, Industrialization and the Marketed Surplus in India: an Essay on the Power of Rural Bias', in David Lehmann, ed., *Agrarian Reform and Agrarian Reformism. Studies of Peru, Chile, China and India* (London, 1974).
Cassels, Walter R. *Cotton: an Account of its Culture in the Bombay Presidency* (Bombay, 1862).
Catanach, I. J. 'Agrarian Disturbances in Nineteenth Century India', *The Indian Economic and Social History Review*, 3, 1, March 1966, pp. 65–84.
  Rural Credit in Western India. Rural Credit and the Co-operative Movement in the Bombay Presidency, 1875–1930 (Berkeley, Los Angeles and London, 1970).
Charlesworth, Neil. *British Rule and the Indian Economy 1800–1914* (London, 1982).
  'The Middle Peasant Thesis and the Roots of Rural Agitation in India, 1914–1947', *Journal of Peasant Studies*, 7, 3, April 1980, pp. 259–80.
  'The Myth of the Deccan Riots of 1875', *Modern Asian Studies*, 6, 4, October 1972, pp. 401–21.
  'Rich Peasants and Poor Peasants in Late Nineteenth Century Maharashtra', in Clive Dewey and A. G. Hopkins, eds., *The Imperial Impact. Studies in the Economic History of Africa and India* (London, 1978), pp. 97–113.
  'The Roots of Rural Agitation in India, 1914–1947: a Reply to Hardiman', *Journal of Peasant Studies*, 9, 4, July 1982, pp. 266–76.
  'The Russian Stratification Debate and India', *Modern Asian Studies*, 13, 1, February 1979, pp. 61–95.
  'Trends in the Agricultural Performance of an Indian Province: the Bombay Presidency, 1900–1920', in K. N. Chaudhuri and Clive Dewey, eds., *Economy and Society. Essays in Indian Economic and Social History* (New Delhi, 1979), pp. 113–40.
Chaudhuri, S. B. *Civil Disturbances during British Rule in India, 1765–1857* (Calcutta, 1955).
Connell, A. K. *Discontent and Danger in India* (London, 1880).
Copland, Ian. 'The Maharaja of Kolhapur and the Non-Brahmin Movement, 1902–1910', *Modern Asian Studies*, 7, 2, April 1973, pp. 209–25.
Dandekar, V. M. and Khudanpur, G. J. *The Working of the Bombay Tenancy Act, 1948. A Report of Investigation* (Poona, 1957).
Desai, Mahadev. *The Story of Bardoli* (Ahmedabad, 1929).
Dewey, Clive. 'Patwari and Chaukidar: Subordinate Officials and the Reliability of India's Agricultural Statistics', in Clive Dewey and A. G. Hopkins, eds., *The Imperial Impact. Studies in the Economic History of Africa and India* (London, 1978), pp. 280–314.
Divekar, V. D. 'Regional Economy 1757–1857. 3. Western India', in Dharma Kumar, ed., *The Cambridge Economic History of India, Vol. 2* (Cambridge, 1983).

Dore, R. P. *Land Reform in Japan* (London, 1959).
Dutt, Romesh C. *The Economic History of India in the Victorian Age* (London, 1903).
Farmer, B. H. *Agricultural Colonization in India since Independence* (London, 1974).
Frykenberg, R. E. *Guntur District, 1788–1848: a History of Local Influence and Central Authority in South India* (Oxford, 1965).
Fukazawa, Hiroshi. 'Agrarian Relations: Western India', in Dharma Kumar, ed., *The Cambridge Economic History of India, Vol. 2* (Cambridge, 1983).
  'Land and Peasants in the Eighteenth Century Maratha Kingdom', *Hitotsubashi Journal of Economics*, 6, 1, June 1965, pp. 32–61.
Gallagher, John. 'Congress in Decline: Bengal 1930 to 1939' in John Gallagher, Gordon Johnson and Anil Seal, eds., *Locality, Province and Nation. Essays on Indian Politics, 1870–1940* (Cambridge, 1973).
Gandhi, M. K. *The Collected Works of Mahatma Gandhi* (New Delhi, 1958).
Gathorne-Hardy, Alfred E. *Gathorne Hardy, First Earl of Cranbrook. A Memoir* (London, 1910).
Geertz, Clifford. *Agricultural Involution. The Process of Ecological Change in Indonesia* (Berkeley and Los Angeles, 1968).
Goodman, David and Redclift, Michael. *From Peasant to Proletarian. Capitalist Development and Agrarian Transitions* (Oxford, 1981).
Gordon, R. G. *The Bombay Survey and Settlement Manual* (Bombay, 1917).
Green, H. *The Deccan Ryots and their Land Tenure* (Bombay, 1852).
Grigg, D. B. *Population Growth and Agrarian Change. An Historical Perspective* (Cambridge, 1980).
Guha, Amalendu. 'Raw Cotton of Western India: 1750–1850', *The Indian Economic and Social History Review*, 9, 1, March 1972, pp. 1–42.
Gupte, K. S. *The Bombay Tenancy Problem* (Poona, 1938).
  The Deccan Agriculturists' Relief Act (Act XVII of 1879) as Modified up to 31 March 1928 (Poona, 1928).
Habib, Irfan. 'Potentialities of Capitalistic Development in the Economy of Mughal India', *Journal of Economic History*, 29, 1, March 1969, pp. 32–78.
Hanley, Susan and Yamamura, Kozo. 'Population Trends and Economic Growth in Pre-industrial Japan', in D. V. Glass and Roger Revelle, eds., *Population and Social Change* (London, 1972), pp. 451–99.
Hardiman, David, 'The Crisis of the Lesser Patidars: Peasant Agitations in Kheda District, Gujarat, 1917–34', in D. A. Low, ed., *Congress and the Raj. Facets of the Indian Struggle, 1917–1947* (London, 1977), pp. 47–75.
  'Politicisation and Agitation among Dominant Peasants in Early Twentieth Century India: Some Notes', *Economic and Political Weekly*, 9, 28 February 1976, pp. 367–70.
  'The Roots of Rural Agitation in India, 1914–1947: a Rejoinder to Charlesworth', *Journal of Peasant Studies*, 8, 3, April 1981.
Harnetty, Peter, 'Cotton Exports and Indian Agriculture, 1861–1870', *Economic History Review*, 24, 3, 1971, pp. 414–29.

Heston, Alan, W. 'National Income', in Dharma Kumar, ed., *The Cambridge Economic History of India, Vol. 2* (Cambridge, 1983), pp. 376–462.

'Official Yields per Acre in India, 1886–1947: Some Questions of Interpretation', *The Indian Economic and Social History Review*, 10, 4, December 1973, pp. 303–32.

Hobsbawm, E. J. *Bandits* (London, 1969).

Holderness, B. A. 'Credit in English Rural Society before the Nineteenth Century, with Special Reference to the Period 1605–1720', *The Agricultural History Review*, 24, 1976.

Hunter, William Wilson, *Bombay 1885–90: a Study in Indian Administration* (London, 1892).

Jacquemont, Victor. *Voyage dans l'Inde pendant les années 1828 à 1832* (Paris, 1841).

Jhirad, J. F. M. 'The Khandesh Survey Riots of 1852: Government Policy and Rural Society in Western India', *Journal of the Royal Asiatic Society*, 1968, Parts 3 and 4, pp. 151–65.

Johnson, Gordon. *Provincial Politics and Indian Nationalism. Bombay and the Indian National Congress 1880–1915* (Cambridge, 1973).

Keatinge, G. *Agricultural Progress in Western India* (London, 1921).

Rural Economy in the Bombay Deccan (London, 1912).

Keer, Dhananjay. *Mahatma Jotirao Phooley. Father of our Social Revolution* (Bombay, 1964).

Kessinger, Tom G. 'The Peasant Farm in North India, 1848–1968', *Explorations in Economic History*, 12, 1975, pp. 303–23.

Vilyatpur, 1848–1968 (Berkeley, Los Angeles and London, 1974).

Klein, Ira. 'Population and Agriculture in Northern India, 1872–1971', *Modern Asian Studies*, 8, 2, April 1974, pp. 191–216.

'Utilitarianism and Agrarian Progress in Western India', *Economic History Review*, 18, 3, 1965, pp. 576–97.

Knight, J. B. *Sugar Cane, its Cultivation and Gul Manufacture* (Bombay, 1914).

Koh, S. J. *Stages of Industrial Development in Asia* (Philadelphia, 1966).

Kulkarni, A. R. 'Village Life in the Deccan in the Seventeenth Century', *The Indian Economic and Social History Review*, 4, 1, March 1967, pp. 38–52.

Kulkarni, L. B. *Improvement of Grazing Areas in the Bombay Presidency* (Poona, 1923).

Kumar, Dharma. *Land and Caste in South India. Agricultural Labour in the Madras Presidency during the Nineteenth Century* (Cambridge, 1965).

'Landownership and Inequality in Madras Presidency: 1853–54 to 1946–47', *The Indian Economic and Social History Review*, 12, 3, July–September 1975, pp. 229–61.

Kumar, Ravinder. 'The Deccan Riots of 1875', *Journal of Asian Studies*, 24, 4, August 1965, pp. 613–35.

'The Rise of the Rich Peasants in Western India', in D. A. Low, ed., *Soundings in Modern South Asian History* (London, 1968), pp. 25–58.

Western India in the Nineteenth Century. A Study in the Social History of Maharashtra (London and Toronto, 1968).
Latthe, A. B. *Memoirs of His Highness Shri Shahu Chhatrapati, Maharaja of Kolhapur* (Bombay, 1924).
Lockwood, W. W. *The Economic Development of Japan* (expanded ed., Princeton, 1968).
McAlpin, Michelle B. 'Death, Famine and Risk: the Changing Impact of Crop Failures in Western India, 1870–1920', *Journal of Economic History*, 39, 1, March 1979, pp. 143–57.
'The Effects of Expansion of Markets on Rural Income Distribution in Nineteenth Century India', *Explorations in Economic History*, 12, 1975, pp. 289–302.
MacKay, Alexander. *Western India. Reports Addressed to the Chambers of Commerce of Manchester, Liverpool, Blackburn and Glasgow* (London, 1853).
McLane, John R. 'Revenue Farming and the Zamindari System in Eighteenth Century Bengal', in R. E. Frykenberg, ed., *Land Tenure and Peasant in South Asia* (New Delhi, 1977).
Macpherson, W. J. 'Investment in Indian Railways, 1845–1875', *Economic History Review*, 8, 2, 1955–6, pp. 177–86.
Mandlik, N. V., ed., *Writings and Speeches of the Late Honourable Rao Saheb Vishvanath Narayan Mandlik* (Bombay, 1896).
Mann, Harold H. *Economic Progress of the Rural Areas of the Bombay Presidency 1911–1922* (Bombay, 1924).
Fodder Crops of Western India (Bombay, 1921).
Land and Labour in a Deccan Village. No. 1. (London and Bombay, 1917).
'The Mahars of a Deccan Village', in Daniel Thorner, ed., *The Social Framework of Agriculture* (London, 1968), pp. 73–81.
Mann, Harold H. and Kanitkar, N. V. *Land and Labour in a Deccan Village. No. 2* (London and Bombay, 1921).
Metcalf, Thomas R. 'From Raja to Landlord: the Oudh Talukdars, 1850–1870', in R. E. Frykenberg, ed., *Land Control and Social Structure in Indian History* (Madison, Milwaukee and London, 1969), pp. 123–41.
Mishra, Satish Chandra. 'Commercialisation, Peasant Differentiation and Merchant Capital in Late Nineteenth Century Bombay and Punjab', *Journal of Peasant Studies*, 10, 1, October 1982, pp. 3–51.
Moore, Barrington, Jr., *Social Origins of Dictatorship and Democracy. Lord and Peasant in the Making of the Modern World* (London, 1967).
Morris, Morris D. *The Emergence of an Industrial Labour Force in India. A Study of the Bombay Cotton Mills, 1854–1947* (Berkeley and Los Angeles, 1965).
'Towards a Reinterpretation of Nineteenth Century Indian Economic History', *The Indian Economic and Social History Review*, 5, 1, March 1968, pp. 1–15.
Musgrave, Peter. 'Landlords and Lords of the Land: Estate Management and Social Control in Uttar Pradesh, 1860–1920', *Modern Asian Studies*, 6, 3, July 1972, pp. 257–75.

'Rural Credit and Rural Society in the United Provinces, 1860–1920', in Clive Dewey and A. G. Hopkins, eds., *The Imperial Impact. Studies in the Economic History of Africa and India* (London, 1978), pp. 216–32.

Myers, Ramon H. *The Chinese Peasant Economy: Agricultural Development in Hopei and Shantung, 1890–1949* (Cambridge, Mass., 1970).

'Socio-economic Change in Villages of Manchuria during the Ch'ing and Republican Periods: Some Preliminary Findings', *Modern Asian Studies*, 10, 4, October 1976, pp. 591–620.

Natarajan, L. 'Maratha Uprising: 1875', in A. R. Desai, ed., *Peasant Struggles in India* (New Delhi, 1979), pp. 159–69.

Neale, Walter C. *Economic Change in Rural India: Land Tenure and Reform in Uttar Pradesh, 1800–1955* (New Haven and London, 1962).

Nightingale, Florence. 'The People of India', *The Nineteenth Century*, 4, July–December 1878, pp. 193–221.

Pandit, Dhairyabala. 'The Myths around Sub-division and Fragmentation of Holdings: a Few Case Histories', *The Indian Economic and Social History Review*, 6, 2, June 1969, pp. 151–63.

Patel, G. D. *The Land Problem of Reorganised Bombay State* (Bombay, 1957).

Patil, P. C. *Studies in the Cost of Production of Crops in the Deccan No. 1. Crops in the Neighbourhood of Poona* (Poona, 1928).

Pedder, W. G. 'Famine and Debt in India', *The Nineteenth Century*, 2, August–December 1877, pp. 177–97.

Perlin, Frank. 'Of White Whale and Countrymen in the Eighteenth Century Maratha Deccan: Extended Class Relations, Rights and the Problem of Rural Autonomy under the Old Regime', *Journal of Peasant Studies*, 5, 2, January 1978, pp. 172–237.

Pocock, David F. *Kanbi and Patidar. A Study of the Patidar Community of Gujarat* (Oxford, 1972).

Postan, M. M. *Medieval Trade and Finance* (Cambridge, 1973).

Potter, Jack M. *Capitalism and the Chinese Peasant: Social and Economic Change in a Hong Kong Village* (Berkeley, 1968).

Ray, Rajat. 'The Crisis of Bengal Agriculture 1870–1927 – the Dynamics of Immobility', *The Indian Economic and Social History Review*, 10, 3, September 1973, pp. 244–79.

Ray, Rajat and Ratna. 'The Dynamics of Continuity in Rural Bengal under the British Imperium', *The Indian Economic and Social History Review*, 10, 2, May 1973.

Reeves, P. D. 'Landlords and Party Politics in the United Provinces, 1934–37', in D. A. Low, ed., *Soundings in Modern South Asian History* (London, 1968), pp. 261–93.

Robert, Bruce, 'Economic Change and Agricultural Organization in "Dry" South India 1890–1940: a Reinterpretation', *Modern Asian Studies*, 17, 1, February 1983, pp. 59–78.

Roberts, John. 'The Movement of Elites in Western India under Early British Rule', *The Historical Journal*, 14, 2, June 1971, pp. 241–62.

Bibliography

Robinson, Geroid T. *Rural Russia under the Old Regime: a History of the Landlord–Peasant World and Prologue to the Peasant Revolution of 1917* (London, 1932).
Rogers, Alexander. *The Land Revenue of Bombay: a History of its Administration, Rise and Progress* (London, 1892).
Sanyal, Nalinaksha. *The Development of Indian Railways* (Calcutta, 1930).
Saul, S. B. *The Myth of the Great Depression, 1873–1896* (London, 1969).
Selden, Mark. *The Yenan Way in Revolutionary China* (Cambridge, Mass., 1971).
Shanin, Teodor. *The Awkward Class* (Oxford, 1972).
'Socio-economic Mobility and the Rural History of Russia, 1905–1930', *Soviet Studies*, 23, 1971–2, pp. 222–35.
Sheldon, C. D. *The Rise of the Merchant Class in Tokugawa Japan, 1600–1868* (New York, 1958).
Siddiqi, Asiya. *Agrarian Change in a Northern Indian State. Uttar Pradesh 1819–1833* (Oxford, 1973).
Sivaswamy, K. G. *Legislative Protection and the Relief of Agriculturist Debtors in India* (Poona, 1939).
Smith, T. C. *The Agrarian Origins of Modern Japan* (Stanford, 1959).
'Landlords and Rural Capitalists in the Modernisation of Japan', *Journal of Economic History*, 16, 2, June 1956, pp. 165–81.
'The Land Tax in the Tokugawa Period', *Journal of Asian Studies*, 18, 1, November 1958, pp. 3–19.
Spate, O. H. K. *India and Pakistan. A General and Regional Geography* (London, 1954).
Stokes, Eric. 'Agrarian Relations. 1. Northern and Central India', in Dharma Kumar, ed., *The Cambridge Economic History of India, Vol. 2* (Cambridge, 1983), pp. 36–86.
'Dynamism and Enervation in North Indian Agriculture: the Historical Dimension', Ch. 10 of *The Peasant and the Raj. Studies in Agrarian Society and Peasant Rebellion in Colonial India* (Cambridge, 1978).
The English Utilitarians and India (Oxford, 1959).
'The Return of the Peasant to South Asian History', Ch. 12 of *The Peasant and the Raj. Studies in Agrarian Society and Peasant Rebellion in Colonial India* (Cambridge, 1978).
'The Structure of Landholding in Uttar Pradesh, 1860–1948', Ch. 9 of *The Peasant and the Raj. Studies in Agrarian Society and Peasant Rebellion in Colonial India* (Cambridge, 1978).
Strachey, Lytton. *Eminent Victorians* (London, 1967).
Sykes, W. H. 'Special Report on the Statistics of the Four Collectorates of Deccan under the British Government', *Report of the Seventh Meeting of the British Association for the Advancement of Science* (London, 1838), pp. 217–336.
Temple, Sir Richard. *Men and Events of my Time in India* (London, 1882).
Visaria, L. and P. 'Population', in Dharma Kumar, ed., *The Cambridge Economic History of India, Vol. 2* (Cambridge, 1983).
Washbrook, David A. 'Country Politics: Madras 1880–1930', *Modern Asian*

Studies, 7, 3, July 1973, pp. 155–211.
 'Economic Development and Social Stratification in Rural Madras: the "Dry Region", 1878–1929', in Clive Dewey and A. G. Hopkins, eds., *The Imperial Impact. Studies in the Economic History of Africa and India* (London, 1978), pp. 68–82.
  The Emergence of Provincial Politics. The Madras Presidency 1870–1920 (Cambridge, 1976).
 'Law, State and Agrarian Society in India', *Modern Asian Studies*, 15, 3, July 1981, pp. 649–721.
Wedderburn, William. *Agricultural Banks for India* (Bombay, 1882).
  The Indian Raiyat as a Member of the Village Community (London, 1884).
  A Permanent Settlement for the Deccan (Bombay, 1880).
West, Raymond. 'Agrarian Legislation for the Deccan and its Results', *Journal of the Society of Arts*, 41, 1892–3, pp. 705–31.
  The Law and the Land in India (Bombay, 1872).
Wilbur, Elvira M. 'Was Russian Agriculture Really that Impoverished? New Evidence from a Case Study from the "Impoverished Centre" at the End of the Nineteenth Century', *Journal of Economic History*, 43, 1, March 1983, pp. 137–44.
Wolf, Eric. 'On Peasant Rebellions', in Teodor Shanin, ed., *Peasants and Peasant Societies. Selected Readings* (London, 1971), pp. 264–74.
  Peasant Wars of the Twentieth Century (London, 1971).

INDEX

agricultural banks, 118, 241–2
agricultural implements, 72, 78, 212
agricultural indebtedness, 82–94, 129, 136
  British policy towards, 115–24
agricultural statistics, 4–5, 205–6
Agris, 275
Ahmedabad District, 15, 165, 173, 198, 200–1, 261, 265, 272
Ahmedabad talukdars (see talukdars)
Ahmedabad Talukdars' Relief Act, 1862, 64
Ahmednagar District, 12, 13, 107, 110, 112, 178, 196, 198, 255, 257, 265
Akola taluk, Ahmednagar District, 104
Alavi, Hamza, 270, 272
Alibag taluk, Kolaba District, 189, 218, 222
Amalner taluk, Khandesh District, 78, 168
American Civil War, 7, 135–7
Anand taluk, Kaira District, 280
Anderson, F. G. H., 42, 213, 234, 236, 251n., 260–1, 263, 266n.
Anderson, W. C., 45, 60
Ankleshwar taluk, Broach District, 132–3
anti-Brahminism, 276–9
Argyll, Duke of, 116
Attwood, D. W., 183, 186

Badami taluk, Belgaum (later Bijapur) District, 26, 28, 48
Baden-Powell, B. H., 22, 30, 34, 35, 45, 63
badhekaris, 32
Bagalkot taluk, Belgaum (later Bijapur) District, 26, 48, 78, 132–3
bajra, 12–13, 70, 214
Bardoli campaign, 1928, 263, 265–6, 283–91
Bardoli taluk, Surat District, 150, 153, 211, 214–15, 224, 264–5, 280
Baroda, 10
Bates, Crispin, 205
Belgaum District, 14, 48, 198, 279

Belgaum taluk, Belgaum District, 156–7, 175, 181, 189
Bell, R. D., 191
Bengal, 3–4, 5
Bengal Rent Act, 1859, 185
bhagdari system, 35–7, 63
Bhatia, B. M., 157
Bhils, 97, 223, 251
Bhimthari taluk, Poona District, 22–3, 74, 99, 101–2, 103, 111, 134, 140, 141, 171, 252, 255, 259
Bijapur District, 14, 177, 185, 197
Blyn, George, 4, 205–6
Bole, S. K., 275
Borsad taluk, Kaira District, 36, 151, 280–1
Boserup, Esther, 75
Brahmins, 31, 60–2, 103, 223, 257, 277–8
Braudel, Fernand, 70, 92–4
Breman, Jan, 223, 229
Broach District, 15, 63, 125, 138, 179, 214, 223, 251
Broomfield, R. S., 283
Bunkapur taluk, Dharwar District, 18, 22

Caird, James, 127
canals, 135–6, 150–1, 166, 183, 230, 237, 296
Candy, E. T., 58
carts, 72–3, 77, 134, 141, 145–6, 212
caste, 165
Catanach, Ian, 109, 110, 243n., 244n.
cattle, 77–8, 140, 143, 156, 159, 212–14, 295–6
Chandrabhan Bapuji, 106
Chaplin, W. H., 33
Chayanov, A. V., 163, 172, 173n., 182, 228, 285
China, 269, 290, 294n.
Chitpavans, 11
climate (see also rainfall), 159, 204
Colvin, Auckland, 103, 140
Coompta, 80
co-operative credit, 4, 244, 246–7

315

cotton, 14, 15, 71, 79–81, 85, 135–7, 149–50, 157, 208, 210–11, 215–16, 223–4, 227–8, 285–7
cotton trade, 71, 80–1, 90–3, 146
credit, 91–4, 170–4, 252–8
crop experiments, 76

Daskroi taluk, Ahmedabad District, 35, 39
Davidson, A. F., 49–52
Davidson, D., 44
Deccan, 12–13, 19–29, 47–56, 74–5, 141, 202
Deccan Agriculturists' Relief Act, 1879, 9, 99, 122–4, 239–40, 242, 247–8
  effects of, 252–60
Deccan Riots, 1875, 68, 99–115, 270
Deccan Riots Commission, 101, 102, 104, 108, 109, 110, 114, 121–2, 135, 138, 152
depression (inter-war period), 208–9, 225–31, 265–6
Desai, Mahadev, 284
Deshasthas, 11
Deshmukhs, 25, 47
dharekaris, 32
Dharwar District, 14, 21, 48, 53, 71, 79, 89, 214, 244, 279
Dholka taluk, Ahmedabad District, 156–7, 159, 190
Dhond Petha, Poona District, 236
Dowell, Lt., 34
Dracup, A. H., 222
Dunlop, J., 34

Elphinston, A., 50–2
Elphinstone, M., 33, 41

famine, 2, 4, 130–1, 164, 204–5, 208–9, 221–2
famine of 1876–8, 126, 140–1
famine of 1896–1900, 126, 155–9, 173, 178, 196
Famine Commission, 1880, 75, 83, 108, 141, 142, 161, 167
Famine Commission, 1901, 129, 245–6
Ferguson, Sir James, 240, 246
fevers, 156
fodder crops, 213–14, 231
fragmentation of land, 7, 232–8, 248–50, 296
Francis, J. T., 45, 56–7
Frykenberg, R. E., 8

Gadgil, D. R., 124

Gandhi, M. K., 264, 268, 280–2
Geertz, Clifford, 297–9
Godbole, Diwan, 248
Gogha taluk, Ahmedabad District, 38–9, 65, 148, 200
Goldsmid, H. E., 43–6, 101
Gooddine, R. N., 24–5, 27
Goodman, David, 294–5
grazing land, 213–14, 231, 295–6
'Great Depression', 1873–96, 131, 160–1
groundnuts, 151
Guha, Amalendu, 145
Gujarat, 15, 34–40, 62–5, 141, 155–6, 161, 202, 279–83

halis and hali system, 180, 222–3, 229
Haliyal taluk, Kanara District, 219
handicrafts, 298
Hardiman, David, 205, 270, 285
Harnetty, Peter, 136
Harris, Lord, 240, 242, 273
Hart, W., 47, 53
Hatban village, Poona District, 235–6
Haveli taluk, Poona District, 74, 104–5, 132–3, 134, 140, 150–1, 166, 235
hay, 152
Heston, Alan, 266
Hobsbawm, E. J., 98
Holderness, B. A., 107
holdings, size of (*see also* fragmentation, subdivision), 206, 208, 231–2
Honya Kenglia, 98–9
Hope, T. C., 245
houses and house types, 167, 171–2
Hunter, W. W., 47

impartible tenure, 251
Inam Commission, 53–5
inam land, 27–9, 35, 53–6, 182, 186–8, 199–201, 272–3
Indapur taluk, Poona District, 20, 43–4, 104–5, 109, 126, 134, 136–7, 143, 156–7, 198
indebtedness (*see* agricultural indebtedness)
influenza epidemic, 2, 130, 205
inheritance procedure, 232, 234–5
irrigation (*see also* canals, wells), 4, 82, 138, 141–2, 173, 236, 296

Jacomb, H. E., 118
Jacquemont, Victor, 83
Jadhav, B. V., 279
jagirs, 27–8
James, H. E. M., 242
Japan, 3, 190, 203, 210–11, 298

Jategaon Budruk village, Poona District, 294
Java, 297
Jayakar, M. S., 264–5, 285
Joint Report, 1847, 21, 22, 44–7, 116
jowar, 12–13, 70, 131–3, 214
Junnar taluk, Poona District, 170, 175, 176, 235, 277

Kaira (Kheda) District, 15, 74, 99, 151, 171, 179, 198–9, 279–80, 282
Kaira Satyagraha, 1918, 63, 264, 280–3
Kalyan taluk, Thana District, 219
Kamthadi village, Poona District, 234–5
Karad taluk, Satara District, 151, 167, 226
Karjat taluk, Ahmednagar District, 204
Karmala taluk, Sholapur District, 220
Karwar land revenue campaign, 1871, 60–2
Keatinge, G., 211, 219, 232, 235–7, 248–50
Kessinger, Tom G., 5
Khandali village, Kaira District, 282
Khandesh District, 12, 49–52, 74, 83, 86, 97, 248, 251
Khandesh Survey Riots, 1852, 49–52
Khed taluk, Ratnagiri District, 273–4
Khot Act, 1880, 58
Khot Commission, 58
Khoti Settlement Act, 1904, 188, 273
khots and khoti system, 31–4, 56–60, 188–9, 200–1, 273–5, 292
Kimberley, Earl of, 241–2
Klein, Ira, 2, 41, 42n.
Kolaba District, 14, 32–3, 169, 189, 195, 247, 265, 274–6
Kolhapur, 10, 278–9
Kolis, 165, 219–20, 271, 277–8
Konkan, 14–15, 30–4, 56–62, 161, 234, 273–6
Kopergaon taluk, Ahmednagar District, 217
Koregaon taluk, Satara District, 148–9
Kothari, V. R., 279
Kulkarnis, 23, 258
Kumar, Dharma, 5, 21, 180
Kumar, Ravinder, 41, 100–1, 119, 134, 292
Kunbis, 11, 257

labourers, 21, 164, 184–5, 219–24, 225–6, 286–8, 293–4
land availability, 73–5, 111–12, 126–8, 153, 160, 188, 206–7, 213–15, 230–1, 296

land mortgages, 177–9, 257, 259
land organisation (*see* fragmentation, holdings, subdivision)
land reclamation, 30
land revenue assessments, 2, 17–18, 42–3, 109–10, 262–6
Land Revenue Code, 244–5, 250–1
land revenue collection, 85–6, 282, 284
land revenue policy, 8–9, 40–7, 65–9, 187–8, 260–7
land revenue receipts, 43–4, 264–5
land sales, 167–70
land transfer, 5–6, 100–6, 193–9, 256–7
land transfer legislation, 242–4, 246
land values, 95–6, 101, 139, 169–70, 179, 211, 237–8
landlessness, extent of, 174–5, 180–1
Latthe, A. B., 279
Lee-Warner, W., 169, 187
legal system and its implications, 96, 102–4, 117–18, 252–5, 258–9
Limitation Act, 1859, 102
Lingayats, 278
living standards, 129

McAlpin, Michelle B., 223–4
Mackay, Alexander, 78–80, 101
McNeill, James, 244
Maconochie, A. F., 244
Madras Presidency, 5, 158
Mahars, 184
Malis, 165, 183, 186, 257–8
Malsiras taluk, Sholapur District, 143
Malthusian interpretations, 75, 126–7
Man taluk, Satara District, 149
Mandlik, V. N., 59
mangoes, 229
Mann, Harold H., 184, 205, 207, 213, 217, 219, 232–3, 248, 294
Marwaris, 11, 68, 103–9, 113–15, 171, 257
Matar taluk, Kaira District, 217, 281–2
Maval taluk, Poona District, 181, 198
Maxwell, R. M., 283
Mehmedabad taluk, Kaira District, 281–2
migration, 166, 221–2
mirasdars, 22–3
Mishra, S. C., 163, 167
moneylenders, 86–90, 98, 103–9, 113–15, 170–2, 179–80, 253–8
  landholdings of, 104–6, 193–9
  violence against (*see also* Deccan Riots), 98–9
Moore, J. G., 260
Morris, M. D., 1–3, 221

Mountford, L. J., 274
Muir-Mackenzie, J. W. P., 243
munsifs, 253
Musgrave, P. J., 90, 186
myrabolams, 165, 277

Nadiad taluk, Kaira District, 280–1
Naroo Babajee, 26
narwa system, 35–7, 63
Nasik District, 12, 74
Navagam village, Kaira District, 281
Neale, Walter C., 185
Newasseh taluk, Ahmednagar District, 26
Nightingale, Florence, 121
North Kanara District, 14
Northbrook, Lord, 115
Nuwulgond taluk, Dharwar District, 25

Palaj village, Kaira District, 281
Panch Mahals District, 15, 169, 251
Pandit, Dhairyabala, 233
Paranjpye, Dr, 248
Parner taluk, Ahmednagar District, 26, 104, 106, 127, 144, 169, 175, 181
Patan taluk, Satara District, 151
Patels, 23–7, 47–9, 51–2, 67–8
Patidars, 11, 37, 202, 279–83
Patoda taluk, Ahmednagar District, 26
Pedder, W. G., 35–7, 63, 69, 74, 86, 88, 119–21, 140, 200
Peile, J. B., 64
Pelly, J. H., 33
Pen taluk, Kolaba District, 274–6
Phooley, Jotirao, 276–8
Pimpla Soudagar village, Poona District, 232–3
plague, 156
Poona District, 12, 74, 79, 83, 88, 107, 110, 112, 189, 191, 196, 262, 265
population trends, 2, 74–5, 111, 130, 143, 155–7, 205–7, 232–3, 237, 297
ports, 147
Potter, Jack, 298
price trends, 89, 112, 131–3, 135, 137, 158, 204, 208–9, 210–11, 225–6, 264–5
Pringle, R. K., 17–18, 41–3
public works (see also canals, railways), 135–6
Punjab, 243
Purandhar taluk, Poona District, 104–5, 151, 165, 199, 259

Rahuri taluk, Ahmednagar District, 77
railways, impact of, 142, 144–9, 154
rainfall levels and trends, 12, 76, 110–11, 140, 159, 204
Ranade, M. G., 124
Ratnagiri District, 12, 14, 32–4, 56–7, 68, 128, 148, 157, 189, 234, 273
Ratnagiri taluk, Ratnagiri District, 146–7
Ray, Rajat and Ratna, 4n., 162
Reay, Lord, 240, 277
Redclift, Michael, 294–5
Reeves, P. D., 6
rental value, 263
rents and rent levels, 187–92, 217–18, 227, 273–6, 288
Ricardo's rent theory, influence of, 41–2, 44
rice, 14
riots (see also Deccan Riots), 271
roads, 79–80
Robert, Bruce, 163
Russia, 6, 183, 297–8

Sabhasad, 25
Salisbury, Lord, 116
Salsette District, 14, 30
Sandhurst, Lord, 247
Sangamner taluk, Ahmednagar District, 104
Satara District, 12, 75, 106, 143, 145–6, 148–9, 151, 154, 169, 171, 198, 216, 272–3, 278
Satya Shodak Samaj, 277–8
Shanin, Teodor, 6
Shevgaon taluk, Ahmednagar District, 83
Shivaji, 11, 25
Shivram Marwari, 103, 105
Sholapur District, 12, 13, 48, 77, 110, 197–8, 226, 254
Sholapur taluk, Sholapur District, 177
Sind, 242, 248
Sirur taluk, Poona District, 108, 156, 198
soil exhaustion, 237
Southern Maratha Country, 13–14, 19–29, 53, 161
Sowda taluk, Khandesh District, 49–50
Stewart, T. H., 242
Stokes, Eric, 5, 41, 163, 165, 174, 269n., 270
stratification, 134–5, 138, 162–203, 217–25, 230, 238, 257–60, 289–91, 293–4
subdivision of land, 232–8
legislation on, 248–50
sugarcane, 15, 152, 165, 218–19, 223, 227–30
Sullivan, H. E., 150
Surat District, 15, 214, 219, 223

Index

surinjams, 27–8, 54
Sykes, W. H., 21, 22, 23, 24, 28, 180

takavi, 139, 261
talukdars (of Ahmedabad District), 38–40, 64–5, 68–9, 95
talukdars' estates, 183, 200–1
tariff protection, 228
Tasgaon taluk, Satara District, 154, 189
Temple, Richard, 63, 108, 122, 161
tenancy, extent and spread of, 21, 174–92, 226, 288
tenancy legislation, 176, 247, 267
tenant politics, 272–6
Thakurs (of Broach and Kaira Districts), 38, 65
Thakurs Encumbered Estates Act, 1877, 65
Thana District, 14, 60, 195
Thorburn, S. S., 6
Thorner, D. and A., 175–6
Thornton, J., 46
Tipu Sultan, 25
tobacco, 15, 151–2, 223, 228–30
Tukaram, 24
Tularam Karamchand, 106
Tytler, C. Fraser, 117, 118, 120, 121, 123, 125, 139

United Provinces, 5, 186, 190

upris, 22–3
utilitarianism, influence of, 41–2, 44

Valva taluk, Satara District, 145n., 154, 170
Vanias, 11, 68, 103–9, 113–15
Viramgam taluk, Ahmedabad District, 145, 150, 153, 154, 200

Waddington, W., 45, 56, 111
wages, levels and trends, 137, 164, 219–20, 222–4, 227, 230, 294
Washbrook, D. A., 12, 162
Wedderburn, William, 86–7, 96, 118, 139, 241, 245–6, 256
wedding expenditure, 81, 92
weights and measures systems, 91
wells, 134–5, 139–40, 144, 167, 261
West, Raymond, 63, 96, 101, 119–20, 240, 253, 255, 257, 260
Willingdon, Lord, 219, 249, 282
Wingate, George, 18–19, 21, 28, 43–52, 56, 65–6, 71, 73, 74, 77, 83, 85, 88, 89, 98, 101, 116–19, 123, 128, 295
Wodehouse, Philip, 115–16, 119, 122
Wolf, Eric, 270–2, 286, 289
Woodburn, A. F., 177–8, 240, 252–5, 257
Woodward, H., 240, 255–6

Yawul taluk, Khandesh District, 49–50
Young, D., 72